Application of Artificial Intelligence in Land Use and Land Cover Mapping II

Application of Artificial Intelligence in Land Use and Land Cover Mapping II

Guest Editors

Sawaid Abbas
Janet E. Nichol
Faisal M. Qamer
Jianchu Xu

MDPI

Basel • Beijing • Wuhan • Barcelona • Belgrade • Novi Sad • Cluj • Manchester

Guest Editors

Sawaid Abbas
Geographical Information System
University of the Punjab
Lahore
Pakistan

Janet E. Nichol
School of Environment and Life Sciences
University of Portsmouth
Portsmouth
United Kingdom

Faisal M. Qamer
Cryosphere and Water
International Centre for Integrated Mountain Development
Lalitpur
Nepal

Jianchu Xu
Kunming Institute of Botany
Chinese Academy of Sciences
Kunming
China

Editorial Office
MDPI AG
Grosspeteranlage 5
4052 Basel, Switzerland

This is a reprint of the Special Issue, published open access by the journal *Remote Sensing* (ISSN 2072-4292), freely accessible at: www.mdpi.com/journal/remotesensing/special_issues/H6SCCGE052.

For citation purposes, cite each article independently as indicated on the article page online and using the guide below:

Lastname, A.A.; Lastname, B.B. Article Title. *Journal Name* **Year**, *Volume Number*, Page Range.

ISBN 978-3-7258-3992-6 (Hbk)
ISBN 978-3-7258-3991-9 (PDF)
https://doi.org/10.3390/books978-3-7258-3991-9

Contents

About the Editors

Sawaid Abbas

Sawaid Abbas is currently leading the Smart Sensing for Climate and Development initiative at the Centre for Geographical Information System, University of the Punjab, Lahore, Pakistan, where he also serves as an Assistant Professor. As an emerging expert in climate and environmental sciences, his work spans the environment, earth sciences, ecology, the urban climate, forestry, and adaptation strategies by effectively utilizing smart sensing, space technologies, and advanced spatial data analysis. Dr. Abbas's contributions to smart sensing technologies have earned him and his team the Outstanding Achievement in Technology Transfer award from the Faculty of Construction and Environment at Hong Kong Polytechnic University. His collaborations with esteemed global and regional institutions, including AIRCAS, AOGEO, USESCAP, Space4Water, WWF, ICIMOD, UNESCAP, ICRAF, AFCD, and KFBG, have further enriched his research and impact. A dedicated educator and communicator of science, Dr. Abbas actively engages with regional stakeholders to ensure that scientific knowledge is translated into impactful policies and sustainable management practices.

Janet E. Nichol

Janet E. Nichol is an Applied Geographer specialising in remote sensing, geo-informatics, and rnvironmental change. Her main research interests are in the application of remote sensing to landscape change, including agricultural and forest resources, especially in context of global climatic change. She has previously worked in universities in Nigeria, the Republic of Ireland, the United Kingdom, Singapore, and Hong Kong. Her current work is focused on trends in wood fuel resources, specifically farmland trees in northern Nigeria, succession in tropical secondary forests, global greening/browning trends under climate change, and remote sensing of atmospheric aerosols, including fine particulates.

Faisal M. Qamer

Faisal M. Qamer is a Remote Sensing Specialist who leads the Resilient River Basins Intervention at ICIMOD, where he applies science-based digital technology approaches—including advanced AI and machine learning techniques—to address climate and water risks across the Hindu Kush Himalaya region. He has driven the development of regional land cover mapping systems, in-season crop mapping, and rapid yield estimation services, and he is exploring emerging natural language processing methods to automate tailored advisories for diverse stakeholders.

Jianchu Xu

Jianchu Xu is a professor at the Kunming Institute of Botany, Chinese Academy of Sciences. He is an internationally respected leading ethno-ecologist who works in coupled human environmental systems. His research includes the investigation of early warning signals of global change, transboundary water governance, landscape restoration, ecosystem services and their resilience, agriculture, and integrative conservation. Xu holds a PhD in Soil Science from China Agricultural University and has published over 100 papers in high-impact journals including *Nature, Science, PNAS, Conservation Biology, Fungal Diversity,* and *Frontiers in Ecology and Evolution.*

Preface

In today's data-intensive environment, the rapid evolution of Earth observation technologies and the advent of high-performance computing have opened unprecedented avenues for enhancing our understanding of landscape changes and sustainable development. The articles compiled in this volume cover a broad spectrum of topics—from advanced semantic segmentation techniques and object detection in remote sensing images to innovative approaches in mapping paddy rice phenology and assessing the impacts of land cover transformations on surface temperature and carbon emissions. Each paper makes a unique contribution by addressing specific methodological challenges and proposing robust solutions, whether through the design of novel neural network architectures or the deployment of comprehensive regional accuracy assessments.

This collection presents a cohesive exploration of cutting-edge methodologies in land cover classification, emphasizing the integration of deep learning techniques to enhance accuracy and applicability across diverse environmental contexts. A notable contribution is the development of an aggregated backbone network (ABNet), which synergistically combines multiple backbone architectures to improve feature representation for fine-grained land cover classification. This approach demonstrates superior performance in handling complex spatial information inherent in high-resolution remote sensing images. Complementing this, a comprehensive regional accuracy assessment of widely used 30-meter land cover products highlights the influence of environmental heterogeneity on classification. Further studies within this issue apply convolutional neural networks to wildfire mapping, showcasing the potential of deep learning in dynamic environmental monitoring. Additionally, innovative segmentation techniques are employed to analyze China's coastal wetlands, demonstrating the versatility of deep learning models in capturing complex ecological patterns.

Collectively, these contributions accentuate the interdisciplinary relevance and practical impact of integrating advanced computational methods in land cover classification, paving the way for more nuanced and accurate environmental assessments. We extend our heartfelt gratitude to all contributors, peer reviewers, and editorial staff for their unwavering commitment to excellence. It is our hope that the insights presented in this reprint will inspire further innovation and inform both scientific research and practical applications in sustainable development. As we look ahead, the integration of artificial intelligence with remote sensing stands as a beacon for future advancements—an invitation to continually reimagine how we monitor and manage our planet's resources.

Sawaid Abbas, Janet E. Nichol, Faisal M. Qamer, and Jianchu Xu
Guest Editors

remote sensing

MDPI

Article

ABNet: An Aggregated Backbone Network Architecture for Fine Landcover Classification

Bo Si [1], Zhennan Wang [2], Zhoulu Yu [1,*] and Ke Wang [1]

[1] College of Environmental and Resource Sciences, Zhejiang University, Hangzhou 310058, China; bo_si@zju.edu.cn (B.S.); kwang@zju.edu.cn (K.W.)
[2] Zhejiang Shuzhi Space Planning and Design Co., Ltd., Hangzhou 310030, China; 1108130121001@stu.bucea.edu.cn
* Correspondence: yuzl@zju.edu.cn; Tel.: +86-13958123139

Abstract: High-precision landcover classification is a fundamental prerequisite for resource and environmental monitoring and land-use status surveys. Imbued with intricate spatial information and texture features, very high spatial resolution remote sensing images accentuate the divergence between features within the same category, thereby amplifying the complexity of landcover classification. Consequently, semantic segmentation models leveraging deep backbone networks have emerged as stalwarts in landcover classification tasks owing to their adeptness in feature representation. However, the classification efficacy of a solitary backbone network model fluctuates across diverse scenarios and datasets, posing a persistent challenge in the construction or selection of an appropriate backbone network for distinct classification tasks. To elevate the classification performance and bolster the generalization of semantic segmentation models, we propose a novel semantic segmentation network architecture, named the aggregated backbone network (ABNet), for the meticulous landcover classification. ABNet aggregates three prevailing backbone networks (ResNet, HRNet, and VoVNet), distinguished by significant structural disparities, using a same-stage fusion approach. Subsequently, it amalgamates these networks with the Deeplabv3+ head after integrating the convolutional block attention mechanism (CBAM). Notably, this amalgamation harmonizes distinct scale features extracted by the three backbone networks, thus enriching the model's spatial contextual comprehension and expanding its receptive field, thereby facilitating more effective semantic feature extraction across different stages. The convolutional block attention mechanism primarily orchestrates channel adjustments and curtails redundant information within the aggregated feature layers. Ablation experiments demonstrate an enhancement of no less than 3% in the mean intersection over union (mIoU) of ABNet on both the LoveDA and GID15 datasets when compared with a single backbone network model. Furthermore, in contrast to seven classical or state-of-the-art models (UNet, FPN, PSPNet, DANet, CBNet, CCNet, and UPerNet), ABNet evinces excellent segmentation performance across the aforementioned datasets, underscoring the efficiency and robust generalization capabilities of the proposed approach.

Keywords: backbone network; landcover classification; semantic segmentation; aggregated feature

Citation: Si, B.; Wang, Z.; Yu, Z.; Wang, K. ABNet: An Aggregated Backbone Network Architecture for Fine Landcover Classification. *Remote Sens.* **2024**, *16*, 1725. https://doi.org/10.3390/rs16101725

Academic Editors: Sawaid Abbas, Janet E. Nichol, Faisal M. Qamer and Jianchu Xu

Received: 31 March 2024
Revised: 7 May 2024
Accepted: 10 May 2024
Published: 13 May 2024

1. Introduction

Landcover classification is the process of distinguishing different types of naturally occurring landcover through remote sensing image data processing, which provides important geospatial data for decision making and monitoring in the fields of environmental monitoring, urban planning, precision agriculture, resource management, and so on [1,2]. With the rapid development of remote sensing imaging technology, the rapid acquisition of meter-scale remote sensing images provides the possibility of fine landcover classification and mapping [3]. However, the feature differences within high-precision landcover categories become significantly larger, making it more difficult to distinguish surface covers with complex compositional elements [4]. Misidentification of landcover categories will cause a huge waste of remote sensing data resources, and it is difficult to produce effective application value. Therefore, how to realize the fine classification of landcover has become a hot issue to be solved [5].

In recent years, deep learning models for landcover classification based on convolutional neural networks (CNNs) have been widely studied and have shown powerful performance over traditional methods that rely on manual feature extraction and shallow machine learning algorithms [6]. Michael Kampffmeyer et al. [7] first introduced an end-to-end convolutional neural network model for remote sensing imagery landcover mapping. After that, the semantic segmentation model for remote sensing images gradually formed an encoder–decoder architecture containing a backbone network, a neck, and a segmentation head [8–10]. Therefore, the semantic segmentation model based on the CNN as the backbone network has become a mainstream method for studying the high-precision landcover classification problem. However, different backbone networks have different abilities to extract features due to their different convolutional structures, which leads to different classification results [11]. Therefore, how to efficiently apply different backbone networks for landcover classification still deserves further exploration.

To compare the performance of semantic segmentation models based on different backbone networks, scholars have carried out related studies on backbone networks and provided some new inspirations. Marc et al. [12] used two model structures, UNet and Deeplabv3+, in combination with three backbone networks, namely, MobileNet-V3, ResNet-50, and EfficientNet-B4, to test their abilities to segment flooded areas under different environmental conditions and data availability. Results showed that the MobileNet-V3-based UNet model performed the best. This implies that the appropriate backbone network needs to be selected in advance under different datasets and environmental conditions. In a pioneering study, Liu et al. [13] generated a more powerful backbone network (CBNet) by combining ResNet and ResNeXt through a stage-by-stage composite connection strategy of different backbone networks, and the experimental results demonstrated that CBNet possessed a stronger generalization ability than a single backbone network on different target detection datasets. Liang et al. [14] further improved CBNet by a dense connectivity strategy of backbones and assisted supervised training to achieve an increase in classification accuracy, which provides a novel idea for model architecture improvement.

In the pursuit of leveraging a backbone network fusion strategy for the development of a remote sensing semantic segmentation model, this study meticulously selects the following three pre-eminent backbone networks for fusion: ResNet, HRNet, and VoVNet. Notably, ResNet facilitates feature extraction by serially connecting residual convolutional blocks at each network stage, thereby enabling profound feature extraction. In contrast, HRNet fosters feature extraction through parallel connections at each stage of the convolutional blocks, adeptly amalgamating features from multi-scale receptive fields. Furthermore, VoVNet strategically aggregates all preceding convolutional blocks exclusively at the final layer of each stage, thereby amplifying the receptive fields and enhancing feature extraction efficiency. After the multi-scale fusion of the backbone network, feature optimization is meticulously achieved through the integration of a convolutional attention mechanism. The amalgamation with the Deeplabv3+ network architecture gives rise to the aggregated backbone network (ABNet), tailored specifically for the fine-grained classification of landcover.

The main contributions of this paper include the following: (1) An efficient feature extraction strategy using multiple existing backbone networks is proposed. (2) An efficient and effective aggregated backbone network (ABNet) to enhance multi-scale feature extraction and fusion capabilities for remote sensing image semantic segmentation work is proposed. ABNet leverages the Deeplabv3+ head and reconstructs the backbone network through the same-stage fusion of ResNet-50, HRNet-48, and VoVNet-39. (3) Experimental results demonstrate that ABNet achieves comparable performance to the state-of-the-art methods on the LoveDA and GID datasets.

The rest of this paper is organized as follows. Section 2 introduces related research work, mainly reviewing the development of representative backbones and ensemble methods of semantic segmentation in the field of fine landcover classification in recent years. Section 3 elaborates on the structure of the proposed ABNet and details the design ideas of the aggregated backbone network. Section 4 gives the experimental details and results on the LoveDA dataset and Gaofen Image Dataset (GID). The advantages, limitations, and potential improvements of our proposed method will be elaborated on in Section 5. In Section 6, conclusions and future works are provided.

2. Related Work

2.1. Backbones for Remote Sensing Image Semantic Segmentation Work

The pivotal role of backbone networks in feature extraction remains integral to the fabric of semantic segmentation models [15], which is particularly evident in the realm of remote sensing classification, as shown in Table 1. Since the inception of AlexNet in 2012, progressive enhancements in backbone networks have predominantly revolved around augmenting both network depth and width. VGG-16 [16] constitutes a prototypical deep network by increasing the number of network layers. However, the deepening of network layers often precipitates the potential vanishing or explosion of computational gradients. Addressing this concern, He et al. [17] ingeniously introduced the residual connection module to alleviate the convergence challenges encountered by deep networks. Furthermore, Inceptionv3 [18] astutely observed that augmenting the model width effectively bolsters feature extraction through the incorporation of branches with multi-scale convolutional kernels. Beyond the model architecture, the effective interconnection between feature layers emerges as a focal avenue for enhancing backbone networks. In this context, DenseNet [19] forges dense connections between preceding and subsequent layers, harnessing multi-layer features to optimize segmentation performance. In a bid to pare down the parameter count, VoVNet [20] boosts computational efficiency by densely linking previous layer features exclusively to the last layer, drawing inspiration from DenseNet. HRNet [21], on the other hand, intertwines the outputs of multiple resolutions in parallel, engendering a richer, high-resolution representation. Leveraging the neural structure search technique, EfficientNet [22] delves into an exploration of network resolution, depth, and width parameters, rationalizing their configuration to bolster model enhancement efficiency. Meanwhile, Res2Net [23] delineates multi-scale features at a granular level by constructing hierarchical residual connections within the same residual block, sans an expansion in the network depth. These contemporary mainstream backbone networks exhibit distinct model structures and generalization proficiencies tailored to their specific developmental tasks.

Table 1. Representative backbones used in remote sensing classification.

Backbone	First Publication	Features	Application in Remote Sensing
VGG-16 [16]	2014	Deep convolutional network	Topography and geomorphology classification [24]
ResNet-50 [17]	2016	Residual connection	Multispectral image classification [25]
Inceptionv3 [18]	2016	Multi-scale convolutional kernel	Sheltered Vessels classification [26]
DenseNet [19]	2017	Dense connection of all layers	Road extraction [27]
VoVNet [20]	2019	Dense connection of the final layer	Open-pit mine extraction [28]

Table 1. *Cont.*

Backbone	First Publication	Features	Application in Remote Sensing
HRNet [21]	2019	Parallel connection of multi-resolution layers	Landcover classification [29]
EfficientNet [22]	2019	Neural structure search	Mars scene recognition [30]
Res2Net [23]	2021	Hierarchical residual connection	Remote sensing scene recognition [31]

2.2. Research on Fine Landcover Classification with Semantic Segmentation Network

Since the full convolutional neural network (FCN) [32] realizes the pixel-by-pixel semantic segmentation of images, many classical semantic segmentation models based on convolutional neural networks have been developed, such as UNet [33], Deeplabv3+ [34], FPN [35], PSPNet [36], DANet [37], UPerNet [38], and CCNet [39], extensively embraced in remote sensing semantic segmentation applications [40–43]. Remote sensing semantic segmentation models dedicated to landcover classification chiefly fall into distinct categories, including the pixel-based CNN, object-based CNN, graph-based CNN, siamese CNN, and ensembled CNN, with the ensembled CNN model emerging as adept at efficiently addressing complex landcover scenarios [44].

In the pursuit of heightened accuracy, Bigdeli et al. [45] achieved superior classification by optimally amalgamating the classification results of CNN models constructed from diverse sensor datasets. Faced with a dearth of labeled samples, Fan et al. [46] crafted an integrated teacher model to autonomously pick samples from an extensive pool of unlabeled data, culminating in dataset generation. A pivotal notion entails the integration of results generated by multiple backbone network models in a channel-by-channel superposition fashion. Complementing inputs or outputs, scholars have sought to consolidate deep learning models during their training process. Notably, Cao et al. [47] bolstered the classification accuracy of the ISPRS 2-D dataset by harnessing complementarities between diverse models, coalescing to jointly learn the solution domains of multiple semantic segmentation models. Additionally, Ekim et al. [48], premised on the training process of three loss optimization methods, honed the integration of distinct model outputs to fortify the robustness of the landcover classification model.

Cumulatively, these investigations meld the computational attributes of divergent networks to yield refined semantic segmentation results. Yet, the variance in classification outcomes stems from disparities in feature layer outputs across dissimilar model structures, disregarding the potential integration of feature layers encapsulating distinct semantic information. This oversight of structural dissimilarities may precipitate the omission of essential extraction outcomes.

3. Methods

3.1. Aggregated Backbone Network

Deep learning semantic segmentation models based on features extracted from backbone networks have been widely used in the fine landcover classification of remote sensing images. However, for remote sensing datasets with complex feature distributions, it is still challenging to select an appropriate backbone network. In addition, a single backbone network is usually unable to extract effective and relevant features from complex scenes, especially on datasets with diverse classes and large intra-class variations. To overcome this problem, this study selects three backbone networks to construct an aggregated backbone network (ABNet), aggregates the feature layers of different backbone networks by same-stage fusion, then optimizes the merged features by a convolution block attention mechanism, and finally outputs the classification results by the Deeplabv3+ head [34] (as shown in Figure 1). Specifically, ResNet, HRNet, and VoVNet are used in this study and described in Section 3.2. Different backbone networks extract the semantic features of the classified objects in different ways, and the aggregation backbone network utilizes the characteristics of each backbone network and enhances the feature extraction ability by using the same-stage fusion method at different scale stages of feature extraction, which will be described in detail in Section 3.3. The CBAM attention mechanism extracts the features of the different backbone networks aggregated at different stages adaptively and optimally using the channel attention module and the spatial attention module, which will be described in Section 3.4.

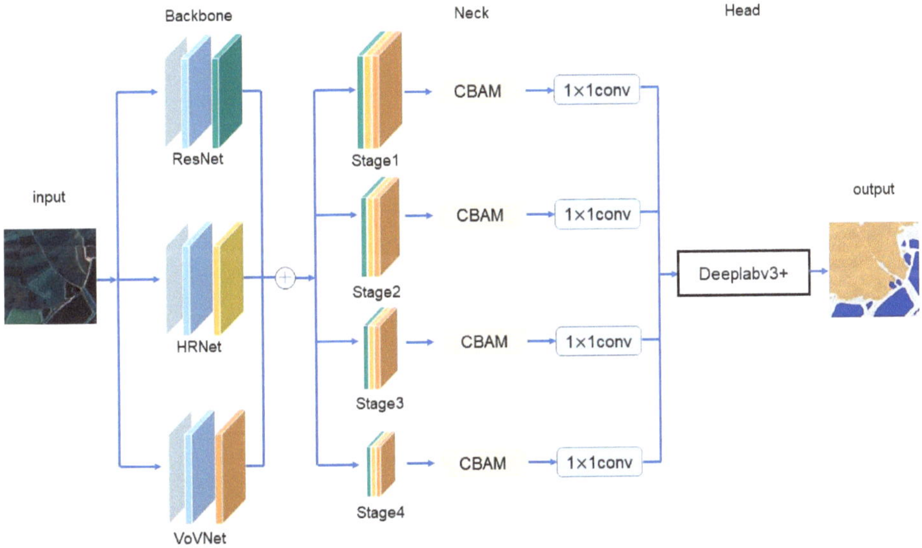

Figure 1. Illustration of the proposed aggregated backbone network (ABNet) architecture for semantic segmentation.

3.2. Basic Backbone Networks

3.2.1. Residual Network (ResNet)

In a conventional convolutional neural network, the output of the layer is used as the input to the neural network in layer (i + 1) during forward propagation. The expression is defined as follows: $xi = Li(xi - 1)$. As the number of neural network layers increases, it leads to the problem of gradient vanishing or gradient explosion. Residual Network (ResNet) adds a jump connection to the traditional convolutional neural network and constructs the residual module by nonlinear transformation and the activation function, whose expression is defined as x$i = Li(xi - 1) + xi - 1$. One of the advantages of ResNet is that the gradient can flow directly from deep to shallow layers, thus avoiding the problem of gradient vanishing and gradient explosion that occurs when the model becomes deeper, and thus it has become the mainstream encoder backbone network.

3.2.2. High-Resolution Network (HRNet)

Typically, deep convolutional neural networks gradually generate features from high to low resolution through cascaded convolutional blocks and downsampling or pooling operations at different stages. However, high-resolution networks employ parallel connections of convolutional streams from high to low resolution, where the output at each resolution is the sum of the already generated resolution in the previous stage, mathematically defined as $R_s^r = f(R_{s-1}^1) + f(R_{s-1}^2) \cdots + f(R_{s-1}^{r-1})$, where s represents the different stages and r represents the different resolutions. Furthermore, as the resolution decreases, information from different resolution features is exchanged repeatedly. HRNet gains richer semantic representation, more precise spatial information, and a stronger feature extraction capability by exchanging information across resolutions repeatedly.

3.2.3. Variety of View Network (VoVNet)

To enhance the fusion of shallow and deep features in the neural networks, a Dense Connected Network (DenseNet) has been proposed. In DenseNet, each subnetwork layer takes the outputs of the previous layers as inputs after concatenating them in the channel dimension, i.e., the input of layer i is the output of layers 1 to $(i-1)$, whose expression is defined as follows: $x_i = L_i([x_0, x_1, \ldots x_{i-1}])$. However, too many dense connections in the intermediate layers leads to the computational inefficiency of DenseNet. VoVNet proposes the strategy of one-shot aggregation, which not only adopts the advantage of DenseNet, i.e., the advantage of representing diverse features with multiple sensory fields, but also overcomes the inefficiency of dense connections. One-shot aggregation is performed only once for all features in the last feature map of each stage, and its expression is defined as $x_s = \sum_{1}^{i-1} L_i([x_0, x_1, \ldots x_{i-1}])$.

While reducing the model complexity, VoVNet still maintains the effective use of global information, and this global context summarization capability improves the interpretation and characterization of fine and complex features.

3.3. *Backbone Ensemble*

The majority of ensemble methods necessitate a voting strategy to compare output results from various models and enhance the final prediction [49]. This research introduces a novel approach by amalgamating multiple existing backbone networks through intricate connections between adjacent backbone networks, thereby creating a more robust composite backbone network known as the aggregated backbone network (ABNet). Through this process, ABNet includes output features from different backbone networks as part of the input features, progressively fusing them for the model and ultimately conducting semantic segmentation via the decoder's feature mapping.

Model diversity implies the significance of training models with distinct architectures or techniques, which has been widely acknowledged within the model ensemble [50]. Hence, this study adopts the same-stage fusion approach to combine the backbones of three different architectures, namely, HRNet, ResNet, and VoVNet. Specifically, HRNet employs the HRNet-48 structure with 4 stages, ResNet employs the ResNet-50 structure with 4 stages, and VoVNet employs the VoVNet-39 structure with 5 stages, as illustrated in Figure 2.

Each stage consists of several convolutional layers with the same-sized feature mappings, except the Stem layer in HRNet-48 and ResNet-50, and apart from the Stem layer and the first stage in VoVNet-39. Furthermore, the downsampling for each stage is accomplished by 3×3 max-pooling with a stride of 2. The specific structures are presented in Table 2. Upon preprocessing an input image of size C through the Stem layer, HRNet-48 and ResNet-50 engage in convolutional operations to produce the first stage output, resulting in a size of $C/4$. Similarly, VoVNet-39 requires convolutional operations in the Stem layer and the first stage to attain an output size of $C/4$, aligning it with the outputs from HRNet-48 and ResNet-50. This process of merging the outputs from various stages of HRNet-48 and ResNet-50 with the subsequent stage output of VoVNet-39 yields a multi-scale aggregated result.

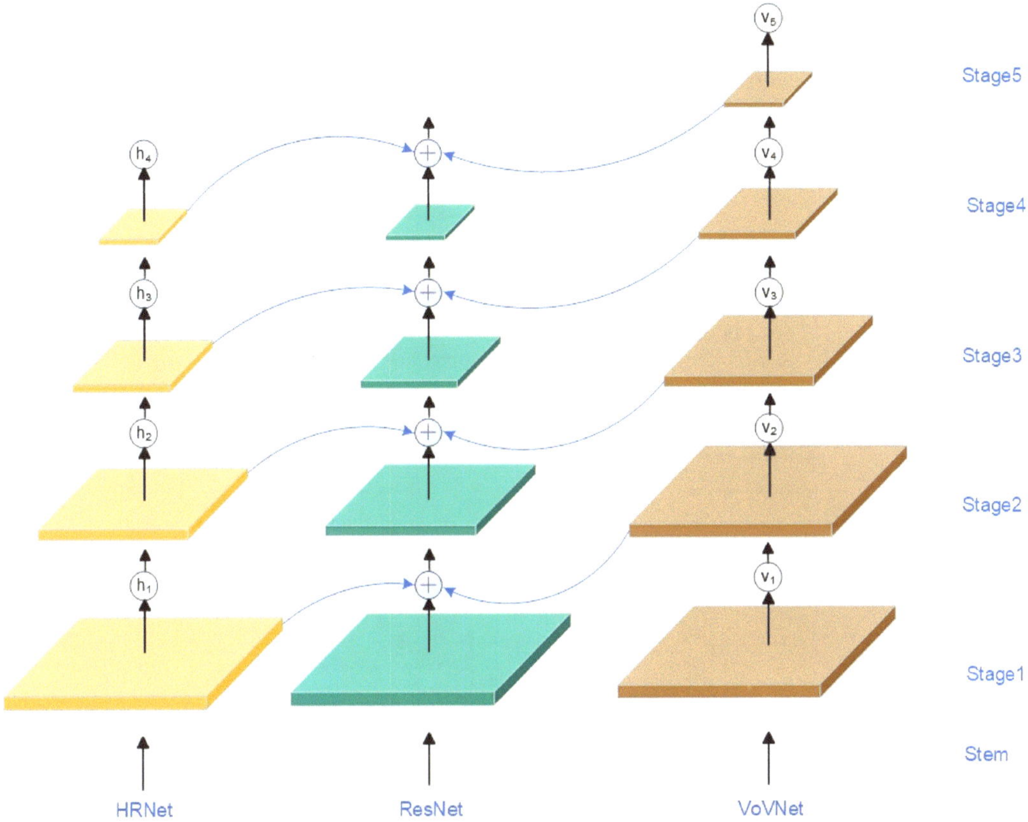

Figure 2. Same-stage fusion method for HRNet, ResNet, and VoVNet.

Table 2. Detailed module structures for each stage of HRNet-48, ResNet-50, and VoVNet-39. The downsampling for each stage is accomplished by 3 × 3 max-pooling with a stride of 2.

Stage	HRNet-48	ResNet-50	VoVNet-39
Stem	3 × 3 conv, 64, stride = 2 3 × 3 conv, 64, stride = 2	7 × 7 conv, 64, stride = 2 3 × 3 max pool, stride = 2	3 × 3 conv, 64, stride = 2
Stage1	[1 × 1conv, 48 3 × 3conv, 48 1 × 1conv, 48] ×4	[1 × 1conv, 64 3 × 3conv, 64 1 × 1conv, 256] ×3	3 × 3 conv, 64, stride = 1 3 × 3 conv, 128, stride = 1
Stage2	[3 × 3conv, 96 3 × 3conv, 96] ×4	[1 × 1conv, 128 3 × 3conv, 128 1 × 1conv, 512] ×4	[3 × 3conv, 128, × 5 concate & 1 × 1conv, 256] ×1
Stage3	[3 × 3conv, 192 3 × 3conv, 192] ×4	[1 × 1conv, 256 3 × 3conv, 256 1 × 1conv, 1024] ×6	[3 × 3conv, 160, × 5 concate & 1 × 1conv, 512] ×1
Stage4	[3 × 3conv, 384 3 × 3conv, 384] ×4	[1 × 1conv, 512 3 × 3conv, 512 1 × 1conv, 2028] ×3	[3 × 3conv, 192, × 5 concate & 1 × 1conv, 768] ×2
Stage5			[3 × 3conv, 224, × 5 concate & 1 × 1conv, 1024] ×2

3.4. *Convolutional Block Attention Module (CBAM)*

To enhance the feature representation after merging multiple backbone networks and mitigating redundant information within the merged feature layer, we propose an optimization method through the utilization of the convolutional block attention model [51] (as shown in Figure 3). This approach aims to focus on essential parts of the feature map while suppressing noise and uncertainty. The convolutional attention model consists of the following two key components: the channel attention and spatial attention models. These models generate the channel attention weights and spatial attention weights, respectively. Subsequently, the attention weights are multiplied by the input feature map to achieve adaptive feature refinement. The specific formula is as follows:

$$F_c = M_c(F) \otimes F \quad (1)$$

$$F_{out} = M_s(F_c) \otimes F_c \quad (2)$$

F represents the input feature map, F_c represents the input feature map after channel attention, M_c represents the channel attention model, M_s represents the spatial attention model, $\otimes$ represents the pixel-by-pixel multiplication, and F_{out} represents the final optimized output.

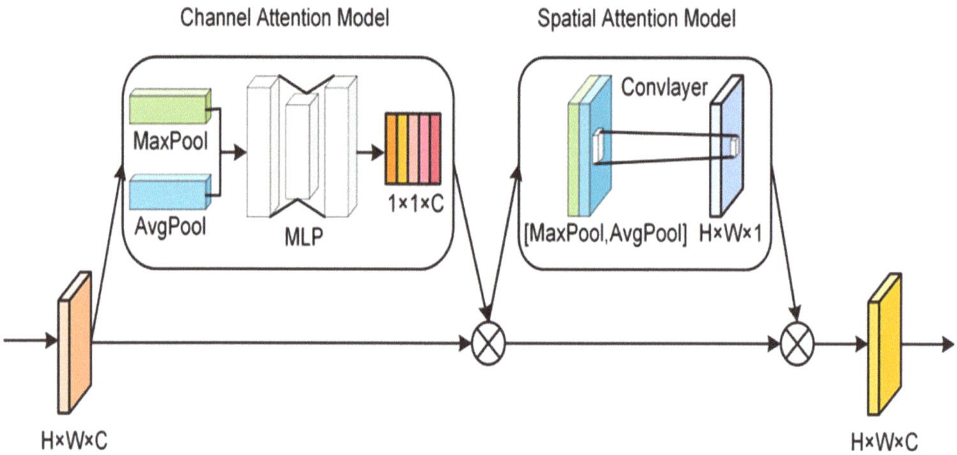

Figure 3. Diagram of the convolutional block attention module (CBAM) structure. H, W, and C represent the height, width, and number of channels of the feature maps. MLP stands for Multi-Layer Perceptron.

4. Experiments and Results

4.1. Datasets

4.1.1. LoveDA Dataset

LoveDA is a well-labeled and challenging landcover classification dataset [52]. The images were collected from Nanjing, Changzhou, and Wuhan cities, covering 18 different administrative districts. LoveDA dataset were acquired by Spaceborne sensors and collected from two main scenarios, urban and rural, containing a total of 2522 training samples and 1669 test samples. Each image contains red, green, and blue bands with a size of 1024 × 1024 pixels and a spatial resolution of 0.3 m. The labels contain the following seven categories: background, building, road, water, barren, forest, and agricultural land (Figure 4). The LoveDA dataset has become one of the most challenging landcover datasets in the field of semantic segmentation for remote sensing due to its multi-scale targets, complex background, and discontinuous category distribution.

Figure 4. Some typical sample images of the LoveDA dataset and their corresponding labels.

4.1.2. Gaofen Image Dataset (GID)

The Gaofen Image Dataset (GID) is a large-scale landcover classification dataset [53] which contains widely distributed Gaofen-2 satellite images. GID contains 150 high-quality GF-2 images acquired from more than 60 different cities in China. The Gaofen image dataset contains a large-scale classification (LSC) dataset and a fine landcover classification (FLC) dataset. To verify the effectiveness of the extracted method, the fine landcover classification set providing 15 categories was selected as the experimental dataset in this study (Figure 5), which is known as the GID15 dataset. The GID15 dataset contains a total of 10 pixel-level labeled images of 7200 × 6800 size, a spatial resolution of 0.81 m, and four spectral bands of IRRGB. This dataset was first cropped into non-overlapping 2100 patches of 512 × 512 size, and then these patches were divided into 1680 training samples and 420 validation samples.

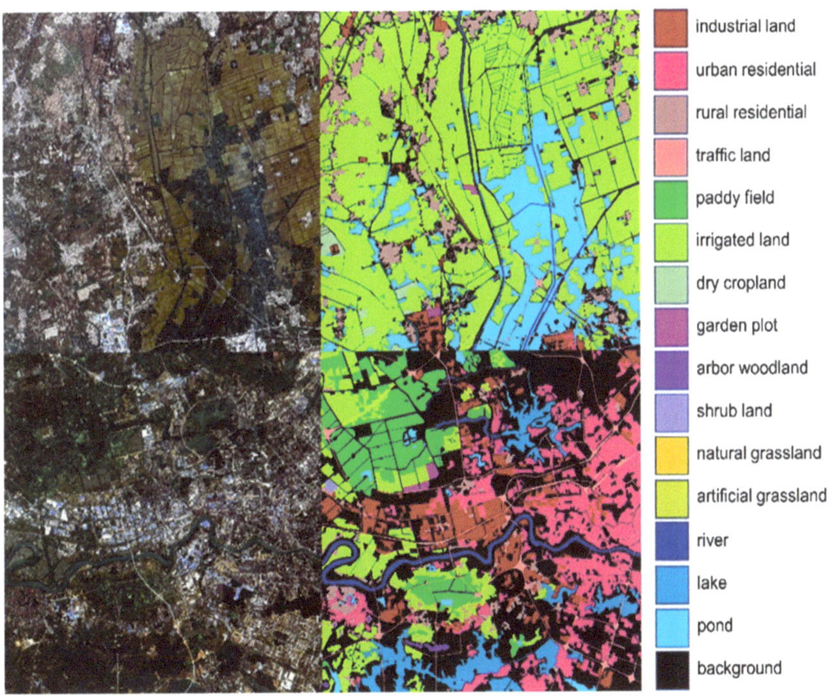

Figure 5. Some typical sample images of the GID15 dataset and their corresponding labels.

4.2. Implementation Details

All experiments were implemented by the Python 3.8 and Pytorch 1.10.0 framework and trained on a 1 NVIDIA RTX A6000 GPU with 48 G of RAM. The batch size was set to eight, the number of working threads was four, and the cross-entropy loss function was used to calculate the loss values for all networks [54]. The optimizer used an SGD with a momentum value of 0.9 and weight decay of 0.0005. The learning rate was initialized with 0.01 and scheduled by the poly strategy. In addition to expanding the dataset, all training samples performed image enhancement operations including a random horizontal flip, vertical flip, and random scaling in the range of [0.5, 0.75, 1.0, 1.25, 1.5, 1.75] (all with equal probability). For all backbone networks, we used the models pre-trained on the ImageNet dataset as the initial weight files for the network training.

4.3. Evaluation Metrics

In this study, we use the overall accuracy (OA), intersection over union (IoU), mean intersection over union (mIoU), and mean accuracy (mAcc) to measure the performance of ABNet, as shown in Equations (3)–(6). The OA denotes the overall accuracy of the classification result; the IoU denotes the similarity between the predicted results and the true values; the mIoU denotes the average IoU value across all categories; the mAcc denotes the average classification accuracy across all categories.

$$\mathrm{OA} = \frac{\mathrm{TP} + \mathrm{TN}}{\mathrm{TP} + \mathrm{TN} + \mathrm{FP} + \mathrm{FN}} \tag{3}$$

$$\mathrm{IoU} = \frac{\mathrm{TP}}{\mathrm{TP} + \mathrm{FP} + \mathrm{FN}} \tag{4}$$

$$\mathrm{mIoU} = \frac{1}{\mathrm{N}} \sum_{\mathrm{c}=1}^{\mathrm{N}} \frac{\mathrm{TP_c}}{\mathrm{TP_c} + \mathrm{FP_c} + \mathrm{FN_c}} \tag{5}$$

$$\mathrm{mAcc} = \frac{1}{\mathrm{N}} \sum_{\mathrm{c}=1}^{\mathrm{N}} \frac{\mathrm{TP_c}}{\mathrm{TP_c} + \mathrm{FN_c}} \tag{6}$$

where TP and TN are the positive and negative examples of correct predictions; FP and FN are the positive and negative examples of incorrect predictions.

Furthermore, the following two criteria have been adopted to evaluate the model complexity: (1) floating-point operations (FLOPs), which is the number of floating-point operations accounting for the computational complexity; (2) parameters, which is the number of trainable parameters representing the model size.

4.4. Ablation Studies

To assess the efficacy of both single backbone networks and aggregated backbone networks, we employed the Deeplabv3+ decoding model as a unified segmentation head. Experiments were individually performed utilizing ResNet-50, HRNet-48, VoVNet-39, and the aggregated backbone as the encoder's backbone network. Ablation experiments were executed on both the LoveDA dataset and the GID15 dataset, with the ensuing experimental results expounded upon in Sections 4.4.1 and 4.4.2, respectively.

4.4.1. LoveDA Dataset

Illustrated in Figure 6, the classification accuracy of the aggregated backbone network demonstrates notable improvements in contrast to ResNet-50, HRNet-48, and VoVNet-39. The aggregated backbone network surpasses ResNet-50, HRNet-48, and VoVNet-39 by over 3% in terms of the overall accuracy (OA), mean accuracy (mAcc), and mean intersection over union (mIoU). Consequently, the amalgamation of diverse backbone networks proves to be an effective strategy for enhancing the performance of semantic segmentation models in the context of encoder–decoders.

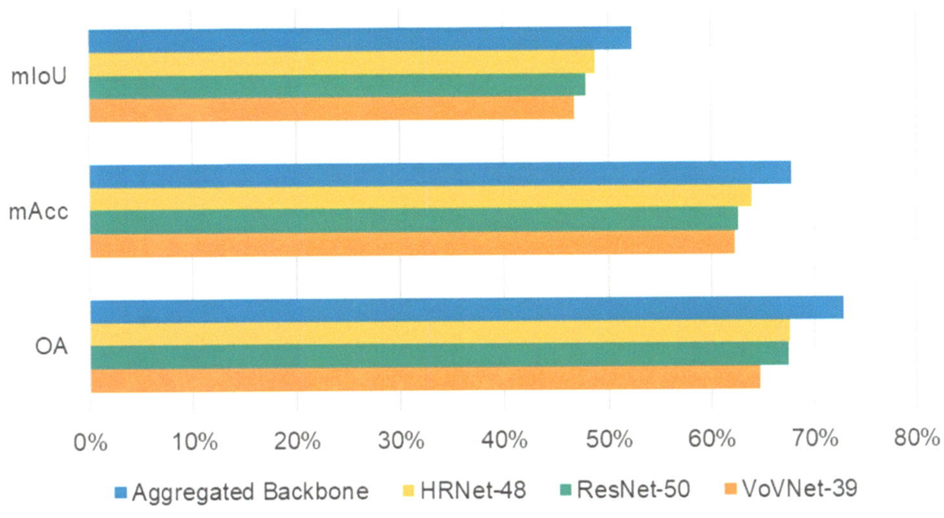

Figure 6. Performance comparison of ABNet with the unaggregated backbones.

As depicted in Table 3, the aggregated backbone network exhibits superior performance over ResNet-50, HRNet-48, and VoVNet-39 across all categories of the LoveDA dataset. Remarkably, ABNet notably enhances the segmentation performance of water bodies and bare land by a minimum of 7.14% and 4.63% in comparison to the other three backbone networks. Moreover, the aggregated backbone network demonstrates varied degrees of improvement across categories, including buildings, roads, forests, agricultural land, and background.

Table 3. IoU (%) for ablation experiments on the LoveDA dataset.

Backbone	Background	Building	Road	Water	Barren	Forest	Agricultural
ResNet-50	52.19	56.03	51.51	62.85	22.4	39.83	50.56
HRNet-48	52.87	59.63	53.01	60.85	29.23	38.58	47.66
VoVNet-39	48.98	62.51	53.51	61.32	18.06	37.48	45.65
ABNet	53.85	63.69	54.06	69.99	33.86	40.23	51.33

To more effectively demonstrate the advantages of the aggregated backbone network in comparison to other backbone networks, we present the visualization results of different backbone networks in Figure 7. In Figure 7a, the aggregated backbone network excels in recognizing complex shapes of bare ground. Moreover, in Figure 7b, the aggregated backbone network demonstrates superior recognition of small-scale water bodies. Figure 7c illustrates the aggregation backbone network's ability to minimize misclassifications of forests, buildings, and roads. While in Figure 7d, the network effectively reduces misclassifications of agricultural land, water bodies, and forests. Combining the observations from Table 3 and Figure 7, it is evident that the aggregated backbone network not only excels in recognizing intricate features that pose challenges to individual backbone networks but also enhances the recognition capabilities across multi-scale targets.

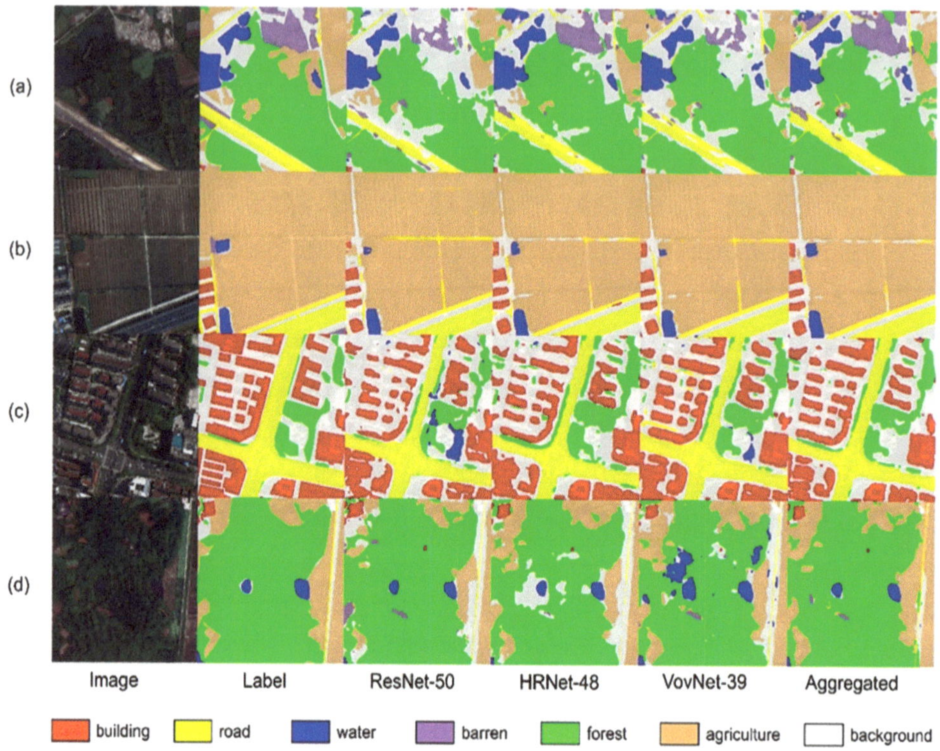

Figure 7. Prediction results of ablation experiments on the LoveDA dataset. (**a–d**) shows the typical case that it is difficult to identify barren, water, building, and agriculture.

4.4.2. GID15 Dataset

To further verify the generalization performance of the aggregated backbone network, we conducted ablation experiments using the GID15 dataset, which encompasses 15 categories. In this study, Table 4 and Figure 8 present the IoU and visualization results from the classification of the GID15 dataset, leveraging ResNet-50, HRNet-48, VoVNet-39, and the aggregated backbone as the backbone network, respectively.

As observed in Table 4, the segmentation accuracy of the aggregated backbone network on the GID15 dataset significantly surpasses that of ResNet-50, HRNet-48, and VoVNet-39, with the mean intersection over union improved by 4.59%, 3.45%, and 3.24%, respectively. Except for rural residential land and artificial grassland, the IoU of segmentation by ABNet exhibits differing degrees of enhancement compared to the other three backbone networks. Specifically, ABNet proves highly effective in classifying shrub land, demonstrating a minimum increase of 6.15% in the IoU. Furthermore, ABNet also enhances the classification accuracy of garden plots, natural grassland, ponds, paddy fields, and dry cropland, with minimum increases of 4.68%, 2.95%, 2.85%, 2.42%, and 2.05%, respectively. Additionally, ABNet improves the segmentation accuracy of the background, industrial land, urban residential land, traffic land, irrigated land, arbor woodland, river, and lake classes to a certain extent.

From Figure 8a–c,e,g, it is evident that our results better preserve the geometric features of the objects, aligning more closely with the manually labeled results and exhibiting more regular shapes. Figure 8d demonstrates that our proposed method reduces misclassifications of small-scale targets in ponds. Furthermore, Figure 8f visually illustrates a substantial improvement in the segmentation accuracy of shrub land, providing compelling evidence for the superiority of ABNet.

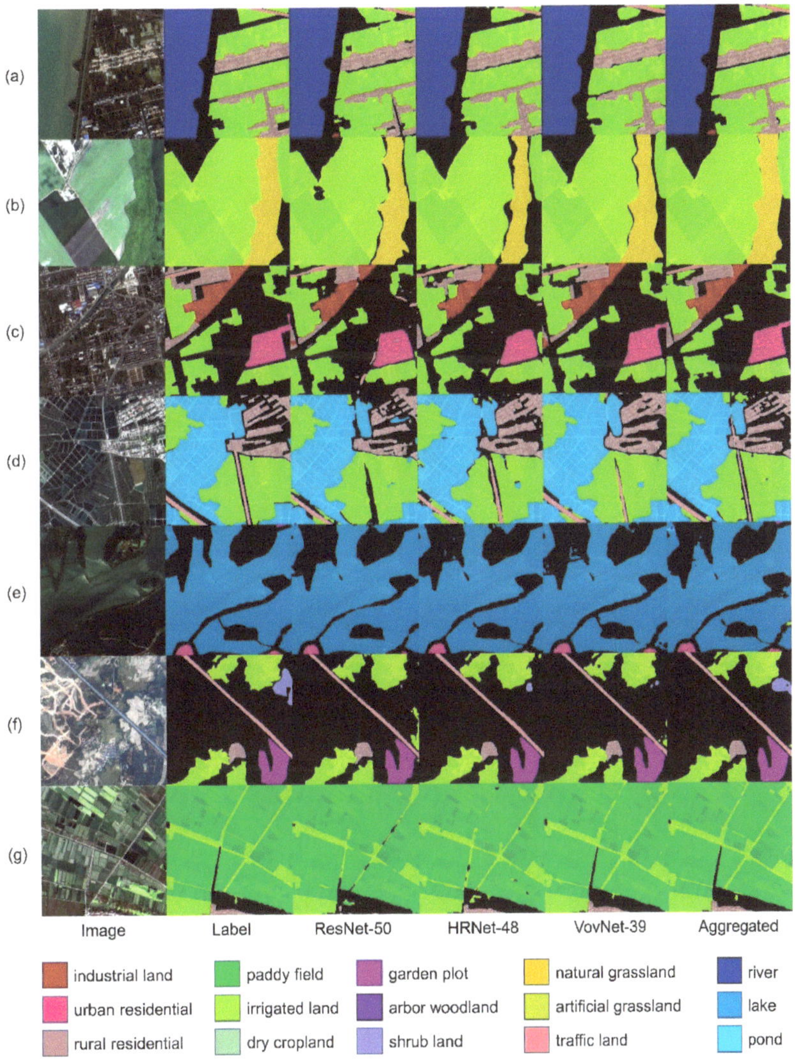

Figure 8. Prediction results of ablation experiments on GID15 dataset. (**a–g**) shows the typical case that it is difficult to identify rural residential, natural grassland, industrial land, pond, lake, shrub land, and paddy field.

Table 4. IoU (%) for ablation experiments on the GID15 dataset.

Class	ResNet-50	HRNet-48	VoVNet-39	ABNet
Background	70.01	69.61	70.47	73.02
Industrial land	59.51	60.03	61.93	63.43
Urban residential	64.34	63.58	66.57	67.38
Rural residential	56.03	54.18	58.09	57.81
Traffic land	59.93	63.86	63.9	64.05
Paddy field	63.07	65.8	64.52	68.22
Irrigated land	71.37	71.29	71.24	73.27
Dry cropland	59.01	56.68	58.76	61.06
Garden plot	31.38	39.54	24.02	44.22
Arbor woodland	78.35	78.64	75.52	79.39
Shrub land	13.67	23.15	21.67	29.3
Natural grassland	61.94	62.54	64.06	67.01
Artificial grassland	24.37	24.53	32.62	30.03
River	79.85	78.81	80.48	82.7
Lake	70.99	67.99	69.79	71.69
Pond	62.29	64.01	63.99	66.86
mIoU	57.88	59.02	59.23	62.47

4.5. Comparing with the State-of-the-Art

To showcase the superior performance of our method, a series of experiments were conducted to compare it with other state-of-the-art semantic segmentation models. This study selected mainstream single backbone network models and a composite backbone network model for experiments with equivalent training configurations on the LoveDA and GID15 datasets. The single backbone network models included UNet, FPN, PSPNet, DANet, CCNet, and UPerNet utilizing the widely used ResNet-50 as the backbone network for experimentation. CBNet, as a representative of composite backbone network models, involved the composite connection of ResNet and Res2Net as the backbone network for experimentation. UNet is a classic encoder–decoder semantic segmentation model; FPN, PSPNet, and UPerNet represent feature pyramid methods; DANet and CCNet represent attention mechanism methods; and CBNet represents an approach to convergence of multiple backbone networks.

4.5.1. LoveDA Dataset

Table 5 illustrates the accuracy results obtained from the different models after training on the LoveDA dataset. Whether considering the OA, mIoU, or mAcc, the segmentation accuracy of ABNet outperforms the other models. In comparison to UNet, the mIoU demonstrates a 7.33% improvement, and when compared with CCNet, there is a 1.51% increase in the mIoU. The mIoU of FPN, PSPNet, DANet, and UPerNet all surpass that of UNet, but are lower than that of CCNet.

Table 5. Accuracy results of different models on the LoveDA dataset (%).

Model	Backbone	OA	mIoU	mAcc
UNet	ResNet-50	63.50	45.10	61.76
FPN	ResNet-50	67.88	49.76	62.99
PSPNet	ResNet-50	68.82	50.53	62.08
DANet	ResNet-50	68.19	48.73	60.13
CBNet	CB-ResNet50	68.26	50.28	62.39
UPerNet	ResNet-50	68.72	50.36	62.03
CCNet	ResNet-50	68.87	50.92	63.72
ABNet	Aggregated Backbone	72.75	52.43	67.82

From Figure 9a, it is evident that ABNet significantly extracts the barren feature compared to the other models, showcasing that features that are challenging to distinguish by a single backbone network can be more effectively identified by the composite backbone network. In Figure 9b, it is noticeable that ABNet is less adept than CCNet and UPerNet in extracting small-sized water bodies. However, as depicted in Figure 9c,d, ABNet delineates agricultural land more effectively than CCNet and UPerNet. This shows that the proposed model can effectively improve the task of semantic segmentation of remote sensing images by fusing the structure of different backbone networks.

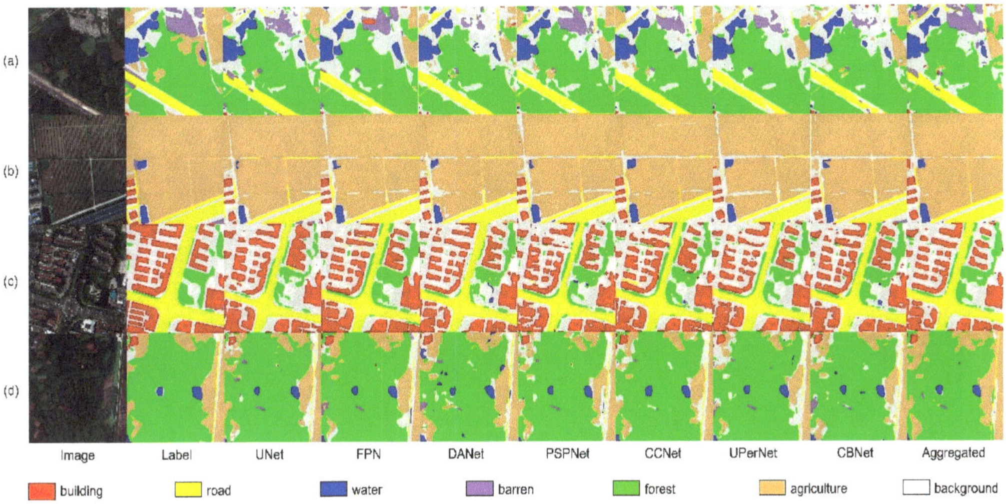

Figure 9. Comparison of prediction results of different state-of-the-art models on the LoveDA dataset. (**a**–**d**) shows the typical case that it is difficult to identify barren, water, building, and agriculture.

4.5.2. GID15 Dataset

In this study, we conducted experiments on the GID15 dataset, which contains finer categories, to evaluate the efficacy of ABNet. Table 6 demonstrates the segmentation accuracy obtained after training various models on the GID15 dataset. Across metrics such as the overall accuracy (OA), mean intersection over union (mIoU), and mean accuracy (mAcc), ABNet consistently outperforms the other models. Notably, in comparison to UNet, the mIoU is improved by 7.58%, while, compared with CBNet, there is a 1.20% enhancement in the mIoU. Additionally, the mIoU values of FPN, PSPNet, DANet, CCNet, and UPerNet surpass that of UNet, although they are lower than that of CBNet.

Table 6. Accuracy results of different models on the GID15 dataset (%).

Model	Backbone	OA	mIoU	mAcc
UNet	ResNet-50	81.33	54.89	65.39
FPN	ResNet-50	81.74	59.66	67.77
PSPNet	ResNet-50	82.05	60.78	68.48
DANet	ResNet-50	82.19	60.84	68.83
CCNet	ResNet-50	82.29	60.46	69.09
UPerNet	ResNet-50	82.27	61.09	69.54
CBNet	CB-ResNet50	82.83	61.27	68.92
ABNet	Aggregated Backbone	83.27	62.47	72.74

Moreover, in Figure 10, we provide a comprehensive comparison result. Specifically, Figure 10a emphasizes the superior capability of ABNet in correctly recognizing small-sized features, a task in which the other models encounter challenges. Furthermore, Figure 10b demonstrates ABNet's precise classification of water body edges, surpassing the performance of the other models. Similarly, in Figure 10c, the result indicates that ABNet excels in extracting shrub land, implying its effectiveness in classifying intricate feature targets. Finally, Figure 10d illustrates the consistency of ABNet in extracting the contour boundaries of rice fields, showcasing its ability to achieve superior segmentation performance, particularly for finer feature classes.

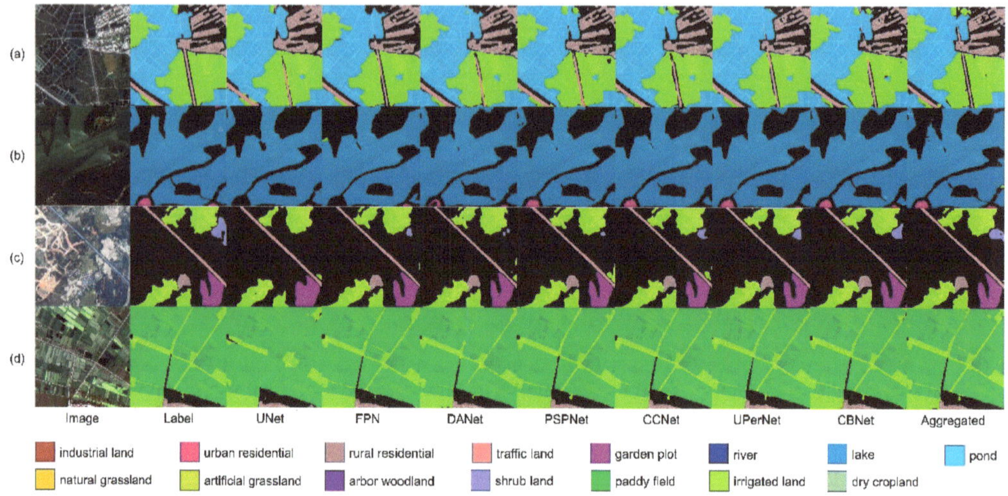

Figure 10. Comparison of prediction results of different state-of-the-art models on the GID15 dataset. (**a**) shows the typical case that it is difficult to identify pond in small-scale feature. (**b**–**d**) shows the typical case that it is difficult to identify lake, shrub land, and paddy field.

5. Discussion

5.1. Advantages and Limitations

5.1.1. Advantages

ResNet utilizes residual connections in convolutional blocks to ensure the stability of neural network convergence at each stage. However, it lacks feature fusion within and between stages of the backbone network. In contrast, VoVNet achieves feature integration within each stage through the one-shot aggregation of each convolutional block in the backbone network. Nonetheless, it does not fuse features between different stages. Meanwhile, HRNet achieves the feature fusion of outputs from different stages through parallel concatenation. However, as the number of stages increases, the output features of each stage become more complex after multi-scale feature fusion, leading to some loss of original information within each stage. The aggregation backbone network (ABNet) uniquely combines the deep feature characterization capabilities of ResNet with the structural features of HRNet's string–parallel combination and VoVNet's global one-shot aggregation. This integration significantly expands the receptive field of the convolutional blocks at each stage, thereby enhancing the model's capacity to leverage spatial context information [55]. As a result, ABNet consistently enhances the classification accuracy across all classes within the LovaDA and GID15 datasets. Furthermore, owing to the multi-stage output structure of ABNet, it efficiently generates merged features of varying scale sizes, which efficiently utilizes the semantic information from both deep and shallow features and improves the classification accuracy of the model for small-scale targets [56]. Additionally, to mitigate information redundancy within the merged features, the convolutional block attention model (CBAM) optimizes the multi-scale merged features across both spatial and channel dimensions. This optimization enhances the distinctive characterizing abilities of the features at each stage [57]. Consequently, when applied to landcover semantic segmentation tasks, this approach yields superior segmentation results and bolsters the model's capability to discern features that pose challenges to a single backbone network's recognition ability.

5.1.2. Limitations

Despite the proven efficacy of ABNet on landcover datasets, its substantial model parameters and higher computational time complexities present notable considerations. To comprehensively assess the practicality of these models, this study quantified the parameters and floating-point operations (FLOPs) for ABNet and comparable models using a sample input image size of 1024 $\times$ 1024 [58]. As illustrated in Table 7, the parameters and FLOPs for ABNet notably surpass those of other models, with ABNet exceeding the second-largest CBNet model by 100.55 M and 0.33 T FLOPs. The abundance of model parameters and computational cost contributes to heightened challenges in model convergence, especially when computational resources are constrained. Consequently, the suitability of ABNet for complex scene segmentation tasks may diminish under limited computational resources.

Table 7. Comparison with other networks on parameters and FLOPs.

Method	Parameters (M)	FLOPs (T)
Deeplabv3plus_Res-50	41.22	0.71
Deeplabv3plus_HRNet-48	68.59	1.01
Deeplabv3plus_VoVNet-39	34.27	0.88
DANet	47.93	0.96
PSPNet	46.61	0.72
CCNet	47.46	0.84
UPerNet	64.04	0.95
CBNet	69.71	1.08
ABNet	170.26	1.41

5.2. *Potential Improvements*

To further optimize ABNet, this study explores three potential directions for enhancing its performance. Firstly, without compromising the model performance, reducing the number of model parameters can be achieved by decreasing the number of layers in the backbone network or minimizing the involvement of the backbone network models in fusion. Secondly, integrating advanced decoders, such as the feature pyramid network (FPN), can facilitate a more effective fusion of deep and shallow features, thereby enhancing classification performance. Lastly, enhancing the fusion of different backbone networks presents an opportunity for improvement. In addition to merging the output layers of the same stage, combining the outputs of adjacent stages of diverse backbone networks through upsampling or downsampling operations can promote the deep fusion of deep and shallow features, fostering further performance enhancements.

6. Conclusions

This paper introduced the aggregated backbone network (ABNet), a novel convolutional neural network designed for fine-grained landcover classification. ABNet leverages the Deeplabv3+ head and reconstructs the backbone network through the same-stage fusion of ResNet-50, HRNet-48, and VoVNet-39. The aggregated backbone network yields merged features in multi-scale sizes, which are further optimized using the convolutional block attention mechanism.

On the LoveDA and GID15 datasets, the aggregated backbone network demonstrates superior classification accuracy compared to a single backbone network, outperforming the overall accuracy (OA), mean intersection over union (mIoU), and mean accuracy (mAcc). Specifically, the mIoU derived from the aggregated backbone network surpasses that of the single backbone network by at least 3%. Notably, ABNet enhances the classification accuracy of diverse land classes and improves the discernment of challenging classes often difficult for a single backbone network to recognize. Moreover, when compared with popular semantic segmentation models (UNet, FPN, PSPNet, DANet, CBNet, CCNet, and UPerNet), the aggregated backbone network exhibits leading performance, affirming the efficacy of the proposed method. Although the ABNet proposed in this paper exhibits excellent segmentation performance on the LovaDA and GID15 datasets, the model has no advantages in terms of the parameters and FLOPs of the algorithm.

Consequently, this approach presents an effective strategy to enhance the classification performance of remote sensing deep learning models. Researchers can further explore and customize the aggregated structure by integrating different types and quantities of backbone networks to tackle more intricate landcover classification challenges. Meanwhile, scholars can investigate how to reduce the parameters of the integrated model while maintaining efficient classification performance.

Author Contributions: Conceptualization, B.S.; methodology, B.S.; software, B.S.; validation, B.S., formal analysis, B.S.; investigation, B.S.; writing—original draft preparation, B.S.; writing—review and editing, B.S. and Z.Y.; visualization, B.S. and Z.W.; supervision, K.W. All authors have read and agreed to the published version of the manuscript.

Funding: This research received no external funding.

Data Availability Statement: We analyzed publicly available datasets in this study. Data are freely available here: (1) LoveDA dataset: https://github.com/Junjue-Wang/LoveDA (accessed on 14 March 2023) (2) GID Dataset: http://www.captain-whu.com/repository.html (accessed on 14 March 2023).

Acknowledgments: Thanks to Ziran Ye for suggesting revisions to this paper.

Conflicts of Interest: The authors declare no conflicts of interest.

References

1. Wang, Y.Z.; Sun, Y.H.; Cao, X.Y.; Wang, Y.H.; Zhang, W.K.; Cheng, X.L. A review of regional and Global scale Land Use/Land Cover (LULC) mapping products generated from satellite remote sensing. *ISPRS J. Photogramm. Remote Sens.* **2023**, *206*, 311–334. [CrossRef]
2. Su, Y.; Qian, K.; Lin, L.; Wang, K.; Guan, T.; Gan, M.Y. Identifying the driving forces of non-grain production expansion in rural China and its implications for policies on cultivated land protection. *Land Use Policy* **2020**, *92*, 104435. [CrossRef]
3. Boguszewski, A.; Batorski, D.; Ziemba-Jankowska, N.; Dziedzic, T.; Zambrzycka, A. LandCover.ai: Dataset for Automatic Mapping of Buildings, Woodlands, Water and Roads from Aerial Imagery. In Proceedings of the IEEE/CVF Conference on Computer Vision and Pattern Recognition (CVPR), Nashville, TN, USA, 19–25 June 2021; pp. 1102–1110.
4. Tong, X.Y.; Xia, G.S.; Zhu, X.X. Enabling country-scale land cover mapping with meter-resolution satellite imagery. *ISPRS J. Photogramm. Remote Sens.* **2023**, *196*, 178–196. [CrossRef] [PubMed]
5. Sertel, E.; Ekim, B.; Osgouei, P.E.; Kabadayi, M.E. Land Use and Land Cover Mapping Using Deep Learning Based Segmentation Approaches and VHR Worldview-3 Images. *Remote Sens.* **2022**, *14*, 4558. [CrossRef]
6. Ma, L.; Liu, Y.; Zhang, X.L.; Ye, Y.X.; Yin, G.F.; Johnson, B.A. Deep learning in remote sensing applications: A meta-analysis and review. *ISPRS J. Photogramm. Remote Sens.* **2019**, *152*, 166–177. [CrossRef]
7. Kampffmeyer, M.; Salberg, A.B.; Jenssen, R. Semantic Segmentation of Small Objects and Modeling of Uncertainty in Urban Remote Sensing Images Using Deep Convolutional Neural Networks. In Proceedings of the 29th IEEE Conference on Computer Vision and Pattern Recognition (CVPR), Las Vegas, NV, USA, 26 June–1 July 2016; pp. 680–688.
8. Ye, Z.R.; Fu, Y.Y.; Gan, M.Y.; Deng, J.S.; Comber, A.; Wang, K. Building Extraction from Very High Resolution Aerial Imagery Using Joint Attention Deep Neural Network. *Remote Sens.* **2019**, *11*, 2970. [CrossRef]
9. Fu, Y.Y.; Liu, K.K.; Shen, Z.Q.; Deng, J.S.; Gan, M.Y.; Liu, X.G.; Lu, D.M.; Wang, K. Mapping Impervious Surfaces in Town-Rural Transition Belts Using China's GF-2 Imagery and Object-Based Deep CNNs. *Remote Sens.* **2019**, *11*, 280. [CrossRef]
10. Zhang, D.J.; Pan, Y.Z.; Zhang, J.S.; Hu, T.G.; Zhao, J.H.; Li, N.; Chen, Q. A generalized approach based on convolutional neural networks for large area cropland mapping at very high resolution. *Remote Sens. Environ.* **2020**, *247*, 111912. [CrossRef]
11. Kumar, D.G.; Chaudhari, S. Comparison of Deep Learning Backbone Frameworks for Remote Sensing Image Classification. In Proceedings of the IEEE International Geoscience and Remote Sensing Symposium (IGARSS), Kuala Lumpur, Malaysia, 17–22 July 2022; pp. 7763–7766.
12. Wieland, M.; Martinis, S.; Kiefl, R.; Gstaiger, V. Semantic segmentation of water bodies in very high-resolution satellite and aerial images. *Remote Sens. Environ.* **2023**, *287*, 113452. [CrossRef]

13. Liu, Y.D.; Wang, Y.T.; Wang, S.W.; Liang, T.T.; Zhao, Q.J.; Tang, Z.; Ling, H.B. CBNet: A Novel Composite Backbone Network Architecture for Object Detection. In Proceedings of the 34th AAAI Conference on Artificial Intelligence/32nd Innovative Applications of Artificial Intelligence Conference/10th AAAI Symposium on Educational Advances in Artificial Intelligence, New York, NY, USA, 7–12 February 2020; pp. 11653–11660.
14. Liang, T.T.; Chu, X.J.; Liu, Y.D.; Wang, Y.T.; Tang, Z.; Chu, W.; Chen, J.D.; Ling, H.B. CBNet: A Composite Backbone Network Architecture for Object Detection. *IEEE Trans. Image Process.* **2022**, *31*, 6893–6906. [CrossRef]
15. Elharrouss, O.; Akbari, Y.; Almaadeed, N.; Al-Maadeed, S. Backbones-Review: Feature Extraction Networks for Deep Learning and Deep Reinforcement Learning Approaches. *arXiv* **2022**, arXiv:2206.08016.
16. Simonyan, K.; Zisserman, A. Very Deep Convolutional Networks for Large-Scale Image Recognition. *arXiv* **2014**, arXiv:1409.1556.
17. He, K.; Zhang, X.; Ren, S.; Sun, J. Deep Residual Learning for Image Recognition. In Proceedings of the 2016 IEEE Conference on Computer Vision and Pattern Recognition (CVPR), Las Vegas, NV, USA, 27–30 June 2016; pp. 770–778.
18. Szegedy, C.; Vanhoucke, V.; Ioffe, S.; Shlens, J.; Wojna, Z. Rethinking the Inception Architecture for Computer Vision. In Proceedings of the 2016 IEEE Conference on Computer Vision and Pattern Recognition (CVPR), Seattle, WA, USA, 27–30 June 2016; pp. 2818–2826.
19. Huang, G.; Liu, Z.; van der Maaten, L.; Weinberger, K.Q. Densely Connected Convolutional Networks. In Proceedings of the 2017 IEEE Conference on Computer Vision and Pattern Recognition, Honolulu, HI, USA, 21–26 July 2017; pp. 4700–4708.
20. Lee, Y.W.; Hwang, J.W.; Lee, S.; Bae, Y.; Park, J. An Energy and GPU-Computation Efficient Backbone Network for Real-Time Object Detection. In Proceedings of the 32nd IEEE/CVF Conference on Computer Vision and Pattern Recognition (CVPR), Long Beach, CA, USA, 16–20 June 2019; pp. 752–760.
21. Sun, K.; Xiao, B.; Liu, D.; Wang, J.D. Deep High-Resolution Representation Learning for Human Pose Estimation. In Proceedings of the 32nd IEEE/CVF Conference on Computer Vision and Pattern Recognition (CVPR), Long Beach, CA, USA, 16–20 June 2019; pp. 5686–5696.
22. Tan, M.X.; Le, Q.V. EfficientNet: Rethinking Model Scaling for Convolutional Neural Networks. In Proceedings of the 36th International Conference on Machine Learning (ICML), Long Beach, CA, USA, 9–15 June 2019.
23. Gao, S.H.; Cheng, M.M.; Zhao, K.; Zhang, X.Y.; Yang, M.H.; Torr, P. Res2Net: A New Multi-Scale Backbone Architecture. *IEEE Trans. Pattern Anal. Mach. Intell.* **2021**, *43*, 652–662. [CrossRef] [PubMed]
24. Ye, M.; Ruiwen, N.; Chang, Z.; He, G.; Tianli, H.; Shijun, L.; Yu, S.; Tong, Z.; Ying, G. A Lightweight Model of VGG-16 for Remote Sensing Image Classification. *IEEE J. Sel. Top. Appl. Earth Obs. Remote Sens.* **2021**, *14*, 6916–6922. [CrossRef]
25. Tao, C.X.; Meng, Y.Z.; Li, J.J.; Yang, B.B.; Hu, F.M.; Li, Y.X.; Cui, C.L.; Zhang, W. MSNet: Multispectral semantic segmentation network for remote sensing images. *GIScience Remote Sens.* **2022**, *59*, 1177–1198. [CrossRef]
26. Liu, K.; Yu, S.T.; Liu, S.D. An Improved InceptionV3 Network for Obscured Ship Classification in Remote Sensing Images. *IEEE J. Sel. Top. Appl. Earth Obs. Remote Sens.* **2020**, *13*, 4738–4747. [CrossRef]
27. Xu, Y.Y.; Xie, Z.; Feng, Y.X.; Chen, Z.L. Road Extraction from High-Resolution Remote Sensing Imagery Using Deep Learning. *Remote Sens.* **2018**, *10*, 1461. [CrossRef]
28. Zhao, L.R.; Niu, R.Q.; Li, B.Q.; Chen, T.; Wang, Y.Y. Application of Improved Instance Segmentation Algorithm Based on VoVNet-v2 in Open-Pit Mines Remote Sensing Pre-Survey. *Remote Sens.* **2022**, *14*, 2626. [CrossRef]
29. Guo, S.C.; Yang, Q.; Xiang, S.M.; Wang, P.F.; Wang, X.Z. Dynamic High-Resolution Network for Semantic Segmentation in Remote-Sensing Images. *Remote Sens.* **2023**, *15*, 2293. [CrossRef]
30. Hu, S.; Liu, J.; Kang, Z.W. DeepLabV3+/Efficientnet Hybrid Network-Based Scene Area Judgment for the Mars Unmanned Vehicle System. *Sensors* **2021**, *21*, 8136. [CrossRef]
31. Das, A.; Chandran, S. Transfer Learning with Res2Net for Remote Sensing Scene Classification. In Proceedings of the 11th International Conference on Cloud Computing, Data Science and Engineering (Confluence), Uttar Pradesh, India, 28–29 January 2021; pp. 796–801.
32. Long, J.; Shelhamer, E.; Darrell, T. Fully Convolutional Networks for Semantic Segmentation. In Proceedings of the IEEE Conference on Computer Vision and Pattern Recognition (CVPR), Boston, MA, USA, 7–12 June 2015; pp. 3431–3440.
33. Ronneberger, O.; Fischer, P.; Brox, T. U-Net: Convolutional Networks for Biomedical Image Segmentation. In Proceedings of the 18th International Conference on Medical Image Computing and Computer-Assisted Intervention (MICCAI), Munich, Germany, 5–9 October 2015; pp. 234–241.
34. Chen, L.C.E.; Zhu, Y.K.; Papandreou, G.; Schroff, F.; Adam, H. Encoder-Decoder with Atrous Separable Convolution for Semantic Image Segmentation. In Proceedings of the 15th European Conference on Computer Vision (ECCV), Munich, Germany, 8–14 September 2018; pp. 833–851.
35. Lin, T.Y.; Dollár, P.; Girshick, R.; He, K.M.; Hariharan, B.; Belongie, S. Feature Pyramid Networks for Object Detection. In Proceedings of the 30th IEEE/CVF Conference on Computer Vision and Pattern Recognition (CVPR), Honolulu, HI, USA, 21–26 July 2017; pp. 936–944.
36. Zhao, H.S.; Shi, J.P.; Qi, X.J.; Wang, X.G.; Jia, J.Y. Pyramid Scene Parsing Network. In Proceedings of the 30th IEEE/CVF Conference on Computer Vision and Pattern Recognition (CVPR), Honolulu, HI, USA, 21–26 July 2017; pp. 6230–6239.
37. Fu, J.; Liu, J.; Tian, H.J.; Li, Y.; Bao, Y.J.; Fang, Z.W.; Lu, H.Q. Dual Attention Network for Scene Segmentation. In Proceedings of the 32nd IEEE/CVF Conference on Computer Vision and Pattern Recognition (CVPR), Long Beach, CA, USA, 16–20 June 2019; pp. 3141–3149.

38. Xiao, T.T.; Liu, Y.C.; Zhou, B.L.; Jiang, Y.N.; Sun, J. Unified Perceptual Parsing for Scene Understanding. In Proceedings of the 15th European Conference on Computer Vision (ECCV), Munich, Germany, 8–14 September 2018; pp. 432–448.
39. Huang, Z.L.; Wang, X.G.; Wei, Y.C.; Huang, L.C.; Shi, H.; Liu, W.Y.; Huang, T.S. CCNet: Criss-Cross Attention for Semantic Segmentation. *IEEE Trans. Pattern Anal. Mach. Intell.* **2023**, *45*, 6896–6908. [CrossRef]
40. Liang, C.B.; Xiao, B.H.; Cheng, B.; Dong, Y.Y. XANet: An Efficient Remote Sensing Image Segmentation Model Using Element-Wise Attention Enhancement and Multi-Scale Attention Fusion. *Remote Sens.* **2023**, *15*, 236. [CrossRef]
41. Wang, D.; Yang, R.H.; Liu, H.H.; He, H.Q.; Tan, J.X.; Li, S.D.; Qiao, Y.C.; Tang, K.Q.; Wang, X. HFENet: Hierarchical Feature Extraction Network for Accurate Landcover Classification. *Remote Sens.* **2022**, *14*, 4244. [CrossRef]
42. Chen, C.; Zhao, H.L.; Cui, W.; He, X. Dual Crisscross Attention Module for Road Extraction from Remote Sensing Images. *Sensors* **2021**, *21*, 6873. [CrossRef] [PubMed]
43. Ye, Z.R.; Si, B.; Lin, Y.; Zheng, Q.M.; Zhou, R.; Huang, L.; Wang, K. Mapping and Discriminating Rural Settlements Using Gaofen-2 Images and a Fully Convolutional Network. *Sensors* **2020**, *20*, 6062. [CrossRef]
44. Kotaridis, I.; Lazaridou, M. Cnns in land cover mapping with remote sensing imagery: A review and meta-analysis. *Int. J. Remote Sens.* **2023**, *44*, 5896–5935. [CrossRef]
45. Bigdeli, B.; Pahlavani, P.; Amirkolaee, H.A. An ensemble deep learning method as data fusion system for remote sensing multisensor classification. *Appl. Soft Comput.* **2021**, *110*, 107563. [CrossRef]
46. Fan, R.Y.; Feng, R.Y.; Wang, L.Z.; Yan, J.N.; Zhang, X.H. Semi-MCNN: A Semisupervised Multi-CNN Ensemble Learning Method for Urban Land Cover Classification Using Submeter HRRS Images. *IEEE J. Sel. Top. Appl. Earth Obs. Remote Sens.* **2020**, *13*, 4973–4987. [CrossRef]
47. Cao, Y.; Huo, C.L.; Xu, N.; Zhang, X.; Xiang, S.M.; Pan, C.H. HENet: Head-Level Ensemble Network for Very High Resolution Remote Sensing Images Semantic Segmentation. *IEEE Geosci. Remote Sens. Lett.* **2022**, *19*, 6506005. [CrossRef]
48. Ekim, B.; Sertel, E. Deep neural network ensembles for remote sensing land cover and land use classification. *Int. J. Digit. Earth* **2021**, *14*, 1868–1881. [CrossRef]
49. Mao, M.; Zhang, B.; Doermann, D.; Guo, J.; Han, S.; Feng, Y.; Wang, X.; Ding, E. Probabilistic Ranking-Aware Ensembles for Enhanced Object Detections. *arXiv* **2021**, arXiv:2105.03139.
50. Chen, M.H.; Fu, J.L.; Ling, H.B. One-Shot Neural Ensemble Architecture Search by Diversity-Guided Search Space Shrinking. In Proceedings of the IEEE/CVF Conference on Computer Vision and Pattern Recognition (CVPR), Nashville, TN, USA, 19–25 June 2021; pp. 16525–16534.
51. Woo, S.; Park, J.; Lee, J.-Y.; Kweon, I.S. CBAM: Convolutional Block Attention Module. *arXiv* **2018**, arXiv:1807.06521.
52. Wang, J.; Zheng, Z.; Ma, A.; Lu, X.; Zhong, Y. LoveDA: A remote sensing land-cover dataset for domain adaptive semantic segmentation. *arXiv* **2021**, arXiv:2110.08733.
53. Tong, X.Y.; Xia, G.S.; Lu, Q.K.; Shen, H.F.; Li, S.Y.; You, S.C.; Zhang, L.P. Land-cover classification with high-resolution remote sensing images using transferable deep models. *Remote Sens. Environ.* **2020**, *237*, 111322. [CrossRef]
54. Xia, L.; Zhao, F.; Chen, J.; Yu, L.; Lu, M.; Yu, Q.Y.; Liang, S.F.; Fan, L.L.; Sun, X.; Wu, S.R.; et al. A full resolution deep learning network for paddy rice mapping using Landsat data. *ISPRS J. Photogramm. Remote Sens.* **2022**, *194*, 91–107. [CrossRef]
55. Qiang, J.; Liu, W.J.; Li, X.X.; Guan, P.; Du, Y.L.; Liu, B.; Xiao, G.L. Detection of citrus pests in double backbone network based on single shot multibox detector. *Comput. Electron. Agric.* **2023**, *212*, 108158. [CrossRef]
56. Neupane, B.; Horanont, T.; Aryal, J. Deep Learning-Based Semantic Segmentation of Urban Features in Satellite Images: A Review and Meta-Analysis. *Remote Sens.* **2021**, *13*, 808. [CrossRef]
57. Cui, Z.Y.; Li, Q.; Cao, Z.J.; Liu, N.Y. Dense Attention Pyramid Networks for Multi-Scale Ship Detection in SAR Images. *IEEE Trans. Geosci. Remote Sens.* **2019**, *57*, 8983–8997. [CrossRef]
58. Zhang, Y.H.; Lu, H.Y.; Ma, G.Y.; Zhao, H.J.; Xie, D.L.; Geng, S.T.; Tian, W.; Sian, K. MU-Net: Embedding MixFormer into Unet to Extract Water Bodies from Remote Sensing Images. *Remote Sens.* **2023**, *15*, 3559. [CrossRef]

remote sensing

MDPI

Article

Double Branch Parallel Network for Segmentation of Buildings and Waters in Remote Sensing Images

Jing Chen [1], Min Xia [1,*], Dehao Wang [1] and Haifeng Lin [2]

[1] Collaborative Innovation Center on Atmospheric Environment and Equipment Technology, B-DAT, Nanjing University of Information Science and Technology, Nanjing 210044, China
[2] College of Information Science and Technology, Nanjing Forestry University, Nanjing 210000, China
* Correspondence: xiamin@nuist.edu.cn

Abstract: The segmentation algorithm for buildings and waters is extremely important for the efficient planning and utilization of land resources. The temporal and space range of remote sensing pictures is growing. Due to the generic convolutional neural network's (CNN) insensitivity to the spatial position information in remote sensing images, certain location and edge details can be lost, leading to a low level of segmentation accuracy. This research suggests a double-branch parallel interactive network to address these issues, fully using the interactivity of global information in a Swin Transformer network, and integrating CNN to capture deeper information. Then, by building a cross-scale multi-level fusion module, the model can combine features gathered using convolutional neural networks with features derived using Swin Transformer, successfully extracting the semantic information of spatial information and context. Then, an up-sampling module for multi-scale fusion is suggested. It employs the output high-level feature information to direct the low-level feature information and recover the high-resolution pixel-level features. According to experimental results, the proposed networks maximizes the benefits of the two models and increases the precision of semantic segmentation of buildings and waters.

Keywords: double branch; CNN; semantic segmentation; buildings and waters; deep learning

check for updates

Citation: Chen, J.; Xia, M.; Wang, D.; Lin, H. Double Branch Parallel Network for Segmentation of Buildings and Waters in Remote Sensing Images. *Remote Sens.* **2023**, *15*, 1536. https://doi.org/10.3390/rs15061536

Academic Editors: Sawaid Abbas, Janet E. Nichol, Faisal M. Qamer and Jianchu Xu

Received: 13 February 2023
Revised: 4 March 2023
Accepted: 9 March 2023
Published: 11 March 2023

1. Introduction

Semantic segmentation is very important in many domains such as unmanned driving, land use, ecological environment monitoring, disaster monitoring and agricultural monitoring. Identifying building and water area types from remote sensing images can provide an efficient technical approach for regional map updating, land planning, risk management [1] and regional economic development forecasting. The resolution of remote sensing images is increasing as modern computer vision and aerospace technology advance quickly, including space, spectrum and time [2]. A highly effective and affordable method for mapping a large area is to use remote sensing techniques [3]. The textural features and spatial structure properties of ground objects can be clearly expressed in high-resolution remote sensing photographs [4–6]. The development of remote sensing image technology is of great significance to the promotion of semantic segmentation tasks.

The three primary approaches used to segment traditional remote sensing images are the threshold, clustering, and maximum likelihood methods. The maximum likelihood method calculates the maximum likelihood discriminant function of each category by training set data, then substitutes the value of each pixel into the calculation, and finally evaluates the reliability of the classification outcomes. The maximum likelihood method has a high requirement for the training set, which can easily lead to very poor estimation results. The threshold method mainly defines the regional attribution of different targets in the image by threshold, but it is sensitive to the noise of the image, and in remote sensing images with a highly complicated background, the gray difference is not obvious and the overlap of different target gray values is not obvious. The appropriate threshold is not easy to find. The clustering rule is to classify the pixels of the image, and some traditional clustering algorithms do not consider the spatial information, which can easily cause a lack of segmentation in image information and a decrease in segmentation accuracy, and the work to be performed in clustering analysis is more complicated. There are also some image segmentation methods such as the segmentation method based on the genetic algorithm, region segmentation and edge segmentation [7]. The disadvantage of region segmentation is that it is easy to cause excessive segmentation of images. The edge segmentation method cannot obtain a better regional structure, and there is a contradiction between accuracy and noise immunity [8]. Usually, the edge segmentation method is combined with the region segmentation method to obtain a better segmentation effect. In the segmentation method based on the genetic algorithm, it is difficult to determine the crossover probability and mutation probability, and the selection of fitness function is more difficult. Recently, some machine learning methods including decision tree, support vector machine, random forest and artificial shallow neural network have been determined unsuitable for massive quantities of data. Recently, fully supervised models have had great success in this area, however the lack of annotated data will result in significant performance loss [9]. In summary, on the one hand, the traditional image segmentation method is limited by the high-resolution remote sensing image spectrum. On the other hand, it has limited ability to process massive data, and there are some problems such as poor segmentation effect and poor generalization ability.

High-resolution remote sensing images have complex features and are difficult to classify. At present, some deep learning methods have performed well in natural image segmentation tasks. If they are directly applied to remote sensing image segmentation, there will be some challenges: 1. Data volume and quality: different from natural images, the data volume of remote sensing images is usually small, and it is affected by clouds, shadows, noise and other factors, resulting in unstable image quality, which may lead to overfitting or underfitting of deep learning algorithms, thereby affecting classification accuracy. 2. The multi-scale problem: remote sensing images usually involve multiple scales and resolutions. At different scales, the object in the image has different appearance and semantic information. Therefore, in the task of remote sensing image segmentation, multi-scale information needs to be processed and fused to capture global and local information, so that the model can classify the targets in the image more accurately. 3. Category imbalance: remote sensing images usually involve multiple categories, some of which may have few samples. This category imbalance may lead to bias in the model, that is, the model tends to predict common categories in training, and the prediction of rare categories is inaccurate. 4. Spatial and temporal issues: in remote sensing images, temporal and spatial information are closely related. For example, images of the same area may be taken at different times, and the location, shape, and number of targets may also change. Therefore, spatio-temporal information needs to be considered when processing remote sensing images in order to better capture change information [1,10–16]. The deep learning method can extract more and deeper feature information, which is suitable for the classification of high-resolution images [17–20]. Long et al. [21] proposed a fully-connected convolutional neural network (FCN), which migrates the feature extraction layer to the segmentation task and updates these parameters through fine-tuning [22]. At the same time, in order to achieve more precise segmentation findings, a unique structure is created to mix shallow semantic information and high-level semantic information. Chen et al. [23] proposed a semantic pixel-level segmentation network (SegNet). SegNet is a lightweight neural network that uses an upsampling layer to increase the resolution of the segmentation results to the same as the input image. In addition, SegNet uses an encoder-decoder structure with skip connections and deconvolution layers to better preserve detail and semantic information. Zhao et al. [24] proposed a pyramid network structure (PSPNet). PSPNet uses dilated convolution for the convolution of the basic ResNet [25], and in the entire encoder coding part, the features remain at the same resolution after the initial pooling layer. Auxiliary loss is introduced in training to help ResNet learning, and the Spatial Pyrmid Pooling module is introduced to integrate semantic information in different regions to obtain better global semantic information. In the same year, a multi-path refinement network was suggested for high-resolution images segmentation (RefineNet) [26]. It is also based on making full use of shallow and deep feature information, cutting the image input, and then extracting the features, respectively. Sun et al. [27] proposed a high-resolution segmentation network (HRNet). All throughout the procedure, HRNet maintains a high-resolution representation. It achieves the purpose of strong semantic information and accurate location information by connecting high-resolution and low-resolution convolutional flow branches in parallel and continuously interacting with information between different branches. Pang et al. [8] proposed a lightweight building and water segmentation network (SGBNet), in which the channel pooling attention module extracts features through two different global pooling modules. While improving the segmentation accuracy, the network is also more efficient and lightweight. In conclusion, CNNs are usually composed of convolutional layers, pooling layers, and fully-connected layers. For semantic segmentation tasks, the encoder-decoder structure is generally used. The encoder is responsible for encoding the input image information into low-dimensional features. The decoder is responsible for mapping the low-dimensional features back to the original image size and outputting the classification results of each pixel. For each application field, the appropriate network structure and training strategy can be designed according to the specific situation to obtain the best performance.

Recently, Transformer [28] has shown great value in some fields. Transformer is the first transduction model that calculates its input and output representations purely using self-attention, and does not require convolutional network or a sequence-aligned recurrent neural network. According to the conventional model, ByteNet grows logarithmically, ConvS2S increases linearly, and the number of operations needed to correlate signals at two random input or output sites increases with distance. Learning the reliance between far-off places is made more difficult as a result. In Transformer, this is reduced to a fixed number of operations. ViT is proposed as a transformer for large-scale image recognition [29]. The core process of ViT includes four parts: image block processing, image block embedding and location coding, Transformer encoder and MLP classification processing. In the task of migrating to small-scale data sets, it can achieve better performance than CNN, and successfully converts visual problems into natural language processing problems. Liu et al. [30] proposed a Transformer structure with moving windows (Swin Transformer). Swin Transformer designs a shift window and performs self-attention calculation in the shift window. The global information is fully interactive, which brings greater efficiency and performance. At the same time, the operation of the moving window can make the adjacent two windows interact with each other, thus achieving the ability of global modeling in disguise and achieving good results in the segmentation task. Based on Swin Transformer and UNet [31], swin-UNet [32] is proposed. It is based on the Swin Transformer module and constructs a symmetric encoder-decoder structure with skip connections to perform corresponding pixel-level segmentation prediction. Lu et al. [22] proposed a bilateral branch model based on the traditional Transformer, using a strip convolution module in the encoder. The information gathered by the two branches helps to guide each other and produce more accurate segmentation renderings. In conclusion, the Transformer model is a neural network structure based on a self-attention mechanism, which is usually composed of an encoder and decoder. For the semantic segmentation task, the encoder part of the Transformer model can be used as a feature extractor to extract the high-dimensional feature representation of the input image and input it into the decoder for pixel-level classification. CNN can only model local features, while Transformer can model global information. Combining them, a Transformer module can be introduced into CNN to capture longer-range contextual information. This method can improve the edge and detail information of the object in the image segmentation task.

High-resolution remote sensing photos' intricate spectral and spatial texture information not only enhances the table's finely detailed features, but also makes semantic segmentation tasks more challenging. Due to the large difference in the size of various types of features, neural networks need to effectively extract features of ground objects from different angles. For example, the shooting angle and distance of some images, the light intensity, the complexity of the terrain, including the diversity of landforms, the difference between the size of urban waters and natural lakes, the density and diversity of urban buildings and the sparseness of rural houses. More thorough criteria for the model of the semantic segmentation task are put forth by these issues. Due to the lack of global information interaction and the single calculation method, the traditional algorithm will lead to a lot of noise when predicting, and the detection and recognition ability of some areas similar to the pixels of non-target areas is insufficient. For some edge details, the loss of information is large, and on the whole, it is prone to misjudgment. Convolutional neural network-based approaches have difficulty learning explicit global and long-term semantic information relationships because of the intrinsic constraints of convolution processes. This research introduces a novel double-branch parallel image segmentation network in response to the shortcomings of the existing approaches. The network is based on Swin Transformer and CNN. In the stage of feature coding, the designed cross-scale multi-level fusion module is used to connect the two branches, and the comprehensive semantic information and spatial semantic information are extracted using CNN and Swin Transformer. The multi-scale fusion module designed by us guides feature information extracted by double branches to each other, giving full play to the characteristics of Swin Transformer's global information interaction, and making up for the errors in judgment brought on by a lack of global information and long-term semantic information interaction of CNN. During the feature decoding stage, the designed multi-scale fusion module is utilized to fuse the high-level feature information in the coding stage and the low-level feature information extracted by CNN, and the high-level feature information is used to direct low-level feature information and upsample step-by-step. Through the joint action of several modules, our network has significantly increased the segmentation precision.

2. Methods

At present, convolutional neural networks have a constrained receptive field and it is difficult to capture global information [33]. The Swin Transformer network adds a mobile sliding window to better capture global feature information and perform global information interaction. The convolutional neural network has translation invariance and global correlation, while these characteristics in the Transformer network structure are insufficient. Taking into account the aforementioned factors, we propose a parallel combination structure of both Swin Transformer and CNN. The segmentation accuracy and generalization ability of the model in segmentation tasks are greatly improved. In addition, it can effectively identify houses, waters and backgrounds in building and water tasks, and the segmented boundary details are more delicate. Figure 1 depicts the parallel network's general design, which is mostly made up of encoders and decoders. Figures 2 depict the detailed layout of each module. For a given image, first enter the encoder, enter the Swin Transformer and CNN, extract the features information, and effectively fuse the features extracted by feature fusion module designed in this paper, and pass the fusion parameters into the Swin Transformer for further feature extraction. We use Swin Transformer as a branch in the research. Compared with the traditional Transformer module, Swin Transformer designs a moving window and performs self-attention calculation in the moving window. The global information is fully interactive, which brings greater efficiency and performance. At the same time, through the operation of the moving window, the feature information interaction between the two adjacent windows can be realized, thus achieving the ability of global modeling in disguise. Through the coding stage, highly detailed information is obtained and the global information is fully understood. We propose a step-by-step fusion upsampling module in the decoding stage. The feature information obtained through the convolutional network and the feature information obtained through the encoder are upsampled step-by-step through the multi-scale fusion module. Through the sufficient interaction of global information and the guidance of low-level feature information, some disadvantages are mitigated. The model's performance is significantly enhanced by the four fusion upsampling modules, which gradually combine feature information from high level to low level.

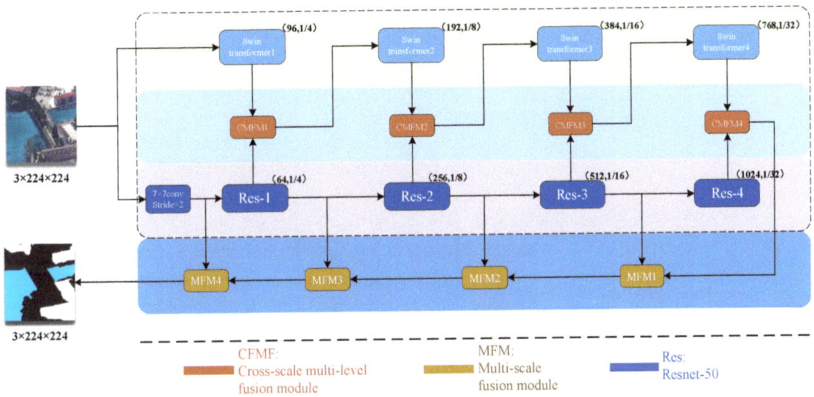

Figure 1. The overall structure of double branch parallel network.

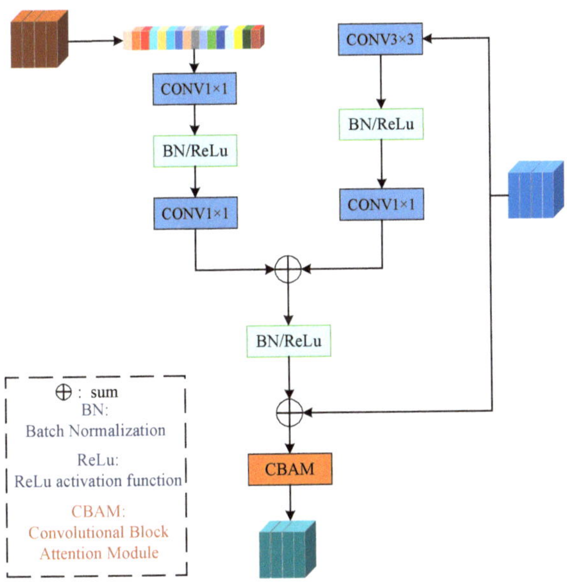

Figure 2. CMFM module structure diagram.

2.1. Overall Structure

The article uses the parallel structure of Swin Transformer and Resnet50 convolution network to draw different levels of information of images. Swin Transformer not only has dynamic attention to focus areas, adding a moving window, but also has a global receptive field and better generalization performance. CNN with Resnet50 as the backbone has two characteristics: local perception and parameter sharing. Local perception refers to the CNN's proposal that each neuron just needs to sense the local pixels in the image rather than all of them, and that this local information may subsequently be combined at a higher level to access all of the image's characterization information. To enhance the performance of the model, we design related modules to fully exploit the advantages of both.

2.2. Cross-Scale Multi-Level Fusion Module

To better improve the accuracy and predictive performance of models in buildings and waters segmentation tasks, we propose a CNN structure Resnet50 and Swin Transformer parallel network structure, but if we simply combine the two structures, we find that the effect of the model is not obvious, which does not meet our task requirements. Considering that two branches output different levels of characteristic information, in order to make full use of the advantages of double branch parallel network, we design a cross-scale multi-level fusion module (CMFM).

As shown in Figure 2, the fusion module we designed here has two branches. It is assumed that the size of the feature map f_1 output by the Resnet50 branch is $C_1 \times H \times W$, and the size of the feature map f_2 output by the Swin Transformer branch is $C_2 \times H \times W$, where C_1 and C_2 represent the number of channels. The f_1 is first passed through a global average pooling layer, which reduces the number of parameters. Global average pooling can better reflect the global information and avoid overfitting. On the other hand, it combines global spatial information and has stronger spatial conversion ability for input images. The next step is to go through a 1×1 convolutional layer, then through the BN and ReLu functions. Finally, the number of channels is changed from C_1 to C_2 through a 1×1 convolutional layer to obtain the other side. We will first process a 3×3 convolutional layer, then the BN and ReLu functions, and finally through a 1×1 convolutional layer. After that, f_{out1} is obtained by adding the $C_2 \times 1 \times 1$ and $C_2 \times H \times W$ size after two 1×1 convolutions, and then f_{out2} is obtained by a BN and ReLu activation function again. After that, we add the $C_2 \times H \times W$ size f_2 to it by a residual operation to obtain f_{out3}. In this paper, we obtain the output f_{out3} and then process the convolutional block attention module (CBAM) to finally obtain the output Y [34]. The above-mentioned process's calculation formula may be represented as:

$$f_1' = f^{1\times1}\left(\delta\left(f^{1\times1} g_a(f_1)\right)\right), \tag{1}$$

$$f_2' = f^{1\times1}\left(\delta\left(f^{3\times3}(f_2)\right)\right), \tag{2}$$

$$Y = CBAM(\delta(f_1' + f_2') + f_2), \tag{3}$$

$f^{1\times1}(\cdot)$ denotes a convolution operation with a convolution kernel of 1×1, $f^{3\times3}(\cdot)$ denotes a convolution operation with a convolution kernel of 3×3, $\delta(\cdot)$ denotes ReLu activation. The ReLu function is sparse, making the sparse model to more effectively extract pertinent features and match training data. The specific calculation formula is shown in Formula (4).

$$\delta(x) = \max(0, x), \tag{4}$$

where x denotes the input of the ReLu function. $g_a(\cdot)$ in Formula (1) denotes one-dimensional average pooling, it calculates an average of all pixels of the feature map of each output channel, and can well suppress overfitting. Here, it changes the feature map of $C_1 \times H \times W$ to the size of $C_1 \times 1 \times 1$. It compresses spatial feature information into channel dimension, and integrates the global spatial information so that the global feature information can be fully utilized. The specific calculation formula is shown in Formula (5).

$$g_a = \frac{1}{H \times W} \sum_{i=1}^{H \times W} p(i), \tag{5}$$

where $p(i)$ represents the pixel value at the i-th position of the feature map. The calculation formula of $CBAM(\cdot)$ in Equation (3) is as follows (assuming the input is F):

$$\begin{aligned} M_c(F) &= \sigma(MLP(AvgPool(F)) + MLP(MaxPool(F))) \\ &= \sigma\left(W_1\left(W_0\left(F^c_{avg}\right)\right) + W_1(W_0(F^c_{max}))\right), \end{aligned} \quad (6)$$

$$\begin{aligned} M_s(F) &= \sigma(f^{7\times7}([AvgPool(F); MaxPool(F)])) \\ &= \sigma\left(f^{7\times7}\left(\left[F^s_{avg}; F^s_{max}\right]\right)\right), \end{aligned} \quad (7)$$

$$F' = M_c(F) \otimes F, \quad (8)$$

$$CBAM = F'' = M_s(F') \otimes F'. \quad (9)$$

In the above formula, the weights W_0 and W_1 of MLP are shared. $AvgPool(\cdot)$ and $MaxPool(\cdot)$ denote average pooling and maximum pooling operations. $M_c(\cdot)$ represents channel attention operation. $M_s(\cdot)$ denotes spatial attention operation. $f^{7\times7}(\cdot)$ denotes a convolution operation with a convolution kernel of 7×7. $\otimes$ represents tensor matrix multiplication. $\sigma(\cdot)$ represents the sigmoid activation function.The sigmoid function can play the role of normalization. Its calculation formula is as follows:

$$\sigma(x) = \frac{1}{1 + e^{-x}}, \quad (10)$$

where x represents the input of the *sigmoid* function.

Figure 3 shows the actual effect of our CMFM module. Among them, (a) is the original image, (b) is its label, (c) and (d) are the feature heat maps of the backbone model without CMFM module and with CMFM module, respectively. The feature heat map demonstrates that some places that were originally concerned or had relatively low attention. Following the addition of the CMFM module to the backbone network, the pixels in these regions—the red portion of the feature heat map—were given more weight by the network. This proves that the designed module is effective.

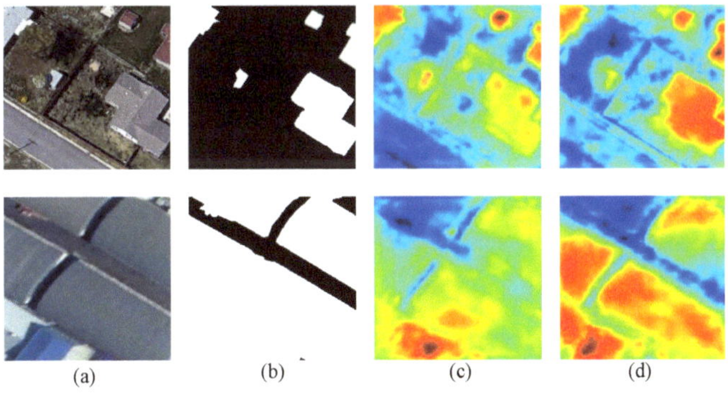

Figure 3. Characteristic heat map comparison of CMFM modules, (**a**) is the original image, (**b**) is its label, (**c**,**d**) are the effect diagrams without module and with module, respectively.

2.3. Multi-Scale Fusion Module

In the decoding stage, if only a simple and crude upsampling recovery is output, it is bound to lose more information, resulting in poor performance of the model; there will be a misjudgment of the situation. To take full advantage of the global interactivity of the Swin Transformer in our backbone network branch, and for high-level semantic features to be used to direct low-level semantic features, we designed a fusion module similar to CMFM. The feature representation of different scales can be captured by guiding at different scales. Figure 4 is a multi-scale fusion module (MFM).

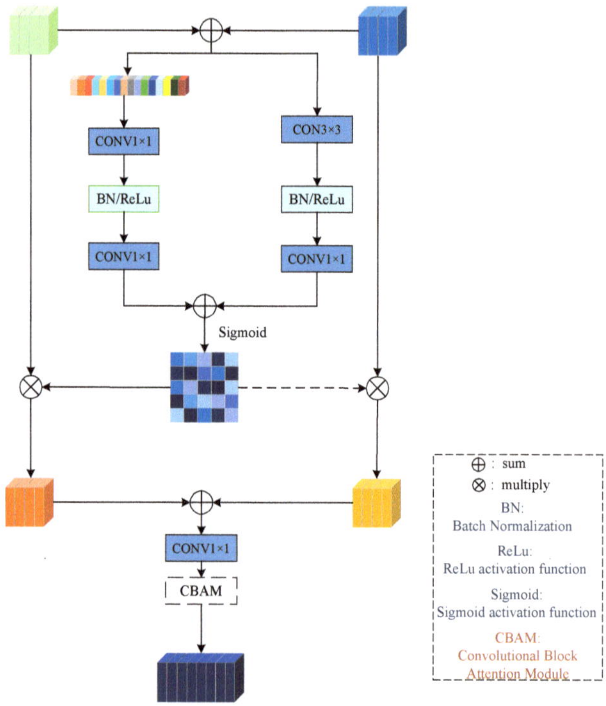

Figure 4. MFM module structure diagram.

Two feature maps with the same scale size make up the module's input, whose structure is shown in Figure 4. We assume that the two inputs are X_1 and X_2, respectively, and the size is $C \times H \times W$, where C denotes the number of channels of the feature map, H and W denote the height and width of the feature map. First, we add the two inputs to obtain X_3, and then parallel out of the two branches X_{31}, X_{32}; X_{31} through the global average pooling, and then through a convolution kernel of 1×1 two-dimensional convolution, then through BN and ReLu function, and finally through a convolution kernel of 1×1 convolution operation to obtain the output X'_{31}. On the other side we direct X_{32} through a convolution kernel for 3×3 two-dimensional convolution operation, the activation functions for BN and ReLu also follow. At last, the output X'_{32} is obtained by convolution operation with convolution kernel of 1×1. Then we add the output of the two to obtain the output X_4, which is activated by sigmoid to obtain the weight s. Since the weight s obtained after sigmoid activation is distributed between (0,1), here we use s and $(1-s)$ as weight coefficients to weight X_1 and X_2, respectively to obtain X_{1out} and X_{2out}, and then add them. To obtain the number of channels matching the next stage, the final output Y is obtained by changing the channel through a two-dimensional convolution with a convolution kernel of 1×1 (the dotted boxed CBAM module is only used in the last MFM module, so Formula (15) is only used in the last MFM. The calculation formula has been described above, so it is not repeated here). The calculation formula of the above-mentioned process can be represented as:

$$X'_{31} = f^{1\times1}\left(\delta\left(f^{1\times1}(g_a(X_1 + X_2))\right)\right), \tag{11}$$

$$X'_{32} = f^{1\times1}\left(\delta\left(f^{3\times3}(X_1 + X_2)\right)\right), \tag{12}$$

$$X_4 = X'_{31} + X'_{32}, \tag{13}$$

$$Y = f^{1\times1}((\sigma(X_4) \otimes X_1 + (1 - \sigma(X_4)) \otimes X_2)), \quad (14)$$

$$Y = CBAM(Y), \quad (15)$$

where $g(\cdot)$ denotes global average pooling, $f^{3\times3}(\cdot)$ denotes a 2D convolution with a 3×3 convolution kernel, $f^{1\times1}(\cdot)$ represents a 2D convolution with a 1×1 convolution kernel, $\delta(\cdot)$ includes Batch Normalization and ReLU function activation, $\otimes$ represents tensor matrix multiplication, $\sigma(\cdot)$ represents the sigmoid activation function.

The above modules together constitute our algorithm network. The network adopts a two-branch parallel method. The convolution branch extracts detailed information to obtain low-level semantic information, and the Swin Transfomer branch extracts contextual information to obtain high-level semantic information. In the coding stage, the CMFM module completely interacts with the contextual information by utilizing the benefits of the two branch networks. In the decoding stage, the process gives full play to the advantages of MFM module, step-by-step fusion recovery, and makes the segmentation boundary more delicate, where in the case of complex backgrounds it can better identify waters and buildings.

3. Results

3.1. Building and Water Dataset

In order to test its effectiveness in the semantic segmentation job of buildings and waters, this paper created a buidings and waters dataset to train and validate the model. In comparison to some other datasets, the dataset created in this experiment has a large spatial span, more angles, and a complex background due to the low angle of view, which necessitates more complex algorithms. The dataset comprises of 10,000 pairs of Google Earth photos divided into the following categories: a riverfront residential complex in China, a private villa in North America, and so on. After that, the photos were divided into 224×224 images, and the data was enhanced on these images. There are three different kinds of strategies: the 50% horizontal flip, the 50% vertical flip, and the 10% random spin, for example. The enlarged dataset can be improved, but it can also raise model training process interference and improve the model's generalizability. Architecture, water, and background are the three object categories that have been manually assigned to these pictures. Figure 5 displays the sliced image together with its label. A single type of image was eliminated, and the remaining photos were then randomly split into an 8:2 training and validation set. The dataset created in this experiment contains a rich variety of backgrounds, which meet the experimental requirements and ensure that the experimental segmentation accuracy will not be biased due to the single dataset type.

The following characteristics are present in the dataset: (1) of the objects in the dataset selected in this paper, such as vehicles, some containers and some similar buildings, some of them have large differences in color, as shown in Figure 5d, and some of the buildings have a great similarity to the surrounding background color, as shown in Figure 5e, so this puts forward higher requirements for the proposed model detection ability. (2) In this dataset, we selected more coverage scenarios to better and more comprehensively test the performance of our model. (3) Because the remote sensing satellite is different in angle and spatial distance when shooting, the difficulty of segmentation is increased to a certain extent. (4) In some dense buildings, some high-rise buildings' shadows cast on nearby low-rise structures causes a certain degree of interference to the segmentation of the model.

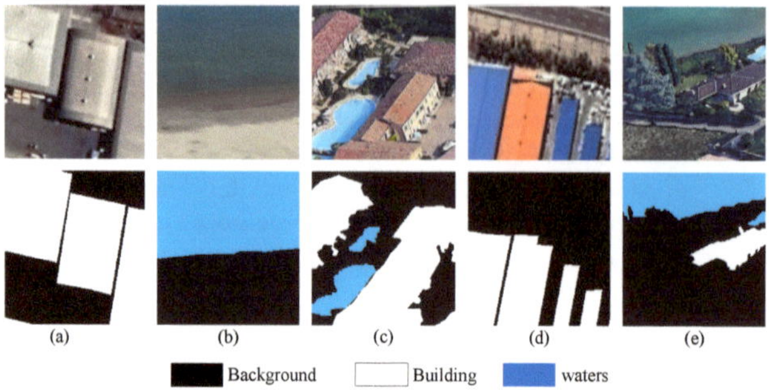

Figure 5. Partial display of the land cover dataset. (**a–e**) are the display of different remote sensing images and corresponding labels, respectively.

3.2. Waters Dataset

This paper selects the waters dataset for verification in order to more thoroughly reflect the network's ability to cope with edge characteristics, e.g., water area, to better validate the generalizability of the network model it proposes. In this experiment, we select China HJ-1A (HJ-1B) multi-spectral environmental remote sensing satellite images as the required dataset. Additionally, the model's capacity for generalization is examined.

Three visible and near-infrared spectral bands from the HJ-1A (HJ-1B) satellite charge coupled device camera are used in the water dataset. In order to effectively utilize the waters information, we selected a mix of bands 1, 2, and 4 to produce a three-channel waters image. To prevent over-fitting and ensure the accuracy of experimental segmentation accuracy, we cut the original image into 256 × 256 images and randomly flip, rotate, and scale the experimental image. Here, we created 8000 waters datasets and divided them into training sets and verification sets according to the 8:2 ratio column. The water dataset is a binary dataset, that is, the model identifies two semantic categories of waters and background. Figure 6 displays the cropped image and its labels.

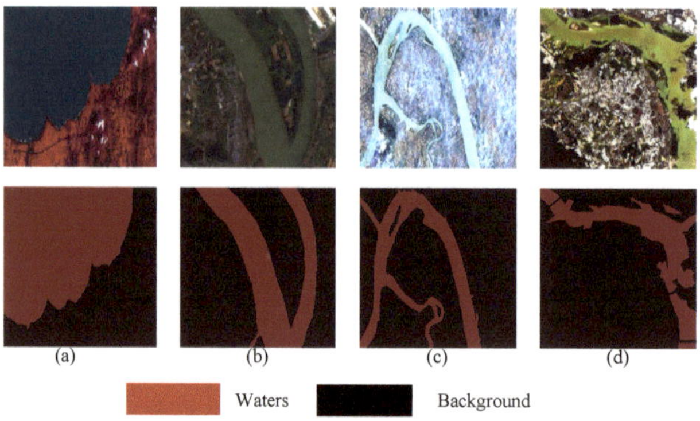

Figure 6. Partial representation of the waters dataset. (**a–d**) are the display of different remote sensing images and corresponding labels, respectively.

3.3. Inria Dataset

In order to further verify the generalization ability of the proposed model, we selected the Inria Aerial Image Labeling Dataset. It is a public data set for computer vision and machine learning research developed by the French National Institute of Digital and Automation (Inria). The dataset contains a set of aerial images taken from high altitude, covering some cities in southern France, and the ground coverage types include buildings, trees and roads. Its resolution is 5000 × 5000 pixels. Here, we cut it into 256 × 256 images and divided them into a training set and validation set according to the ratio of 8:2. As shown in Figure 7, part of the data set is displayed.

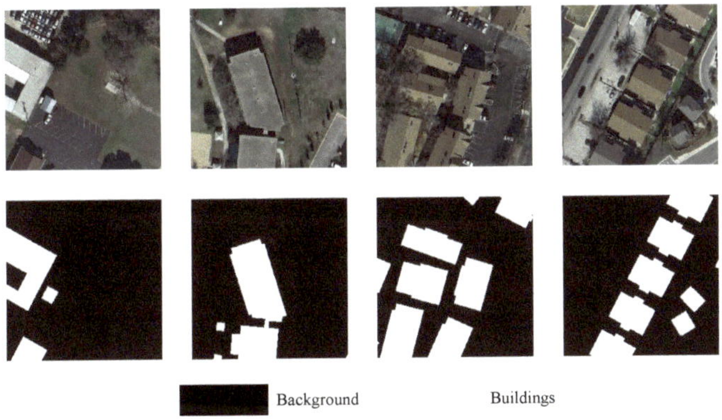

Figure 7. Partial representation of the Inria dataset.

A machine with an NVIDIA RTX3080 graphics card was used to carry out all the experiments in this study. The operating system adopted in this experiment is Windows 10. The construction of the experimental model in this paper is based on the deep learning framework of pytorch (2017). In terms of optimizer, this paper uses an adaptive moment estimation optimizer, it combines the advantages of both momentum and RMSProp optimization algorithms, the first-order moment estimation of the gradient and the second-order moment estimation, comprehensive consideration, and then calculates the update step. The number of iterations is set to 300 in all of the experiments in this work because, according to experimental observation, most experiments tend to converge after 200 iterations. The loss function used in the experiment is BCEWITH-LogitsLoss. Due to the physical memory limitations of the computer's graphics GPU, the experiment's batch size was set to 8. The experimental index is an important reference for evaluating the effect of the model. Here, we use the mean pixel accuracy (MPA), pixel accuracy (PA) and mean intersection over union (MIOU) on the union set as evaluation indicators. The following are the MPA, PA, and MIOU formulae.

$$MPA = \frac{1}{k}\sum_{i=0}^{k} \frac{p_{i,j}}{\sum_{j=0}^{k} p_{i,j}}, \tag{16}$$

$$PA = \frac{\sum_{i=0}^{k} p_{i,j}}{\sum_{i=0}^{k}\sum_{j=0}^{k} p_{i,j}}, \tag{17}$$

$$MIOU = \frac{1}{k+1}\sum_{i=0}^{k} \frac{p_{i,j}}{\sum_{j=0}^{k} p_{i,j} + \sum_{j=0}^{k} p_{j,i} - p_{i,i}}, \tag{18}$$

where k denotes the class of object segmentation (excluding background), $P_{i,i}$ shows the real number, $P_{i,j}$ represents the number of pixels that belong to category i but are predicted to be j.

3.4. Ablation Experiment

We initially use Resnet50 as the backbone network, and then upsample each layer and connect them for output. Then we add the Swin Transformer branch in parallel based on Resnet50, and each layer of the two branches is added and up-sampled step by step for addition and output. Then, to verify the efficiency of the models and modules created in this research, we gradually add each module to the model. Here, the model's primary evaluation metric is MIOU. The ablation experiments are shown in Table 1

Table 1. Results of ablation experiments in land cover datasets.

Method	MIOU (%)
Resnet	82.59
Resnet + Swin Transformer	84.40
Resnet + Swin Transformer + CMFM	85.53
Resnet + Swin Transformer + CMFM + MFM	87.86

(1) To more effectively extract the image's feature information, extract more scale spatial location information, and fully carry out global information interaction, we first simply add the Swin Transformer branch in parallel to the Resnet50 branch. Through the experimental results, we find that the MIOU value can be increased to 84.40% when the two branches extract features in parallel. (2) Two branches are added to the connection module in parallel. For two branches in parallel, if each layer is simply added, there will be some lost spatial and semantic position information. To fully extract the feature information and improve how the high-level feature information directs the underlying feature information, the overall recognition ability of the model and the processing of some detailed features are improved. Based on this, this paper designs a cross-scale multi-level fusion module. It is found that the MOIU value of the model was increased to 85.53% after adding the CMFM module. (3) Ablation (MFM) for high-low information mutual guidance fusion module: since the shape and size of some rivers are not constant in the building and water segmentation task, some of the previous methods for river boundary treatment are not delicate enough. To restore the characteristic information of the edge of the river including some buildings, in this paper, we integrate low-level feature information extracted through convolutional network and high-level feature information through the Swin Transformer, utilizing the high-level feature information to direct the low-level feature information. We processed the global feature pooling in the module, once again fully incorporating global feature information interaction, improving the accuracy of the category area identification and model performance. Finally, the MIOU value of our model reached 87.86%.

3.5. Contrast Experiment

We contrast our model with some excellent models for building and water segmentation in this part, such as DABNet [35], FCN8s, PSPNet, DeeplabV3plus [36], Shufflenetv2 [37], BisenetV2 [38], Dual-branch [22], and so on. The Deeplab series mainly uses dilated convolution and pyramid pooling ASPP [25]. By using different dilated convolutions on a given feature layer, it can effectively resample and construct convolution kernels of different receptive fields to obtain information of multi-scale objects. The Bisenet series fuses the extracted deep feature information and spatial information through a spatial branch and a semantic branch, and supervises model training through an auxiliary loss function. FCN solves the issue of semantic-level picture segmentation, as the fully-connected layer of a traditional CNN is changed to a convolutional layer, which classifies images at the pixel level [39]. The pyramid pooling module, which is the primary component of PSPNet, may gather contextual information from several regions to boost global information acquisition capacity [23]. The Shufflenet series uses grouping convolution to group different features of the input layer, and then uses different convolution kernels to convolve each group, thereby reducing the amount of convolution calculation, mainly playing a lightweight effect. DABNet proposes a deep non-decomposable bottleneck module, which effectively uses asymmetric convolution kernel dilated convolution to construct the bottleneck layer, generates sufficient acceptance domain, intensively uses context information, and greatly reduces parameters.

Table 2 shows that SegNet has the worst segmentation effect, and the MIOU and MPA values are only 80.06% and 89.06%. In general convolutional neural networks, PSPNet (backbone adopts Resnet50) network has the highest segmentation accuracy, where the MIOU value and MPA value can reach 86.61% and 92.88%, respectively. At the same time, the double-branch parallel network designed in this paper is compared with the network swin UNet that improves the swin Transformer, where MPA is 90.00% and 94.11%, and MIOU is 81.14% and 87.86%. Compared with these algorithms, our double-branch parallel network achieves the optimal value in three indicators. Comparing our approach to other models, it can be shown that it significantly improves segmentation, and our model has a strong pertinence for semantic segmentation tasks of buildings and waters. None of the semantic segmentation networks in the table pre-loaded the training weight, and the training requirement parameters were set uniformly to ensure the fairness of the comparison experiment.

Table 2. Experimental results compared to other algorithms.

Method	PA (%)	MPA (%)	MIOU (%)
SegNet [40]	89.88	89.06	80.06
UNet [31]	92.96	92.19	86.27
PAN [41]	92.78	93.22	85.78
HRNet [27]	92.37	92.68	84.78
DFNet [42]	91.74	90.31	84.42
DeepLabV3+ [36]	92.83	93.42	86.18
BiSeNetV2 [38]	91.19	91.13	82.81
DANet [43]	92.16	92.50	84.25
DABNet [35]	92.61	92.33	85.38
ShuffleNetV2 [37]	90.63	89.75	81.08
PSPNet [24]	93.40	92.88	86.61
FCN8s [21]	92.66	92.81	85.91
SwinUNet [32]	89.68	90.00	81.14
Dual-branch [22]	92.80	93.25	86.02
MFANet [12]	93.02	93.73	86.62
Ours	93.64	94.11	87.86

Figure 8 compares the prediction maps of some networks. We comprehensively verify our network by comparing the prediction effect maps of seven remote sensing images. Figure 8i is the label graph. Figure 8b–g are the experimental comparison results, Figure 8h is the proposed network prediction rendering. The comparison shows that the effect map predicted by the network model in this research is more accurate in detecting the buildings and waters, and there is no omission overall. This is due to the fact that our double-branch parallel structure fully exploits its own benefits. On the basis of the global information interaction of Swin Transformer, the two modules we designed are added to the global average pooling again. With the cooperation of these modules, the global information interaction is continuously carried out. In addition, by using high-level feature information to direct low-level feature information, some edge feature information of the segmented target are repaired. Finally, we find that the effect prediction graph of the double-branch parallel network model is the closest to the label graph, and our prediction accuracy is the best.

Figure 8a is a remote sensing image, Figure 8b–h represent DeepLabV3, FCN-8s, DABNet, PAN, PSPNet, UNet and the paper designed a double-branch parallel network, Figure 8i is the label graph. To show the superiority of our network model over other networks more directly, the red box is used to mark some areas for more intuitive comparison. It is clear that the double-branch parallel network designed in this research reduces the occurrence of misjudgment and missed judgment by grasping the global information and processing some details. The processing of some edge information is also relatively good, and the segmentation accuracy is improved.

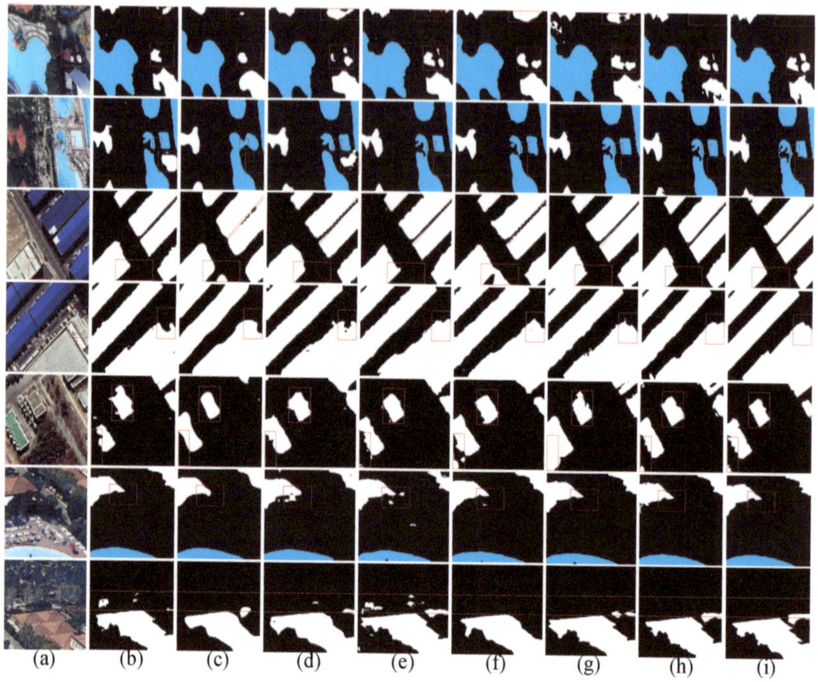

Figure 8. Prediction comparison of different algorithms on land cover dataset. (**a**) Original image; (**b**) DeepLabV3; (**c**) FCN-8s; (**d**) DABNet; (**e**) PAN; (**f**) PSPNet; (**g**) UNet; (**h**) our model; (**i**) Label.

3.6. Generalization Experiment

3.6.1. Waters Dataset

Considering the generalization ability of the double-branch parallel network model, we chose to conduct experiments on the water dataset. Compared with the building and water dataset, the water dataset can detect the segmentation ability of our model in more complex background environments. In this research, we more thoroughly demonstrate the superiority of the network using experimental comparison using several network models on the water dataset, which increases the diversity of trials.

Here, our model is contrasted with some land cover neural networks, such as ESP-NetV2, SegNet, DeepLabV3+, PSPNet, FCN8s and other traditional land segmentation networks. This article also compares some of the latest improved networks on Transformer, such as PVT (Pyramid Vision Transformer), VIT (Vision Transformer), CVT, conformer. The experiment was carried out under the same conditions. The segmentation accuracy is shown in Table 3. It is evident from the table that the segmentation accuracy of the model utilized in this investigation, which came in at 96.38%, is the highest.

Figure 9 compares the predicted representations of various models from the waters dataset. Among them, Figure 9a is the remote sensing image, Figure 9b–h represent CVT, DeepLabV3+, DFN, FCN-8s, SegNet, ShuffleV2 and the double-branch parallel network designed in this paper, Figure 9i is the label graph. Here, some areas are marked with yellow boxes for more intuitive comparison. From the figure, we can see that our network can still detect rivers well under different complex background conditions. Although other models can also detect the river, in some small tributaries there will always be some missing parts, and our model can be a good detector of small tributaries with the information. This is because the two branches of our model can fully extract spatial feature information and detail feature information, under the action of CMFM module, the global feature information is fully interacted, which can better grasp the global information and accurately detect the location of the river. Finally, the MFM module guides the low-level feature information by using the advanced feature information, making up for the lack of some feature information.

Table 3. Compare experimental results with different models on a water dataset.

Method	PA (%)	MPA (%)	MIOU (%)
Conformer [44]	96.81	96.19	91.65
SegNet [40]	97.48	97.04	93.35
ESPNetV2 [45]	98.23	98.09	95.27
GhostNet [46]	96.80	96.32	91.61
DeeplabV3+ [36]	98.14	97.50	95.08
DFNet [47]	98.11	97.82	94.96
CVT [48]	97.34	96.61	93.03
ShuffleNetV2 [37]	98.17	98.02	95.10
FCN8s [21]	97.45	97.01	93.28
Dual-branch [22]	98.28	98.13	95.39
PVT [49]	97.88	97.75	94.34
VIT [29]	87.81	85.51	70.66
MFANet [12]	98.25	97.96	95.33
Ours	98.65	98.39	96.38

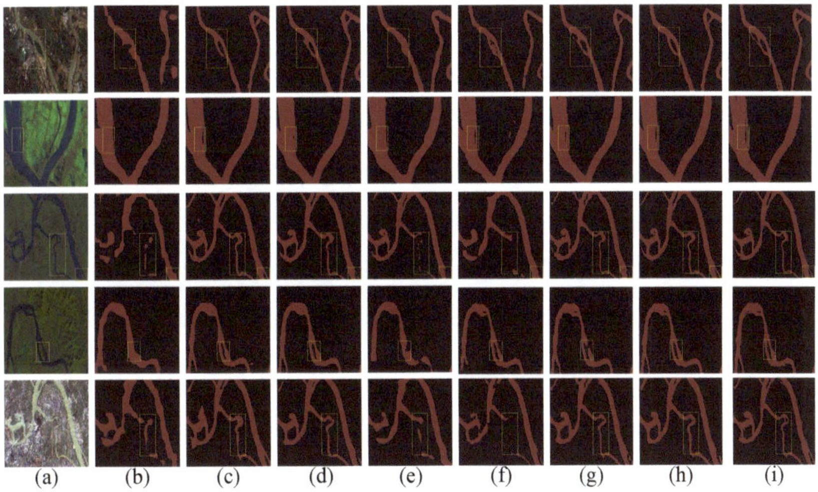

Figure 9. Comparison of prediction effects of different models in water dataset. (**a**) Original image; (**b**) CVT; (**c**) DeepLabV3+; (**d**) DFN; (**e**) FCN-8s; (**f**) SegNet; (**g**) ShuffleV2; (**h**) our model; (i) Label.

3.6.2. Inria Dataset

The main task of the experimental model in this paper is to segment the remote sensing images of buildings and waters. After the generalization experiment results on the waters dataset, our model has a better segmentation effect on the waters. In order to further fully reflect the ability of our model, we conducted generalization experiments on the Inria dataset and compared it with some land segmentation network models, as shown in Table 4.

Table 4. Compare experimental results with different models on Inria dataset.

Method	PA (%)	MPA (%)	MIOU (%)
DFNet [47]	93.95	91.20	82.98
CVT [48]	90.91	86.34	75.72
GhostNet [46]	91.40	87.01	76.93
MFANet [12]	94.47	92.04	84.28
SegNet [40]	94.21	91.20	83.78
HRNet [27]	93.89	90.08	82.85
DeeplabV3+ [36]	94.33	91.37	84.09
FCN8s [36]	93.21	89.31	81.49
SGBNet [8]	94.12	91.52	83.36
PVT [49]	93.95	91.26	82.48
Dual-branch [22]	94.38	92.16	83.96
ShuffleNetV2 [37]	94.22	91.34	83.75
OCRNet [50]	94.77	92.27	85.13
Ours	95.27	92.66	86.19

From the table, we can see that the model designed in this paper has achieved the best in all three indicators. The data show that the generalization ability of the model is strong and persuasive. Figure 10 shows the comparison between our model and other models. From the figure, we can see that the model designed in this paper still achieves good results on the public data set Inria. Since our two-branch model complements the advantages of CNN and swin Transformer through the designed fusion module, on the prediction effect diagram, we can see that our model has achieved good results in both the processing of edge details and the problem of misjudgment, which proves that our model has strong generalization ability.

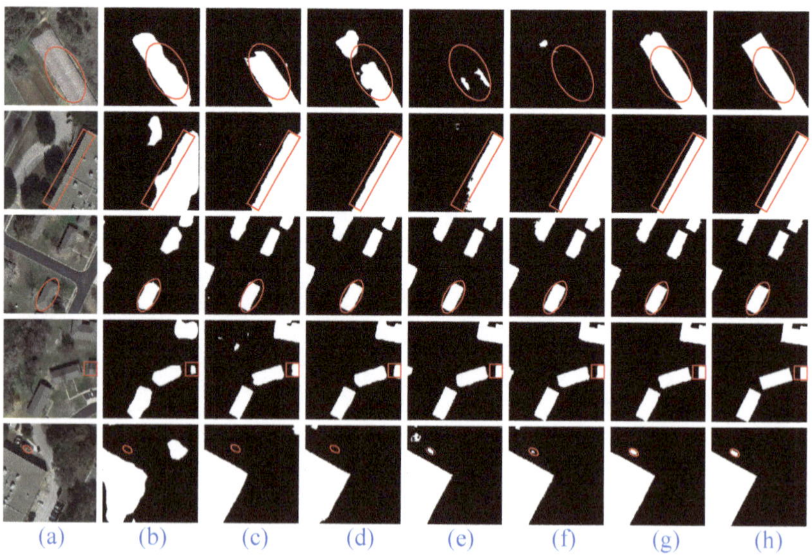

Figure 10. Comparison of prediction effects of different models in Inria dataset. (**a**) Original image; (**b**) CVT; (**c**) DeepLabV3+; (**d**) DFN; (**e**) Dual-branch; (**f**) SegNet; (**g**) our model; (**h**) Label.

4. Discussion

4.1. About the Model

The network is based on Swin Transformer and CNN. In the stage of feature coding, the designed cross-scale multi-level fusion module is used to connect the two branches, and the comprehensive semantic information and spatial semantic information are extracted using CNN and Swin Transformer. The multi-scale fusion module designed by us guides feature information extracted by double branches to each other, giving full play to the characteristics of Swin Transformer's global information interaction, and making up for the judgment errors brought on by a lack of global information and long-term semantic information interaction of CNN. During the feature decoding stage, the designed multi-scale fusion module is utilized to fuse the high-level feature information in the coding stage and the low-level feature information extracted by CNN, and the high-level feature information is used to direct low-level feature information and upsample step-by-step. Through the joint action of several modules, our network has significantly increased the segmentation precision. The following are this paper's main contributions:

1. A double-branch parallel network of Swin Transformer and CNN is proposed. The two network structures extract feature information separately and aggregate the extracted feature information, which can better improve the accuracy and generalization of segmentation. Swin Transformer makes up for the deficiency of the limited receptive field of convolutional neural network (CNN) and can better perform global information interaction; in addition, CNN can make up for the lack of translation in the variance of Transformer.
2. Considering the difference of feature information extracted from two branches, a cross-scale bilateral feature aggregation module is proposed. This method can effectively aggregate different levels of feature information and guide each other, so that more feature information can be globally interacted. It effectively reduces the occurrence of misjudgment. In the upsampling stage, an aggregation module is also proposed, which fully utilizes high-level semantic information to direct low-level semantic information, and recovers high-resolution pixel-level feature information and edge feature information.

4.2. About the Experiment

In order to verify the ability of our model, this study conducted comparative experiments and generalization experiments on the building water dataset and the water dataset and the public Inria dataset. In the comparative experiment, our model is superior to other classical network models in the three indicators. The PA value, MPA value and MIOU value reached 93.64%, 94.11% and 87.86%, respectively. In the prediction effect diagram, the model designed in this paper is compared with other networks regarding the problem of misjudgment, as well as dealing with some edge feature information, which is used to reflect the advantages of our dual branch network. CNN and Swin Transformer give full play to their respective advantages under the action of the fusion module; more feature information is globally interacted, and advanced feature information is used to guide low-level feature information, making edge detail features more delicate. We conducted generalization experiments on the water data set and the Inria data set. Similarly, from the numerical and prediction effect diagrams, our network model is still superior to other models, which proves that our model has better generalization ability.

4.3. Limitations and Future Prospects of the Model

Since both CNN and Transformer models have large computational overhead, our work in the future is to further optimize the structure of the model while ensuring the segmentation accuracy of the model, design more efficient model structure and more effective training strategies to reduce training complexity and training difficulty.

5. Conclusions

In remote sensing images, house waters are important geographical indications. They have important practical significance for land planning, water resources protection planning and geographic mapping. The segmentation task of buildings and waters is also an important part of the land cover segmentation task. Existing image segmentation network models primarily employ CNNs to extract feature information from images. In order to make up for the deficiency of CNN in feature extraction and better fully interact with global semantic information, this paper presents a double branch parallel network structure algorithm for segmentation task. In the coding process of the algorithm, we use Resnet50 and Swin Transformer to extract features for the two branches, obtain rich context information and spatial information, utilize the benefits of the two branches' feature extraction information to the fullest extent, and fuse the feature information extracted by different branches through our fusion module, which provides rich pixel information for the upsampling information recovery. In the process of decoding, we use a fusion module designed to fuse the encoded high-level feature information with the feature information of ResNet50 branch, and use high-level feature information to direct low-level feature information. Upsampling can gradually refine and restore high-resolution images and obtain more spatial details. Compared with some of the current semantic segmentation network models, our model has greatly improved the accuracy of segmentation in buildings and waters. From the performance of different datasets, our model has good anti-interference and recognition capabilities, and can accurately determine the location of waters and houses in complex background environments, while the segmented edges are also more delicate. In the future, in order to enhance the practical applications, we will further optimize the model structure under the assumption of ensuring segmentation precision, and improve the model training speed.

Author Contributions: Conceptualization, J.C. and M.X.; methodology, M.X. and J.C.; software, J.C.; validation, D.W. and H.L.; formal analysis, M.X.; investigation, J.C.; resources, M.X.; data curation, J.C.; writing–original draft preparation, J.C.; writing–review and editing, M.X.; visualization, M.X.; supervision, M.X.; project administration, M.X.; funding acquisition, M.X. All authors have read and agreed to the published version of the manuscript.

Funding: This work is supported in part by the National Natural Science Foundation of PR China (42075130).

Data Availability Statement: The data and the code of this study are available from the corresponding author upon request.

Conflicts of Interest: The authors declare no conflict of interest.

References

1. Shu, Q.; Pan, J.; Zhang, Z.; Wang, M. DPCC-Net: Dual-perspective change contextual network for change detection in high-resolution remote sensing images. *Int. J. Appl. Earth Obs. Geoinf.* **2022**, *112*, 102940. [CrossRef]
2. Song, L.; Xia, M.; Weng, L.; Lin, H.; Qian, M.; Chen, B. Axial Cross Attention Meets CNN: Bi-Branch Fusion Network for Change Detection. *IEEE J. Sel. Top. Appl. Earth Obs. Remote Sens.* **2023**, *16*, 32–43. [CrossRef]
3. Yu, Y.; Huang, L.; Lu, W.; Guan, H.; Ma, L.; Jin, S.; Yu, C.; Zhang, Y.; Tang, P.; Liu, Z.; et al. WaterHRNet: A multibranch hierarchical attentive network for water body extraction with remote sensing images. *Int. J. Appl. Earth Obs. Geoinf.* **2022**, *115*, 103103. [CrossRef]
4. Qu, Y.; Xia, M.; Zhang, Y. Strip pooling channel spatial attention network for the segmentation of cloud and cloud shadow. *Comput. Geosci.* **2021**, *157*, 104940. [CrossRef]
5. Lu, C.; Xia, M.; Lin, H. Multi-scale strip pooling feature aggregation network for cloud and cloud shadow segmentation. *Neural Comput. Appl.* **2022**, *34*, 6149–6162. [CrossRef]
6. Hu, K.; Li, M.; Xia, M.; Lin, H. Multi-Scale Feature Aggregation Network for Water Area Segmentation. *Remote Sens.* **2022**, *14*, 206. [CrossRef]
7. Wang, T.; Lan, J.; Han, Z.; Hu, Z.; Huang, Y.; Deng, Y.; Zhang, H.; Wang, J.; Chen, M.; Jiang, H.; et al. O-Net: A novel framework with deep fusion of CNN and transformer for simultaneous segmentation and classification. *Front. Neurosci.* **2022**, *16*, 876065. [CrossRef]
8. Pang, K.; Weng, L.; Zhang, Y.; Liu, J.; Lin, H.; Xia, M. SGBNet: An Ultra Light-weight Network for Real-time Semantic Segmentation of Land Cover. *Int. J. Remote Sens.* **2022**, *43*, 5917–5939. [CrossRef]
9. Chen, J.; Sun, B.; Wang, L.; Fang, B.; Chang, Y.; Li, Y.; Zhang, J.; Lyu, X.; Chen, G. Semi-supervised semantic segmentation framework with pseudo supervisions for land-use/land-cover mapping in coastal areas. *Int. J. Appl. Earth Obs. Geoinf.* **2022**, *112*, 102881. [CrossRef]
10. Miao, S.; Xia, M.; Qian, M.; Zhang, Y.; Liu, J.; Lin, H. Cloud/shadow segmentation based on multi-level feature enhanced network for remote sensing imagery. *Int. J. Remote Sens.* **2022**, *43*, 5940–5960. [CrossRef]

11. Wang, Z.; Xia, M.; Lu, M.; Pan, L.; Liu, J. Parameter Identification in Power Transmission Systems Based on Graph Convolution Network. *IEEE Trans. Power Deliv.* **2022**, *37*, 3155–3163. [CrossRef]
12. Chen, B.; Xia, M.; Huang, J. Mfanet: A multi-level feature aggregation network for semantic segmentation of land cover. *Remote Sens.* **2021**, *13*, 731. [CrossRef]
13. Ma, Z.; Xia, M.; Weng, L.; Lin, H. Local Feature Search Network for Building and Water Segmentation of Remote Sensing Image. *Sustainability* **2023**, *15*, 3034. [CrossRef]
14. Ding, Y.; Zhao, X.; Zhang, Z.; Cai, W.; Yang, N.; Zhan, Y. Semi-supervised locality preserving dense graph neural network with ARMA filters and context-aware learning for hyperspectral image classification. *IEEE Trans. Geosci. Remote Sens.* **2021**, *60*, 1–12. [CrossRef]
15. Zhang, Z.; Ding, Y.; Zhao, X.; Siye, L.; Yang, N.; Cai, Y.; Zhan, Y. Multireceptive field: An adaptive path aggregation graph neural framework for hyperspectral image classification. *Expert Syst. Appl.* **2023**, *217*, 119508. [CrossRef]
16. Ding, Y.; Zhang, Z.; Zhao, X.; Hong, D.; Cai, W.; Yu, C.; Yang, N.; Cai, W. Multi-feature fusion: Graph neural network and CNN combining for hyperspectral image classification. *Neurocomputing* **2022**, *501*, 246–257. [CrossRef]
17. Chen, B.; Xia, M.; Qian, M.; Huang, J. MANet: A multi-level aggregation network for semantic segmentation of high-resolution remote sensing images. *Int. J. Remote Sens.* **2022**, *43*, 5874–5894. [CrossRef]
18. Hu, K.; Weng, C.; Zhang, Y.; Jin, J.; Xia, Q. An Overview of Underwater Vision Enhancement: From Traditional Methods to Recent Deep Learning. *J. Mar. Sci. Eng.* **2022**, *10*, 241. [CrossRef]
19. Hu, K.; Ding, Y.; Jin, J.; Weng, L.; Xia, M. Skeleton Motion Recognition Based on Multi-Scale Deep Spatio-Temporal Features. *Appl. Sci.* **2022**, *12*, 1028. [CrossRef]
20. Ding, Y.; Zhang, Z.; Zhao, X.; Cai, Y.; Li, S.; Deng, B.; Cai, W. Self-supervised locality preserving low-pass graph convolutional embedding for large-scale hyperspectral image clustering. *IEEE Trans. Geosci. Remote Sens.* **2022**, *60*, 1–16. [CrossRef]
21. Long, J.; Shelhamer, E.; Darrell, T. Fully convolutional networks for semantic segmentation. In Proceedings of the IEEE Conference on Computer Vision and Pattern Recognition, Boston, MA, USA, 7–12 June 2015; pp. 3431–3440.
22. Lu, C.; Xia, M.; Qian, M.; Chen, B. Dual-branch Network for Cloud and Cloud Shadow Segmentation. *IEEE Trans. Geosci. Remote Sens.* **2022**, *60*, 1–12. [CrossRef]
23. Chen, L.C.; Papandreou, G.; Kokkinos, I.; Murphy, K.; Yuille, A.L. Deeplab: Semantic image segmentation with deep convolutional nets, atrous convolution, and fully connected crfs. *IEEE Trans. Pattern Anal. Mach. Intell.* **2017**, *40*, 834–848. [CrossRef]
24. Zhao, H.; Shi, J.; Qi, X.; Wang, X.; Jia, J. Pyramid scene parsing network. In Proceedings of the IEEE Conference on Computer Vision and Pattern Recognition, Honolulu, HI, USA, 21–26 July 2017; pp. 2881–2890.
25. He, K.; Zhang, X.; Ren, S.; Sun, J. Deep residual learning for image recognition. In Proceedings of the IEEE Conference on Computer Vision and Pattern Recognition, Las Vegas, NV, USA, 27–30 June 2016; pp. 770–778.
26. Lin, G.; Milan, A.; Shen, C.; Reid, I. Refinenet: Multi-path refinement networks for high-resolution semantic segmentation. In Proceedings of the IEEE Conference on Computer Vision and Pattern Recognition, Honolulu, HI, USA, 21–26 July 2017; pp. 1925–1934.
27. Sun, K.; Xiao, B.; Liu, D.; Wang, J. Deep high-resolution representation learning for human pose estimation. In Proceedings of the IEEE/CVF Conference on Computer Vision and Pattern Recognition, Long Beach, CA, USA, 15–20 June 2019; pp. 5693–5703.
28. Vaswani, A.; Shazeer, N.; Parmar, N.; Uszkoreit, J.; Jones, L.; Gomez, A.N.; Kaiser, Ł.; Polosukhin, I. Attention is all you need. *Adv. Neural Inf. Process. Syst.* **2017**, *30*, 6000–6010.
29. Dosovitskiy, A.; Beyer, L.; Kolesnikov, A.; Weissenborn, D.; Zhai, X.; Unterthiner, T.; Dehghani, M.; Minderer, M.; Heigold, G.; Gelly, S.; et al. An image is worth 16x16 words: Transformers for image recognition at scale. *arXiv* **2020**, arXiv:2010.11929.
30. Chu, X.; Tian, Z.; Wang, Y.; Zhang, B.; Ren, H.; Wei, X.; Xia, H.; Shen, C. Twins: Revisiting the design of spatial attention in vision transformers. *Adv. Neural Inf. Process. Syst.* **2021**, *34*, 9355–9366.
31. Ronneberger, O.; Fischer, P.; Brox, T. U-net: Convolutional networks for biomedical image segmentation. In Proceedings of the International Conference on Medical Image Computing and Computer-Assisted Intervention, Munich, Germany, 5–9 October 2015; Springer: Berlin/Heidelberg, Germany, 2015; pp. 234–241.
32. Cao, H.; Wang, Y.; Chen, J.; Jiang, D.; Zhang, X.; Tian, Q.; Wang, M. Swin-unet: Unet-like pure transformer for medical image segmentation. *arXiv* **2021**, arXiv:2105.05537.
33. Gao, J.; Weng, L.; Xia, M.; Lin, H. MLNet: Multichannel feature fusion lozenge network for land segmentation. *J. Appl. Remote Sens.* **2022**, *16*, 016513. [CrossRef]
34. Woo, S.; Park, J.; Lee, J.Y.; Kweon, I.S. Cbam: Convolutional block attention module. In Proceedings of the European Conference on Computer Vision (ECCV), Munich, Germany, 8–14 September 2018; pp. 3–19.
35. Li, G.; Yun, I.; Kim, J.; Kim, J. Dabnet: Depth-wise asymmetric bottleneck for real-time semantic segmentation. *arXiv* **2019**, arXiv:1907.11357.
36. Chen, L.C.; Zhu, Y.; Papandreou, G.; Schroff, F.; Adam, H. Encoder-decoder with atrous separable convolution for semantic image segmentation. In Proceedings of the European Conference on Computer Vision (ECCV), Munich, Germany, 8–14 September 2018; pp. 801–818.
37. Ma, N.; Zhang, X.; Zheng, H.T.; Sun, J. Shufflenet v2: Practical guidelines for efficient cnn architecture design. In Proceedings of the European Conference on Computer Vision (ECCV), Munich, Germany, 8–14 September 2018; pp. 116–131.

38. Yu, C.; Gao, C.; Wang, J.; Yu, G.; Shen, C.; Sang, N. Bisenet v2: Bilateral network with guided aggregation for real-time semantic segmentation. *Int. J. Comput. Vis.* **2021**, *129*, 3051–3068. [CrossRef]
39. Fang, L.; Liu, J.; Liu, J.; Mao, R. Automatic segmentation and 3d reconstruction of spine based on fcn and marching cubes in ct volumes. In Proceedings of the 2018 10th International Conference on Modelling, Identification and Control (ICMIC), Guiyang, China, 2–4 July 2018; pp. 1–5.
40. Badrinarayanan, V.; Kendall, A.; Cipolla, R. Segnet: A deep convolutional encoder-decoder architecture for image segmentation. *IEEE Trans. Pattern Anal. Mach. Intell.* **2017**, *39*, 2481–2495. [CrossRef]
41. Liu, S.; Qi, L.; Qin, H.; Shi, J.; Jia, J. Path aggregation network for instance segmentation. In Proceedings of the IEEE Conference on Computer Vision and Pattern Recognition, Salt Lake City, UT, USA, 18–22 June 2018; pp. 8759–8768.
42. Jiang, W.; Wu, Y.; Guan, L.; Zhao, J. Dfnet: Semantic segmentation on panoramic images with dynamic loss weights and residual fusion block. In Proceedings of the 2019 International Conference on Robotics and Automation (ICRA), Montreal, QC, Canada, 20–24 May 2019; pp. 5887–5892.
43. Fu, J.; Liu, J.; Tian, H.; Li, Y.; Bao, Y.; Fang, Z.; Lu, H. Dual attention network for scene segmentation. In Proceedings of the IEEE/CVF Conference on Computer Vision and Pattern Recognition, Long Beach, CA, USA, 16–20 June 2019; pp. 3146–3154.
44. Peng, Z.; Huang, W.; Gu, S.; Xie, L.; Wang, Y.; Jiao, J.; Ye, Q. Conformer: Local features coupling global representations for visual recognition. In Proceedings of the IEEE/CVF International Conference on Computer Vision, Montreal, QC, Canada, 10–17 October 2021; pp. 367–376.
45. Mehta, S.; Rastegari, M.; Shapiro, L.; Hajishirzi, H. Espnetv2: A light-weight, power efficient, and general purpose convolutional neural network. In Proceedings of the IEEE/CVF Conference on Computer Vision and Pattern Recognition, Long Beach, CA, USA, 16–20 June 2019; pp. 9190–9200.
46. Han, K.; Wang, Y.; Tian, Q.; Guo, J.; Xu, C.; Xu, C. Ghostnet: More features from cheap operations. In Proceedings of the IEEE/CVF Conference on Computer Vision and Pattern Recognition, Seattle, WA, USA, 14–19 June 2020; pp. 1580–1589.
47. Yu, C.; Wang, J.; Peng, C.; Gao, C.; Yu, G.; Sang, N. Learning a discriminative feature network for semantic segmentation. In Proceedings of the IEEE Conference on Computer Vision and Pattern Recognition, Salt Lake City, UT, USA, 18–22 June 2018; pp. 1857–1866.
48. Wu, H.; Xiao, B.; Codella, N.; Liu, M.; Dai, X.; Yuan, L.; Zhang, L. Cvt: Introducing convolutions to vision transformers. In Proceedings of the IEEE/CVF International Conference on Computer Vision, Montreal, QC, Canada, 10–17 October 2021; pp. 22–31.
49. Wang, W.; Xie, E.; Li, X.; Fan, D.P.; Song, K.; Liang, D.; Lu, T.; Luo, P.; Shao, L. Pyramid vision transformer: A versatile backbone for dense prediction without convolutions. In Proceedings of the IEEE/CVF International Conference on Computer Vision, Montreal, QC, Canada, 10–17 October 2021; pp. 568–578.
50. Yuan, Y.; Chen, X.; Chen, X.; Wang, J. Segmentation transformer: Object-contextual representations for semantic segmentation. *arXiv* **2019**, arXiv:1909.11065.

remote sensing

Article

Scale-Invariant Multi-Level Context Aggregation Network for Weakly Supervised Building Extraction

Jicheng Wang [1,2,†], Xin Yan [3,†], Li Shen [3,*], Tian Lan [3], Xunqiang Gong [2] and Zhilin Li [3]

1 Key Laboratory of Land Resources Evaluation and Monitoring in Southwest China of Ministry of Education, Sichuan Normal University, Chengdu 610068, China
2 Key Laboratory of Mine Environmental Monitoring and Improving around Poyang Lake of Ministry of Natural Resources, East China University of Technology, Nanchang 330013, China
3 Faculty of Geosciences and Environmental Engineering, Southwest Jiaotong University, Chengdu 611756, China
* Correspondence: lishen@swjtu.edu.cn
† These authors contributed equally to this work.

Abstract: Weakly supervised semantic segmentation (WSSS) methods, utilizing only image-level annotations, are gaining popularity for automated building extraction due to their advantages in eliminating the need for costly and time-consuming pixel-level labeling. Class activation maps (CAMs) are crucial for weakly supervised methods to generate pseudo-pixel-level labels for training networks in semantic segmentation. However, CAMs only activate the most discriminative regions, leading to inaccurate and incomplete results. To alleviate this, we propose a scale-invariant multi-level context aggregation network to improve the quality of CAMs in terms of fineness and completeness. The proposed method has integrated two novel modules into a Siamese network: (a) a self-attentive multi-level context aggregation module that generates and attentively aggregates multi-level CAMs to create fine-structured CAMs and (b) a scale-invariant optimization module that cooperates with mutual learning and coarse-to-fine optimization to improve the completeness of CAMs. The results of the experiments on two open building datasets demonstrate that our method achieves new state-of-the-art building extraction results using only image-level labels, producing more complete and accurate CAMs with an IoU of 0.6339 on the WHU dataset and 0.5887 on the Chicago dataset, respectively.

Keywords: building extraction; high-resolution remote sensing image; weakly supervised semantic segmentation; self-attentive aggregation; class activation map

Citation: Wang, J.; Yan, X.; Shen, L.; Lan, T.; Gong, X.; Li, Z. Scale-Invariant Multi-Level Context Aggregation Network for Weakly Supervised Building Extraction. *Remote Sens.* **2023**, *15*, 1432. https://doi.org/10.3390/rs15051432

Academic Editors: Janet E. Nichol, Faisal M. Qamer, Jianchu Xu and Sawaid Abbas

Received: 7 February 2023
Revised: 1 March 2023
Accepted: 2 March 2023
Published: 3 March 2023

1. Introduction

Automatic building extraction from high-resolution images has become an active topic in the field of remote sensing in recent decades. It plays a vital role in a variety of applications, such as urban monitoring [1], population and economic estimation [2], and geospatial database making and updating [3]. Building extraction aims to classify each pixel as building or non-building, which can be regarded as binary semantic segmentation. However, this task is very challenging due to the difficulty in distinguishing between buildings with complex appearances and varying scales in high-resolution images with rich details and intra-class variance characteristics [4].

Convolutional neural networks (CNNs) have gained widespread popularity in various domains in recent years, including computer vision [5], climate change [6], and industrial detection [7,8]. With remarkable success in image segmentation, CNNs have been applied to building extraction from high-resolution remote sensing imagery using networks such as UNet [9], DeeplabV3+ [10], and PSPNet [11]. Researchers have also proposed building-specific approaches based on analyzing the characteristics of buildings, which have shown promising results [12–15]. However, these approaches have limited applications due to the need for a large number of pixel-level annotations, which are time-consuming and labor-intensive to collect. Instead, weakly supervised semantic segmentation (WSSS) tries to alleviate this issue by utilizing weak supervision, such as image-level labels [16], bounding boxes [17], and scribbles [18]. As image-level labels are more readily available than other forms of supervision, this paper focuses on weakly supervised building extraction using image-level labels.

Image-level labels only indicate the presence or absence of buildings in the image without any localization cue, making it challenging for WSSS to achieve compelling results with fully supervised semantic segmentation. Fortunately, a class activation map (CAM) [16] is proposed to perform object localization only using image-level labels. Most advanced WSSS approaches are based on CAM and follow a three-stage learning paradigm: (1) using image-level labels to train a classification network to obtain the initial CAMs; (2) refining the initial CAMs to generate pseudo-pixel-level labels by semantic affinity-based methods [19], dense conditional random fields (CRF) [20] or saliency detection methods [21]; and (3) training a semantic segmentation network with these pseudo-labels. As supervised information is only utilized in the first stage, the key to the WSSS method is generating a promising CAM that is accurately activated on entire objects and not the background. To this point, many methods have been extensively investigated in the field of computer vision [22–26] and have shown effectiveness in processing natural scene images. However, these methods may not be suitable for high-resolution remote sensing images, which often contain vast amounts of visual information, significant spatio–spectral variability, and a much wider field of view [27].

In recent years, the remote sensing community has seen a growing interest in utilizing WSSS techniques for building extraction in high-resolution images. Advances in this area have resulted in several notable contributions, such as the SPMF-Net [28], which incorporates a superpixel-pooling mechanism to enhance the CAM and better preserve the shape and boundary information of buildings. The MSG-SR-Net [29] further advances this line of research by integrating a multi-scale generation strategy to improve the fineness of the CAM. Other methods, such as that of Li et al. [30], have utilized conditional random fields (CRF) to optimize both the CAM and segmentation results. In an effort to learn more building-specific information and encourage the network to perform more accurately, some studies have explored the exploitation of inherent relationships, such as pixel affinity [31] and adversarial information [32], to benefit building representation. Despite the demonstrated effectiveness of these techniques, the quality of the pseudo-labels generated by the CAM remains a critical factor affecting the performance of WSSS for building extraction. As illustrated in Figure 1, the CAMs often only activate the most discriminative regions, making it challenging to generate complete buildings. Furthermore, over-activation and under-activation of the CAMs can result in fuzzy boundaries, presenting opportunities for further improvement in the performance of weakly supervised building extraction.

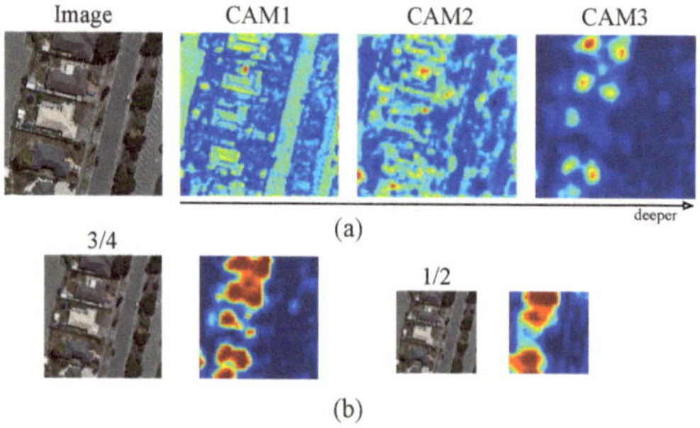

Figure 1. Visualization of CAMs generated by GradCAM++: (**a**) CAMs generated at different levels; (**b**) CAMs generated from different scales.

The limitations in building extraction using weakly supervised methods stem from the absence of localization information in image-level labels, creating a supervision gap between fully and weakly supervised methods. To overcome this challenge, it is crucial to incorporate more spatial information into the weakly supervised method, such as by utilizing the outputs of various layers in a neural network. As depicted in Figure 1, our research findings indicate that the CAM generated by lower-level layers possesses more details, and also more noise, than the CAM produced by high-level layers. By effectively combining CAMs from various levels, the accuracy of the CAMs can be significantly improved. Furthermore, our observations suggest that CAMs generated at different image scales do not always align with the scale variations of the buildings. The CAM generated at a coarser scale tends to highlight the complete area of the buildings but lacks detail at the boundaries, whereas the CAM generated at a finer scale exhibits the opposite trend. This discrepancy can serve as supervisory information to enhance the integrity of the building representations in a CAM.

Based on the above observation and analysis, we present a unified network that aims to enhance the quality of a CAM with regard to building representation. This is achieved through two key improvements to CAMs: (1) multi-level context aggregation for fine-structured refinement and (2) utilization of multi-scale supervision for integrity improvement. To achieve the first improvement, we introduce the self-attentive multi-level context aggregation module (SMCAM), which is based on GradCAM++ [23]. This module generates CAMs from multiple levels while suppressing noise in the network and uses a self-attention mechanism to effectively combine these CAMs, resulting in a more refined representation of building structures. For the second improvement, we propose the scale-invariant optimization module (SIOM) to further improve the integrity of CAMs. This module uses CAMs generated on a coarse scale as pixel-level supervision, guiding the network in learning to improve the integrity of the CAMs, thus ensuring that the CAMs are more aligned with the scale variation of buildings. Therefore, by incorporating the two enhancements, the proposed unified network is capable of generating high-quality CAMs that not only preserve fine structures but also ensure the integrity of the building representation. This leads to more reliable pixel-level training samples that are crucial for the performance of subsequent semantic segmentation steps. The main contributions of this study are summarized as follows:

- A self-attentive method that effectively generates and aggregates multi-level CAMs is proposed to produce fine-structured CAMs;
- A scale-invariant optimization method that incorporates multi-scale supervision is proposed to improve the completeness of CAMs;
- The Siamese network that integrates the above two improvements with designed losses is introduced with the aim of narrowing the supervision gap between fully and weakly supervised building extraction.

The organization of this paper is as follows: In Section 2, we give a comprehensive overview of previous studies on building extraction and weakly supervised semantic segmentation methods. The proposed network and its crucial components, SMCAM and SIOM, are thoroughly explained in Section 3. In Section 4, the performance of the proposed network is evaluated and compared with existing methods on two commonly used building datasets. A thorough analysis and discussion of the results are presented in Section 5, followed by the conclusions and future work in Section 6.

2. Related Works

This section presents a comprehensive overview of the most recent deep learning methods for building extraction from high-resolution remote sensing images. With the rapid advancements in the field, many researchers have proposed new techniques for weakly supervised semantic segmentation and building extraction specifically. Instead of giving a complete assessment of all current approaches in the field, this review only focuses on the related studies.

2.1. Building Extraction

Recently, fully convolutional networks (FCNs) have gained significant attention in the remote sensing community for their ability to perform efficient and accurate building extraction from high-resolution images. As a type of convolutional neural network (CNN), FCNs use a combination of a convolutional encoder and decoder to make dense predictions for every pixel in the input image, which makes them a promising solution for building extraction tasks. Studies have shown that FCNs for building extraction outperform traditional methods that rely on hand-crafted features in terms of accuracy and computational efficiency [12,15,33–42]. Some of these studies modify existing semantic segmentation networks to adapt them to building extraction, such as Schuegraf and Bittner [12], who combine two parallel U-Net-like [9] FCNs to extract the spatial and spectral features, respectively, and then fuse them to extract buildings and Yuan et al. [40], who improve PSPNet [11] by adding a designed feature pooling layer to capture both local and global relationships in building extraction. Others have proposed FCN models for building extraction that are specific to building characteristics. Guo et al. [42] propose a novel coarse-to-fine boundary refinement network (CBR-Net) that accurately extracts buildings. Li et al. [39] develop the CrossGeoNet with a Siamese network and a cross-geolocation attention module to provide the general building representation in different cities. There are also FCN models that utilize auxiliary data, such as digital surface models [36,38], to improve building extraction results.

The aforementioned methods can produce desirable outcomes; however, they heavily rely on large quantities of pixel-level annotations during training. To mitigate this issue, semi-supervised, unsupervised, and weakly supervised methods have emerged as alternatives for building extraction. Existing semi-supervised building extraction methods can be divided into three categories: self-training, generative adversarial network (GAN)-based, and consistency regularization methods [43]. Self-training models are trained with labeled samples, and then predictions for unlabeled samples are utilized as pseudo-labels for supervised training [4,44]. GAN-based methods mainly use limited annotated data to train the generators, which, in turn, generate synthetic annotations for the unannotated training images [45]. Consistency regularization methods can learn the distribution of unlabeled data by detecting the consistency of the output before and after perturbation [46]. However, these methods still rely on pixel-level labels in essence. Moreover, some researchers have attempted unsupervised building extraction [47]. As expected, the task presents significant challenges, and much progress must still be made before it can be accomplished successfully. More studies are focused on weakly supervised building extraction and are based on annotations that are less supervised than pixel-level labels, such as scribbles [48], image-level labels, and bounding boxes. In this paper, the approach taken is a weakly supervised segmentation method that exploits image-level labels, as described in the next subsection.

2.2. Weakly Supervised Semantic Segmentation

Weakly supervised semantic segmentation (WSSS) methods with image-level labels have gained widespread attention in the field of computer vision due to their cost-effectiveness as compared to pixel-level labeling. WSSS methods typically involve localizing objects with class activation maps (CAMs), generating pseudo-labels from CAMs, and training a semantic segmentation network. However, native CAMs [16,23] tend to produce inaccurate pseudo-labels as they only highlight the most discriminative regions, which are often incomplete and rough for objects. Hence, as the key to WSSS, many efforts have been made to improve CAMs for more complete object localization, such as iterative erasing [49], random dropping [50], super-pixel pooling [25], and local-to-global transferring [51]. Other studies have focused on refining CAMs using techniques such as multitask learning [52], region growing [53], pixel affinity [19], and inter-pixel relations [54]. Researchers also believe that the limitations of WSSS methods can be attributed to the supervision gap between the classification and segmentation tasks and have thus proposed methods to introduce auxiliary supervisory information to narrow this gap.

Wang et al. [24] propose SEAM to exploit pixel-level supervision through constraints between various affine changes. Du et al. [22] explore pixel-level supervisory signals with a combination of contrastive learning and consistency regularization. Sub-pixel supervision information is also introduced to compensate for the lack of supervision information [55]. Although these WSSS methods have achieved promising results for handling natural images in the field of computer vision, they may struggle to perform efficiently with remote sensing images as they are not specifically designed for this domain. Generalizing weakly supervised methods to different domains remains a challenge, as has been established by research [27].

Researchers have recently started exploring the potential of weak supervision in building extraction from remote sensing imagery, recognizing the need for methods specifically tailored to building characteristics to improve the accuracy and consistency of building representations in CAMs. Several studies have been conducted to address this challenge and enhance the quality of CAMs. For example, Fu et al. [56] develop WSF-Net for binary segmentation in remote sensing images and address class imbalance through a balanced binary training approach. MSG-SR-Net [29] further improves CAM fineness by integrating a multi-scale generation strategy. Meanwhile, methods such as conditional random fields by Li et al. [30] and superpixel pooling by Chen et al. [28] have been utilized to explore the spatial context and enhance building representation, respectively. The use of semantic affinity [32] and pixel affinity [31] has also been shown to improve the quality of CAMs. In contrast to these methods, we propose a novel WSSS method, based on the observations of CAM (Figure 1), which addresses two crucial aspects: multi-scale and multi-level. By generating and aggregating CAMs at multiple levels and training the images with multi-scale inputs, our method generates auxiliary pixel-level supervised information. This interplay of multi-scale and multi-level results in higher-quality CAMs compared to existing methods. It is noteworthy that MSG-SR-Net also leverages multi-level CAM fusion; however, our approach incorporates a self-attentive mechanism that enables CAMs to automatically calculate their importance during fusion, thus preserving valuable information and suppressing noise.

3. Scale-Invariant Multi-Level Context Aggregation Network

This section provides a description of the methodology and design of the network. Firstly, we define the problem context and describe the process of generating a CAM. Secondly, the overall architecture of the proposed network is presented. Thirdly, we delve into the two modules that are utilized to improve the quality of the CAMs with respect to building characteristics. Finally, we discuss the loss functions that are used in the proposed network.

3.1. Prerequisites

The problem of weakly supervised building extraction can be defined as follows: Given a set of images and corresponding image-level labels, the goal is to learn a model that can predict a pixel-level segmentation mask for the buildings in the image. Specifically, each training image, represented as $\mathbf{I}_{\mathcal{D}} \in \mathbb{R}^{W \times H \times 3}$ in dataset $\mathcal{D}$, is associated with a binary image-level label $y \in \{0, 1\}$, where 1 indicates the presence of buildings in image $\mathbf{I}$ and 0 indicates their absence. Since the image-level labels do not provide any localization information, current methods typically follow a two-stage approach to tackle this task, i.e., first training a classification network to identify regions in the image that correspond to buildings, and then using these regions to generate pseudo-labels for training a semantic segmentation network for building extraction. The majority of existing WSSS approaches rely on CAMs to produce the localization map. The training phase of the task can be formulated as follows:

$$\boldsymbol{M} = \mathcal{F}_{\text{CAM}}(\mathbf{I}|\mathbf{I}_{\mathcal{D}}, \mathbf{y}_{\mathcal{D}}) \in \mathbb{R}^{W \times H \times 2}, \quad \boldsymbol{Y}_{pred} = \mathcal{F}_{\text{SEG}}(\mathbf{I}|\mathbf{I}_{\mathcal{D}}, \mathcal{F}_{pseudo}(\boldsymbol{M})) \in \mathbb{R}^{W \times H \times 2} \tag{1}$$

Here, $\boldsymbol{M}$ represents the CAMs for the image $\mathbf{I}$ generated by the classification network $\mathcal{F}_{\text{CAM}}$, which is trained on the image $\mathbf{I}_{\mathcal{D}}$ and corresponding image-level labels $\mathbf{y}_{\mathcal{D}}$ in dataset $\mathcal{D}$. $\mathcal{F}_{\text{SEG}}$ is an FCN network for semantic segmentation, and $\mathcal{F}_{\text{pseudo}}$ is used to generate the required pseudo-pixel-level labels. The final score map of the buildings is denoted by $\boldsymbol{Y}_{\text{pred}}$. Accurate and complete CAMs are crucial for the success of semantic segmentation, as they significantly influence the performance of the pseudo-labels. Therefore, this paper focuses on improving the generation of CAMs for building extraction.

The proposed network is based on GradCAM++ [23] to improve the quality of CAMs. We chose GradCAM++ due to its two key features. Firstly, it can be loosely considered as a generalized version of the original CAM with an improved ability to localize features. Secondly, it does not require any modifications to the network, preserving the original classification network's learning capabilities. GradCAM++ generates a visual explanation for the specified class label c by using a weighted combination of the positive partial derivatives of the last convolutional layer's feature maps with respect to the class score, which is calculated as:

$$\boldsymbol{M} = relu(\sum_{k} \alpha_k^c \boldsymbol{A}_k) \tag{2}$$

where $\boldsymbol{A}_k$ denotes the k -th channel of the output feature map generated by the last convolutional layer. The rectified linear unit, $relu()$, filters out the features with negative values, where $relu(x) = max(0, x)$. α_k^c denotes the weight of class c corresponding to the k -th channel, which is calculated as follows:

$$\alpha_k^c = \frac{\frac{\partial^2 Y^c}{\partial \boldsymbol{A}_k^2}}{2\frac{\partial^2 Y^c}{\partial \boldsymbol{A}_k^2} + \sum \boldsymbol{A}_k \cdot \frac{\partial^3 Y^c}{\partial \boldsymbol{A}_k{}^3}} \cdot relu(\frac{\partial Y^c}{\partial \boldsymbol{A}_k}) \tag{3}$$

where $\frac{\partial Y^c}{\partial \boldsymbol{A}_k}$, $\frac{\partial^2 Y^c}{\partial \boldsymbol{A}_k^2}$, and $\frac{\partial^3 Y^c}{\partial \boldsymbol{A}_k^3}$ represent the first-order, second-order, and third-order gradients of the prediction score Y^c, respectively. For computational convenience, it is common to make $Y^c = exp(f_c)$; here, f_c is the output score of the classification network.

It should be noted that the CAMs generated by GradCAM++ have the same size as feature map $\boldsymbol{A}$ and are not normalized. To make them more suitable for subsequent processing, we resample them to the size of the input image and normalize them using the following formula:

$$\boldsymbol{M}^* = \frac{\boldsymbol{M} - MIN(\boldsymbol{M})}{MAX(\boldsymbol{M}) - MIN(\boldsymbol{M})} \tag{4}$$

where $MAX()$ represents the maximum value, and $MIN()$ represents the minimum value.

3.2. Overall Network Architecture

We proposed a network for the task of weakly supervised building extraction, as depicted in Figure 2, which leverages multi-level features and multi-scale information to improve the quality of CAMs. It consists of three main components: a Siamese classification network, the self-attentive multi-level context aggregation module (SMCAM), and the scale-invariant optimization module (SIOM). The Siamese network utilizes shared weights to classify two input images with different scales, utilizing the ground-truth image-level labels as the target. Both images are processed simultaneously using the same network architecture, which can be designed based on well-known networks, such as ResNet [57] and VGG [58]. The main purpose of the Siamese network is to generate multi-level features at different scales through training with an image-level labeled dataset. The other two modules are designed to utilize these features to guide the improvement of the CAM. Specifically, SMCAM generates multi-level CAMs based on GradCAM++ and fuses these CAMs through a self-attention mechanism. SIOM further exploits these features to improve the integrity of CAMs by mutual learning between different scale supervisions. Both modules are described in more detail in the subsequent subsections.

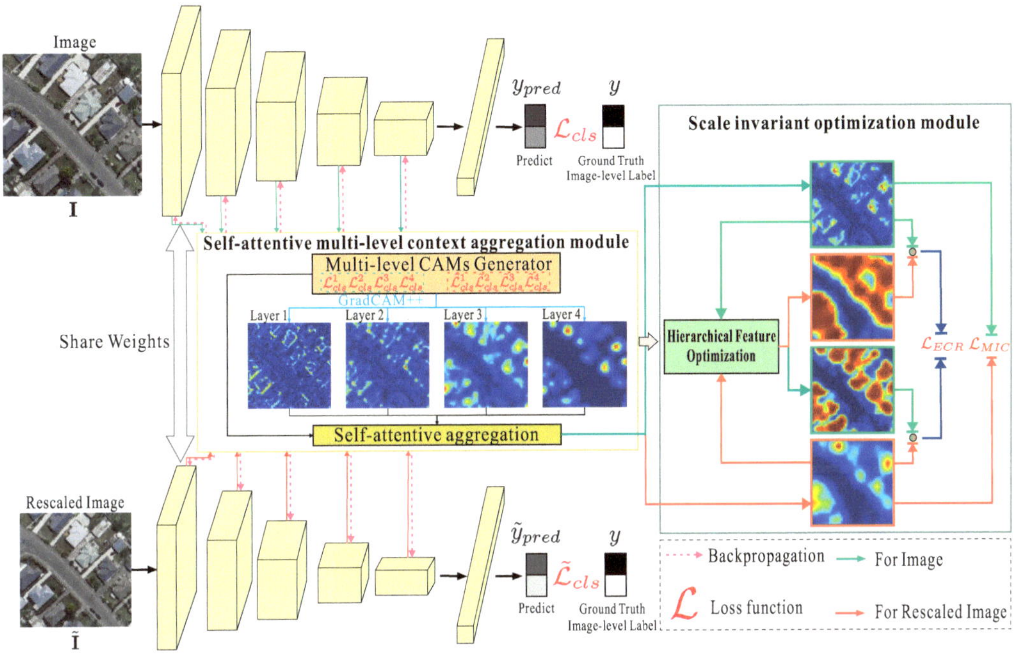

Figure 2. The framework of the proposed network.

It is noteworthy that the proposed network takes two images of different scales as input during the training phase in order to provide multi-scale supervision. During the inference phase, the network only requires a single input image, and the CAM is generated through sequential optimization of SMCAM and SIOM.

3.3. Self-Attentive Multi-Level Context Aggregation Module

The proposed module, SMCAM, is designed to address two major problems in building extraction tasks: the generation of multi-level CAMs and the aggregation of these CAMs. The main aim of SMCAM is to utilize GradCAM++ to generate CAMs at different levels of a deep neural network and to aggregate these CAMs to produce more detailed CAMs. To achieve the first goal, we utilize GradCAM++ to backpropagate gradients from score maps to any nodes of the network (such as the red dot on each ResNet Block in Figure 3). However, it is observed that CAMs generated from low-level features may contain a significant amount of noise, making them challenging to use directly. This noise is mainly due to two factors: disturbance introduced during gradient backpropagation due to the long path and presence of noise in the low-level features themselves. To tackle these issues, SMCAM introduces auxiliary classification branches at each node to improve the semantic depth of low-level features and shorten the gradient backpropagation path. These branches consist of a 1x1 convolutional layer with 1024 output channels, followed by an average pooling layer, which ultimately outputs an image-level score. The auxiliary classification branches enhance the semantic information in low-level features, making the generated CAMs more useful for the building extraction task.

To achieve the second goal of aggregating multi-level CAMs, our proposed module incorporates an attention-based mechanism. Unlike traditional fusion methods, such as averaging and concatenation, which ignore the relative importance of features at different levels, our model aims to learn the varying contributions of each level for every pixel in the CAMs. Specifically, the process starts with passing the feature maps generated by the average pooling layer of each auxiliary branch through the channel average pooling layer, generating a score map for each level, $l \in \{1, ..., L\}$, where each score map has a single channel. The score maps are then upsampled to match the size of the CAMs. Mathematically, let $\boldsymbol{H}_l$ denote the weight score generated at level l. A softmax function is applied across the levels to compute the specific weights, $\mathbf{w}_l$, of the CAMs for each level:

$$\mathbf{w}_l = \frac{exp(\boldsymbol{H}_l)}{\sum_{l=1}^{L} exp(\boldsymbol{H}_l)}. \tag{5}$$

Finally, the CAMs are aggregated into a single map, $\boldsymbol{M}$, through a weighted sum of the score maps of all levels, as calculated by:

$$\boldsymbol{M} = \sum_{l=1}^{L} \mathbf{w}_l \boldsymbol{M}^l, \tag{6}$$

where $\boldsymbol{M}^l$ represents the CAM generated by GradCAM++ at the auxiliary branch of level l, and the weights are determined by the softmax function applied to the weight scores generated at each level.

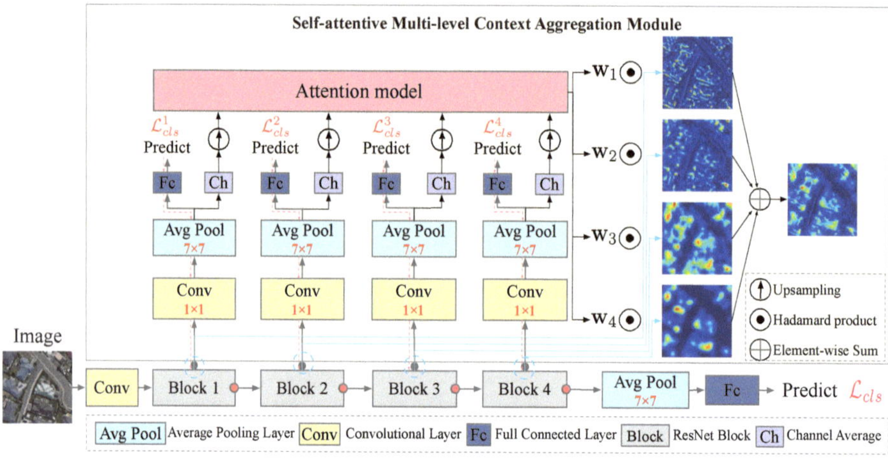

Figure 3. The structure of self-attentive multi-level context aggregation module.

3.4. Scale-Invariant Optimization Module for Improving Integrity of CAMs

Although the CAMs generated by SMCAM utilize multi-level features, there still exists a drawback in that the generated CAMs may not cover the entire building object, as they often struggle to activate the most discriminative regions without adequate pixel-level supervision. One of our key observations is that CAMs generated by the classification network for different scales of input images activate different regions for building objects, providing valuable additional supervisory information. Therefore, we proposed the scale-invariant optimization module (SIOM) to leverage this multi-scale information. As illustrated in Figure 2, SIOM consists of two crucial parts: a mutual learning mechanism of CAMs and hierarchical feature optimization. For the former, a mutual learning mechanism benefits from the architecture of the Siamese network, which consists of two subnetworks with shared parameters that process two different scale images simultaneously and generate a series of CAMs at different scales under two scales with SMCAM. By motivating CAMs of different scales to be similar at multiple levels, the network can learn scale-invariant representations that better capture complete building objects. Therefore, we proposed the multi-level invariant constraint loss $\mathcal{L}_{MIC}$ by extending the equivariant constraint to multiple levels, as presented in the next subsection.

For the latter, although the multi-level invariant constraint loss can provide additional pixel-level supervision at multiple levels, the CAM generation is still limited to the framework of the classification networks. To alleviate this, we further propose a separate learnable branch, or hierarchical feature optimization, for enhancing the completeness of a CAM. In particular, it optimizes a CAM in a progressive manner, leveraging progressively the image and multi-level features generated by the classification network through a three-stage process, as shown in Figure 4. The first two stages utilize self-attention mechanisms to uncover non-local relationships [59] within the multi-level features of the classification network, resulting in improved object integrity. Each stage consists of two convolutional layers with learnable parameters. The optimization process can be mathematically expressed as follows:

$$\boldsymbol{M}' = softmax(\mathbf{x}^T W_1^T W_2 \mathbf{x})\boldsymbol{M}, \tag{7}$$

where $\boldsymbol{M}'$ denotes the optimized CAM. W_1 and W_2 denote the parameters of the first and second convolutional layers, respectively. The utilized features are indicated by $\mathbf{x}$, for stage 1, $\mathbf{x} = [\mathbf{x}_4, Down(\mathbf{x}_2)]$ and for stage 2, $\mathbf{x} = [\mathbf{x}_3, Down(\mathbf{x}_1)]$, where $[]$ and $Down()$ denote the concatenating and downsampling, respectively. This self-attention mechanism essentially uses the relationships between all of the pixels in feature maps to refine the CAM and has been shown to be particularly effective at mining global relationships. However, as the size of the CAM increases, the size of the attention map $W_1^T W_2$ increases exponentially, leading to memory explosions on the graphics card during training; hence, we only use this mechanism in the first two stages. In the third stage, we fully exploit the local information in the image, which is used to enhance the detail stages of the CAM. We use a fixed-parameter pixel adaption convolution [60], and the optimization process can be represented by the following equation:

$$M_i'' = \sum_{j \in \mathcal{N}(i)} D_{i,j} M_{i,j}', \tag{8}$$

where M_i denotes the value at pixel i in the CAM; $\mathcal{N}(i)$ denotes the set of neighboring pixels of pixel i; and $\mathbf{D}$ denotes the affinity map, where $D_{i,j}$ denotes the similarity relationship between pixel i and pixel j, as calculated by the following equation:

$$D_{i,j} = \frac{exp(k(\mathbf{I}_i, \mathbf{I}_j))}{\sum_{j \in \mathcal{N}(i)} exp(k(\mathbf{I}_i, \mathbf{I}_j))}, \tag{9}$$

where $\mathbf{I}_i$ denotes the spectral vector at image pixel i; k is the kernel function, where the Gaussian function is used, i.e., $k(\mathbf{I}_i, \mathbf{I}_j) = exp(-\frac{1}{2}(\mathbf{I}_i - \mathbf{I}_j)^T(\mathbf{I}_i - \mathbf{I}_j))$; and the softmax function is used to normalize. It is important to note that this operation is distinct from the Hadamard product, as indicated by the notation $\odot$ in Figure 4. The parameters of SIOM are learned by comparing optimized and unoptimized CAMs on two scales. The gradient propagation path of this module is detached from the classification network to prevent interference with learning classification representation. By optimizing the hierarchy of non-local and local information, the problem of building edges and integrity can be solved to a large extent, effectively improving the quality of the CAM for building extraction.

It should be noted that the basic architecture of SIOM is inspired by SEAM [24] but has three differences. Firstly, we are based on GradCAM++, which is more generalized than the original CAM applied in SEAM. Secondly, we propose hierarchical feature optimization, which can utilize multi-level features and images to optimize CAM with regard to building extraction. Thirdly, our loss function is specifically designed to address the objective of building extraction. Therefore, the proposed SIOM is more appropriate for weakly supervised building extraction than SEAM.

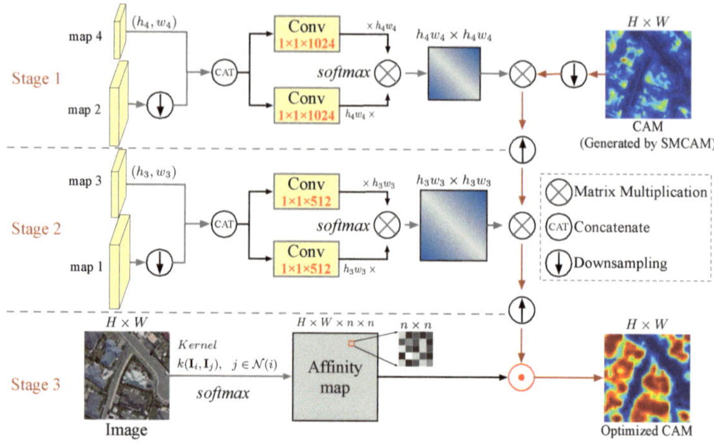

Figure 4. The structure of scale-invariant optimization module.

3.5. The Design of Loss Function

Since only image-level labels are available, the design of the loss function is exceptionally important in order to introduce more supervised information, especially some pixel-level supervisions. In summary, the total loss of the proposed network is defined as follows:

$$\mathcal{L}_{ALL} = \mathcal{L}_{CLS} + \mathcal{L}_{MIC} + \mathcal{L}_{ECR}, \tag{10}$$

where the classification loss is denoted as $\mathcal{L}_{CLS}$, which exploits image-level supervision, and $\mathcal{L}_{MIC}$ is the multi-level invariant constraint loss in SIOM. The equivariant cross regularization loss $\mathcal{L}_{ECR}$ is adopted to train the branch of hierarchical feature optimization.

The binary cross-entropy loss is utilized as the classification loss in the proposed network and is defined as follows:

$$\mathcal{L}_{cls} = -\frac{1}{N}\sum_{n=1}^{N}(y_n log(y_{pred,n}) + (1 - y_n) log(1 - y_{pred,n})), \tag{11}$$

where N is the batch size; y_{pred} denotes the output of the classification network and y_n is the corresponding image-level label. The network is a Siamese network and can generate two scores for the original image $\mathbf{I}$ and the rescaled image $\tilde{\mathbf{I}}$, respectively. In addition, the proposed SMCAM module has multiple output branches for various levels. The classification loss of the network is made up of many sub-losses, which is described below:

$$\mathcal{L}_{CLS} = \mathcal{L}_{cls} + \tilde{\mathcal{L}}_{cls} + \frac{1}{L}\sum_{l=1}^{L}(\mathcal{L}^{l}_{cls} + \tilde{\mathcal{L}}^{l}_{cls}), \tag{12}$$

where $\mathcal{L}_{cls}$ and $\tilde{\mathcal{L}}_{cls}$ are the losses calculated for the Siamese network, while SMCAM at level l has its own losses, represented by $\mathcal{L}^{l}_{cls}$ and $\tilde{\mathcal{L}}^{l}_{cls}$.

The multi-level invariant constraints (MIC) loss is built with a multi-level equivariant metric based on the L1 loss. The aim is to enhance the CAMs generated by different branches of the Siamese network through multiple levels of features, leading to scale-invariant features that result in more complete building representations in the CAMs. The calculation of this loss is described below.

$$\mathcal{L}_{MIC} = \frac{1}{L}\sum_{l=1}^{L}\|\boldsymbol{M}^{l} - up(\tilde{\boldsymbol{M}}^{l})\|_1, \tag{13}$$

where $\|\cdot\|$ denotes the L1-norm, and $up(\cdot)$ denotes the upsampling.

For the equivariant cross regularization loss, we refer to the SEAM [24], which can be represented as:

$$\mathcal{L}_{ECR} = \frac{1}{2}(\|\boldsymbol{M}'' - up(\tilde{\boldsymbol{M}})\|_1 + \|up(\tilde{\boldsymbol{M}}'') - \boldsymbol{M}\|_1), \tag{14}$$

where $\boldsymbol{M}''$ is the CAM optimized by hierarchical feature optimization; $\tilde{\boldsymbol{M}}$ denotes the CAM generated by SMCAM the from rescaled image. This loss can further improve the quality of the CAMs and prevent degeneration.

4. Experiments

4.1. Experimental Datasets

The proposed method is evaluated using two publicly available high-resolution remote sensing datasets, namely the WHU Building Dataset [34] and the Inria Aerial Image Labeling Dataset [61]. The WHU Building Dataset (abbreviated as "WHU dataset") consists of 8189 tiles with 512 × 512 pixels, covering approximately 450 km^2 in the Christchurch area. Each image has three channels (RGB) and a spatial resolution of 0.3 m. These images are divided into 4736, 1036, and 2416 patches with corresponding ground-truth labels, which are then split into training, validation, and test sets. The Inria Aerial Image Labeling Dataset includes images and building labels for 10 cities, with a subset dataset of Chicago (abbreviated as "Chicago dataset") selected for the experiment. It contains 36 color image tiles of 5000 × 5000 pixels with a 0.3 m spatial resolution. All of these images have corresponding building annotations and are divided into a training set of 24 images, a testing set of 8 images, and a validation set of 4 images in this experiment.

Both datasets provide pixel-level labels and are primarily used to test and evaluate fully supervised building extraction methods. For the proposed weakly supervised method, only image-level labels are utilized to train the model. Therefore, pixel-level labels are used to generate image-level annotations for evaluating the method. Following the preprocessing in previous works [29] and [32], the images are cropped into patches with 256 × 256 pixels and a stride of 128 pixels. The image-level labels of these patches are determined by the percentage of building pixels in the pixel-level annotations. Specifically, patches with more than 22 percent of building pixels (about 2/9) are labeled as building, while patches with a percentage of 0, i.e., without any building pixels, are labeled as non-building. The remaining patches with a percentage between 0 and 22 are simply ignored. After processing, the WHU dataset contains 27,879 image patches for training, with 14,316 building patches, 13,563 non-building patches, and 18,364 patches for testing. The Chicago dataset contains 24,736 patches for training, including 12,793 building patches, 12,943 non-building patches, and 11,020 patches for testing. Some processed examples from both datasets are shown in Figure 5. To gain a better understanding of these two datasets, we utilize t-SNE [62] to visualize the feature distributions of both datasets. The features are extracted from the last layer output of a residual neural network [57]. Despite having the same resolution and purpose, the t-SNE visualization (Figure 5b) reveals that the feature distributions of the two datasets are notably distinct.

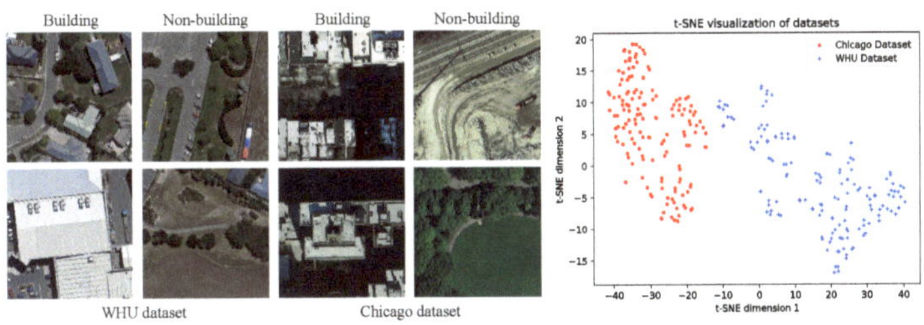

Figure 5. Sample images (**a**) and t-SNE visualization (**b**) for two datasets.

4.2. Experimental Setup

4.2.1. Methods for Comparison

In the experiments, seven WSSS methods are compared with the proposed method for the selected datasets. The main information about these methods, including ours, is summarized as follows:

- CAM-GAP: Zhou et al. [16] propose CAM-GAP for discriminative localization by adding a global average pooling modification to the network. It actually builds a generic localizable deep representation that can be applied to weakly supervised semantic segmentation;
- GradCAM++: Chattopadhay et al. [23] propose GradCAM++ based on gradients without changing the network structure. Its goal is to provide a visual interpretation for CNN-based models and can also be used for WSSS. It can be regarded as the generalization of a CAM;
- WILDCAT: Durand et al. [26] introduce WILDCAT to simultaneously align image regions for spatial invariance and learn strongly localized features for WSSS. It uses a single generic training scheme for classification, object localization, and semantic segmentation;
- SPN: The superpixel pooling network (SPN), proposed by Kwak et al. [25], utilizes superpixel segmentation of the input image as a pooling layout to cooperate with low-level features for semantic segmentation learning and inferring. The network architecture decouples the semantic segmentation task into classification and segmentation, allowing the network to learn class-agnostic shapes prior to the noisy annotations. It achieves outstanding performance on the challenging PASCAL VOC 2012 segmentation benchmark;
- WSF-Net: WSF-Net [56] is proposed for binary segmentation in remote sensing images with the potential to handle class imbalance through a balanced binary training strategy. It introduces a feature fusion strategy to adapt to the characteristics of objects in remote sensing images. The experiments achieve a promising performance for water and cloud extraction;
- MSG-SR-Net: MSG-SR-Net is proposed by Yan et al. [29] for weakly supervised building extraction from high-resolution images. It integrates two modules, i.e., multiscale generation and superpixel refinement, to generate high-quality CAMs so as to provide reliable pixel-level training samples for subsequent semantic segmentation. It achieves excellent performance for building extraction;
- SEAM: The self-supervised equivariant attention mechanism (SEAM) is proposed by Wang et al. [24] for WSSS. It embeds self-supervision into the weakly supervised learning framework through equivariant regularization, which forces CAMs generated from various transformed images to be consistent. It achieved state-of-the-art performance on the PASCAL VOC 2012 dataset;
- Ours–SIOM: The proposed method that only utilized SIOM. The CAM is generated by GradCAM++ at the last convolutional layer;
- Ours–SMCAM: The proposed method that only utilized SMCAM, including mutual learning between different scales ($\mathcal{L}_{MIC}$);
- Ours: The proposed network with SMCAM and SIOM.

As most methods focus on improving the quality of a CAM for target extraction as well as ours, we mainly compare the completeness and fineness of buildings in a CAM.

4.2.2. Evaluation Criteria

In the evaluation of our proposed methods, we employ three commonly used quantitative criteria: the F1 score, overall accuracy (OA), and intersection over union (IoU). The F1 score is computed as the harmonic mean of precision and recall, which are defined as follows:

$$\mathrm{F1} = \frac{2}{recall^{-1} + precision^{-1}}, \tag{15}$$

where *precision* and *recall* are defined as:

$$precision = \frac{TP}{TP+FP}, \quad recall = \frac{TP}{TP+FN}, \tag{16}$$

where TP and TN, respectively, represent a true positive and true negative, while FP is a false positive and FN is a false negative.

The OA is calculated as the ratio of correctly classified instances to the total instances, as represented by the following equation:

$$\text{OA} = \frac{TP+TN}{TP+TN+FP+FN}. \tag{17}$$

The IoU is calculated as the ratio of the area of intersection between the predicted and ground-truth segmentation divided by the area of union between the predicted and ground-truth segmentation and can be calculated as follows:

$$\text{IoU} = \frac{TP}{TP+FP+FN}. \tag{18}$$

The above three metrics are used to quantitatively evaluate the results of the proposed methods, both for CAMs and building extractions. Additionally, comparisons of visualizations are also utilized to evaluate the results.

4.2.3. Implementation Details

The proposed method employs ResNet50 [57] as the backbone of the Siamese network, and the pre-trained weights from ImageNet (provided by PyTorch) are utilized for the backbone initialization. The generation of multi-scale CAMs is depicted in Figure 3 and derived from the blocks of ResNet50. The training process is carried out by using stochastic gradient descent (SGD) with momentum, where a momentum coefficient of 0.9 and weight decay of 0.0005 are set [11,29,31]. In accordance with the approach presented in [11], a poly-like learning rate rule is employed, where the learning rate is defined as base one multiplying $(1 - t/T)^{power}$ with the base set to 0.01, power to 0.9, t denoting the current iteration, and T denoting the maximum iteration. The network is trained with a batch size of 8, over a total of 30 epochs, using 2 GPUs. In the initial 5 epochs, only cross-entropy loss ($\mathcal{L}_{CLS}$) is employed, while the total loss is utilized in the subsequent epochs. Adhering to SEAM ([24]), the proposed method employs online hard example mining (OHEM) on the ECR loss ($\mathcal{L}_{ECR}$), retaining the top 20% of pixel losses. Random data augmentation, including mirroring, rotation, Gaussian blurring, and color jittering, is applied during the training process, and the Siamese structure expands the matched augmentation to the rescaled image. During inference, the shared-weight Siamese network only requires one branch to be restored.

To ensure a fair comparison, all comparison methods are reimplemented based on the ResNet50 [57] backbone, with the exception of SPN, which presents some difficulties in separating its main modules from the classification network. Both SPN and MSG-SR-Net employ the simple linear iterative clustering algorithm [63] to pre-segment the images into approximately 256 superpixels per image. Furthermore, DeepLabV3+ (based on Resnet50) is utilized as the semantic segmentation network $\mathcal{F}_{\text{SEG}}$ for all methods. To train DeepLabV3+, we employ cross-entropy as the loss function along with the same SGD optimizer as mentioned previously and with the same parameters. However, the initial learning rate is set to 0.007 for this process [10]. The training batch size is set to 8, and a total of 20 epochs are trained. For data augmentation, we only apply random mirroring and rotation. All experimental settings are made consistent across methods as much as possible, though some hyperparameters (e.g., learning rate and momentum) specific to each model may have been adjusted to enable efficient convergence during training. It should be noted that the background thresholds of CAMs for each approach differ as well, and one should be chosen with the best F1 score of pseudo-labels after traversing all background threshold options on the validation set.

The proposed method and all comparison methods are implemented on a Linux 5.15 platform, using Python 3.9 and PyTorch for deep learning. The CUDA 11.6 version is utilized for GPU acceleration. The experiments are run on a Linux platform equipped with two Intel Xeon 8-core CPUs @ 2.2 GHz and two NVIDIA RTX 4000 GPUs with 8 GB memory and 2304 shading units.

4.3. Comparison of CAM Results

4.3.1. Results of Chicago Dataset

For the quantitative aspect, Table 1 reports three evaluation criteria of the CAM results in terms of building extraction accuracy. The proposed method yields the highest values of F1, IoU, and OA when compared to other methods, achieving the best pseudo-label accuracy. In particular, CAM-GAP and GradCAM++ obtain pool accuracy due to the fact that they only activate the most discriminative regions. However, WILDCAT achieves the worst results, which demonstrates that this method is not suitable for building extraction in the remote sensing field. Ours–SMCAM still achieves remarkable results with 0.6774 of F1 and 0.8081 of OA, which is slightly behind MSG-SR-NET, which benefits from the superpixel's ability to utilize the low-level features. It also illustrates the effectiveness of the attentive aggregation of multi-scale CAMs since we do not employ the superpixel-like technique. As for ours–SIOM, it achieves slightly better results than SEAM. Both of these two methods utilize the Siamese networks. Overall, the results indicate the effectiveness of our method, which incorporates SMCAM and SIOM for optimal performance.

Table 1. Evaluation of CAMs generated by various methods with IoU (%), OA (%), and F1 score (%) on WHU dataset and Chicago dataset.

Method	WHU Dataset			Chicago Dataset		
	IoU	F1	OA	IoU	F1	OA
CAM-GAP	41.85	59.01	84.56	36.17	53.13	73.72
GradCAM++	42.73	59.88	84.13	35.74	52.66	73.29
WILDCAT	37.11	54.13	85.51	33.61	50.31	71.22
SEAM	51.83	68.27	90.97	46.82	63.78	79.56
SPN	40.79	57.94	85.31	44.76	61.84	75.93
MSG-SR-NET	53.76	69.93	91.90	51.25	67.77	80.98
WSF-Net	46.58	63.56	87.77	45.27	62.33	79.04
Ours-SIOM	51.91	68.34	91.78	47.48	64.39	80.38
Ours-SMCAM	54.38	70.45	92.21	51.22	67.74	80.01
Ours	58.04	73.45	93.29	52.68	69.01	81.53

Five image patches containing buildings with different sizes and shapes are selected for visual comparison, as shown in Figure 6. It can be observed that WILDCAT performed the worst in identifying various sizes of buildings, making it difficult to separate the buildings from the background. As inferred above, CAM-GAP and GradCAM++ only activate the most discriminative parts of buildings (e.g., edges and texture); the difference is that the former is relatively complete and the latter relatively fine. SEAM achieves promising results in terms of building completeness but still has some over-activation and under-activation issues. The CAMs generated by WSF-Net are relatively coarse and not suitable for building extraction, as this method is designed for water and cloud extraction in remote sensing images. Both SPN and MSG-SR-Net benefit from using superpixel for more detailed information but are also limited by the inaccurate activation of non-building regions. In addition, the results of MSG-SR-Net are notably better than SPN for building extraction. The CAMs obtained by the proposed methods are better than those obtained by the other methods. Specifically, ours–SIOM obtained a relatively integral building in

CAMs, while ours–SMCAM contained more detailed information. The proposed method combines the advantages of both modules to obtain the best overall results.

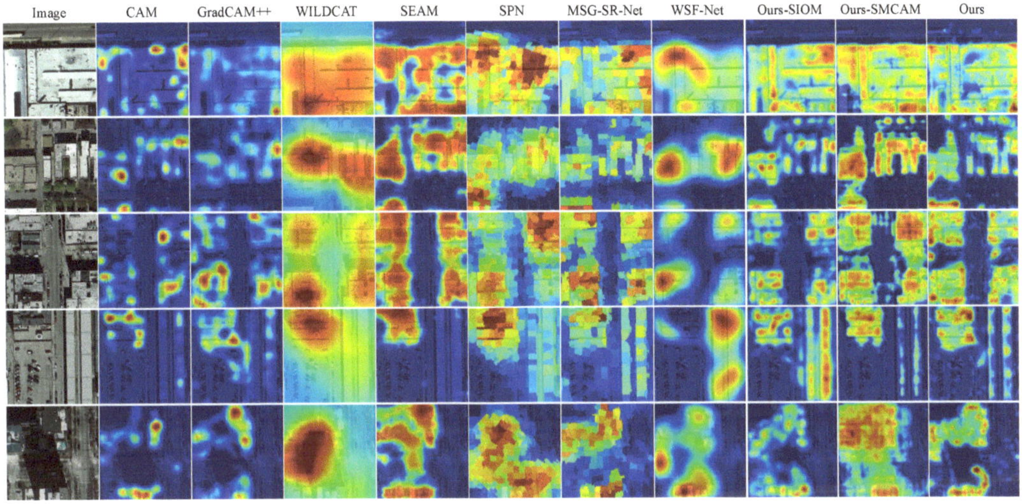

Figure 6. Qualitative comparison of the CAM results obtained by various methods on the Chicago dataset.

4.3.2. Results of WHU Dataset

Table 1 also reports the F1 score, IoU, and OA of different methods on the WHU dataset. It can be observed that: (1) WILDCAT cannot successfully generate a CAM for buildings. Other than that, CAM-GAP, GradCAM++, and WILDCAT have the worst results, and MSG-SR-Net, designed for remote sensing images, achieves remarkable results, as well as the Chicago dataset. (2) The proposed method outperforms all other methods, but with a larger lead than the results of the Chicago dataset. (3) Ours-SMCAM also shows a comparative performance with other methods. This reveals the advantages of the proposed module. (4) Ours-SIOM shows a better performance than SEAM, with a higher F1 score and IoU and an improvement of over one percent in OA, demonstrating its effectiveness in distinguishing buildings from the background.

In order to visually compare the performances of the different methods, five image patches featuring buildings of various sizes and shapes have been selected, as shown in Figure 7. Notably, the quality of images and labels in the WHU dataset is better than that in the Chicago dataset. The images in the WHU dataset are well-calibrated and contain fewer shadows from the buildings. Since buildings and shadows often accompany each other, almost all methods are difficult to distinguish them from one another due to the lack of relevant supervision information. Therefore, both in terms of quality and quantity, the results for the WHU dataset are superior to those of the Chicago dataset. Additionally, the corresponding pixel-level labels are more accurate for producing better image-level labels. Similar to the results for the Chicago dataset, WILDCAT, CAM-GAP, and GradCAM++ still struggle to accurately identify buildings. Specifically, WILDCAT has over-activated too many regions, which overwhelms the buildings. Additionally, CAM-GAP and GradCAM++ still highlight the most distinctive areas. SEAM performs better than the others but still shows over-activation in some areas. The CAMs generated by WSF-Net are coarse for buildings and more suitable for large targets. Although MSG-SR-Net outperforms SPN when utilizing superpixel segmentation, it still results in false negatives along building boundaries. Our proposed methods outperform the other methods, as seen in the results for the Chicago dataset. In particular, ours–SIOM generates more complete building activations and distinguishes buildings from backgrounds more effectively. However, the edges of the buildings remain somewhat blurred. Conversely, ours–SMCAM preserves the structural details and boundaries of the buildings well but may result in some false activations. By combining both modules, the proposed method is able to achieve the best results, as it fuses multi-level and multi-scale features to enhance the quality of CAMs for building extraction. The results confirm the effectiveness of the proposed method for building extraction in remote sensing images.

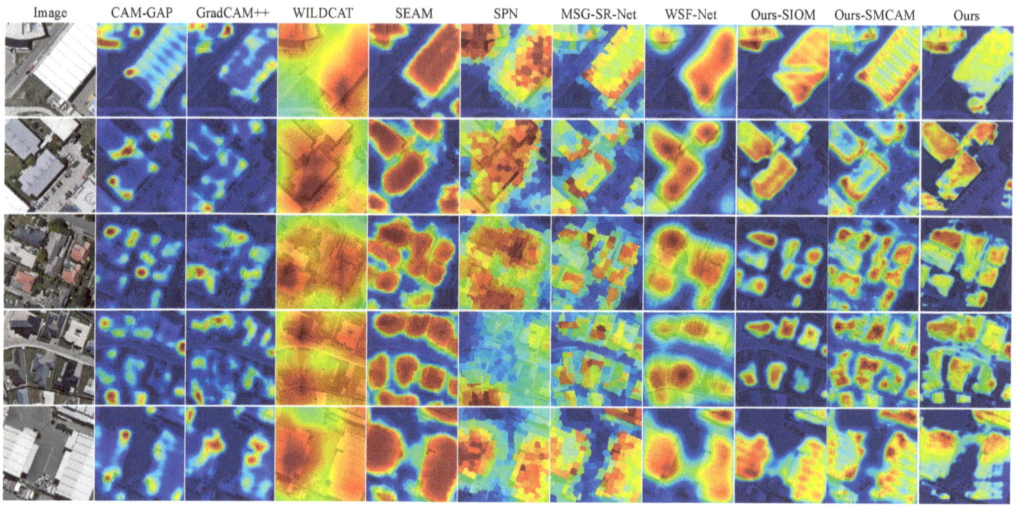

Figure 7. Qualitative comparison of the CAM results obtained by various methods on the WHU dataset.

4.4. Comparison of Building Extraction Results

A CAM can generate building extraction results directly with the use of a suitable threshold, but such results often have significant limitations and are therefore are used commonly as pseudo-labels to train a segmentation network. For a more comprehensive comparison, we trained fully convolutional networks for each method to perform building extraction. Table 2 presents the building extraction results on the two datasets. Firstly, when comparing the CAM results listed in Table 1, almost all methods show improvement in their building extraction results after training the fully convolutional network, with the exception of WILDCAT, which produces a poor CAM that results in ineffective training of the segmentation network. Secondly, the building extraction accuracies of SEAM and MSG-SR-Net are relatively good. Both methods achieve high overall accuracy (OA) scores, demonstrating their capabilities in effectively distinguishing between buildings and non-building objects. Thirdly, when applied to the WHU dataset, our proposed method achieved an F1 score of 0.7759, an OA score of 0.9457, and an IoU score of 0.6339. Similarly, on the Chicago dataset, our method achieved an F1 score of 0.7411, an OA score of 0.8436, and an IoU score of 0.5887. Both outperform the other methods significantly. This highlights the superior effectiveness and robustness of our proposed method for building extraction from remote sensing images, which can be attributed to its ability to produce high-quality CAMs.

The visual comparison of building extraction results from three image patches with varying building densities from each dataset is presented in Figure 8. The results reveal a strong correlation between the quality of the building extraction and the accuracy of the CAM. The higher quality of the annotations and images in the WHU dataset may explain the improved performance compared to the Chicago dataset. Unfortunately, the results of WILDCAT and SPN are below expectations and fail to effectively extract the buildings. On the other hand, CAM, GradCAM++, WSF-Net, and SEAM tend to over-highlight building regions, resulting in coarse outcomes. Although MSG-SR-Net performs better than the aforementioned methods, it still has some inaccuracies and incomplete building extractions. Our proposed method, however, demonstrates the best results, especially on the WHU dataset, where individual buildings are clearly distinguishable and have well-defined edges. This highlights the superior effectiveness of our approach for WSSS building extraction.

Table 2. Evaluation of building extraction results generated by various methods with IoU (%), OA (%), and F1 score (%) on WHU dataset and Chicago dataset.

Method	WHU Dataset			Chicago Dataset		
	IoU	F1	OA	IoU	F1	OA
CAM	45.11	62.17	88.99	40.12	57.27	75.33
GradCAM++	42.19	59.34	87.62	38.30	55.39	74.49
WILDCAT	30.28	46.49	64.33	28.38	44.21	67.11
SEAM	55.44	71.33	91.94	51.27	67.79	81.56
SPN	43.25	60.38	86.6	50.40	67.02	82.35
MSG-SR-NET	57.07	72.67	92.34	57.68	73.16	83.28
WSF-Net	49.60	66.31	91.03	50.15	66.80	80.24
Ours	63.39	77.59	94.57	58.87	74.11	84.36

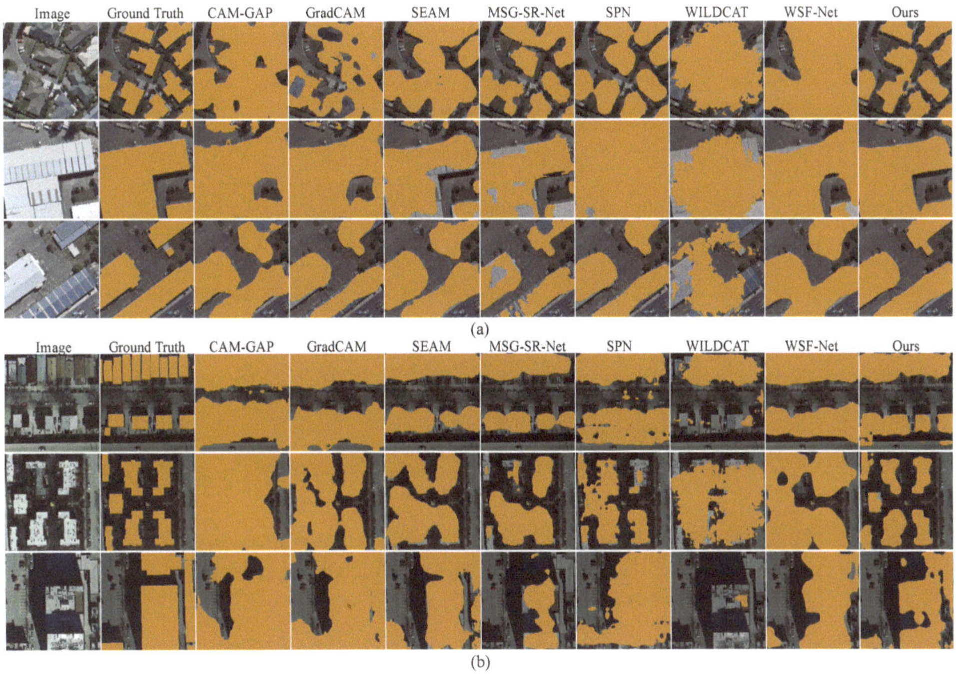

Figure 8. Examples of building extraction results of the proposed method and other comparison methods on two datasets: (**a**) WHU dataset; (**b**) Chicago dataset.

5. Discussion

5.1. The Influence of Auxiliary Branches for Classification Network

The proposed SMCAM introduces auxiliary classification branches that enhance the performance of GradCAM++ by shortening the gradient propagation path and improving the semantic representation of low-level features. This allows for the effective fusion of CAMs at multiple levels. However, the alteration of the original classification network structure and the reliance on image-level annotations for supervised learning raise concerns about the impact on the network's feature learning capabilities. In order to assess this, the classification accuracies of the trunks and branches of the Siamese network are calculated, and the results are presented in Table 3. The results indicate that the accuracies of all branches and trunks are higher than 0.97, demonstrating that the auxiliary branches do not compromise the training accuracy of the classification network. Specifically, branch 1 has slightly lower training and testing accuracies compared to the other branches and trunks, due to its weaker feature map generalization capabilities, yet it still meets the accuracy requirement.

Table 3. The training and testing accuracies of branches and trunks in classification network.

Position	Subnetwork 1 (I)		Subnetwork 2 ($\tilde{I}$)	
	Training Acc.	Testing Acc.	Training Acc.	Testing Acc.
Branch 1	0.971	0.980	0.975	0.972
Branch 2	0.990	0.993	0.991	0.989
Branch 3	0.992	0.995	0.995	0.995
Branch 4	0.993	0.994	0.992	0.995
Trunk	0.990	0.992	0.992	0.989

As shown in Figure 9, the different levels of CAMs obtained from the branching network are also satisfactory, effectively suppressing the noise of low-level CAMs while highlighting the regions related to the buildings. In summary, the auxiliary branches introduced by SMCAM do not negatively impact the performance of the classification network.

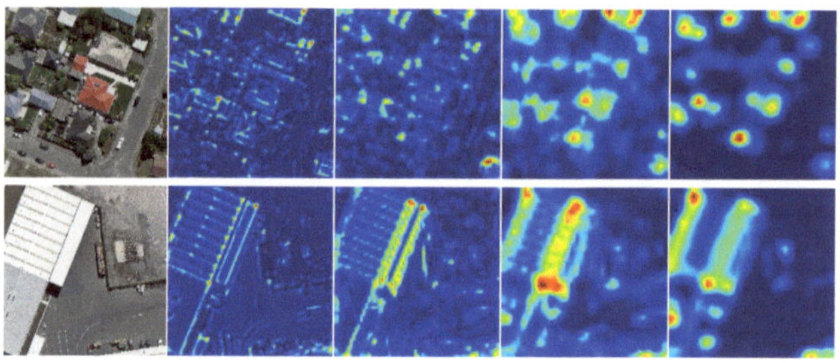

Figure 9. Visulizations of CAMs generated at various levels.

5.2. Comparison of Different Fusion Strategies in SMCAM

The main objective of SMCAM is to attain a fine-structure CAM by aggregating multi-level CAMs generated by GradCAM++. A critical aspect of this challenge is to effectively fuse these CAMs. In this study, we propose a self-attention mechanism for fusing multi-level CAMs by exploiting multi-level features. To evaluate the effectiveness of the proposed self-attention mechanism (denoted as "Attention"), we conduct experiments and compare them with direct addition fusion (denoted as "Addition") in the framework of the proposed method. Additionally, we include the CAM generated from the last convolutional layer (denoted as "Last Conv.") for comparison. The results shown in Table 4 reveal that the self-attention mechanism outperforms Addition, demonstrating its suitability for fusing multi-level CAMs. Additionally, the accuracies of both Attention and Addition are significantly improved compared to Last Conv., implying that utilizing multi-level CAMs enhances the quality of CAMs. The comparison with the results of GradCAM++ listed in Table 1 further highlights the effectiveness of the mutual learning mechanism ($\mathcal{L}_{MIC}$) in the Siamese network in enhancing the quality of the CAM. Furthermore, the improvement in the accuracies of Addition, Attention, and Last Conv. after incorporating SIOM further verifies the effectiveness of SIOM.

Visualizations of the CAMs in this experiment can be seen in Figure 10, with the self-attention method producing more complete and fine-grained CAMs than the others.

To conclude, the proposed self-attention mechanism effectively fuses multi-level CAMs and can be used in conjunction with SIOM to produce better results.

Table 4. The IoU (%), OA (%), and F1 score (%) of different fusion strategies on WHU Dataset.

	IoU	F1	OA
Last Conv.	48.63	65.44	90.01
Addition	51.73	68.19	92.07
Attention	54.38	70.45	92.21
Last Conv. + SIOM (Ours-SIOM)	51.91	68.34	91.78
Addition + SIOM	56.48	72.19	92.98
Attention + SIOM (Ours)	58.04	73.45	93.29

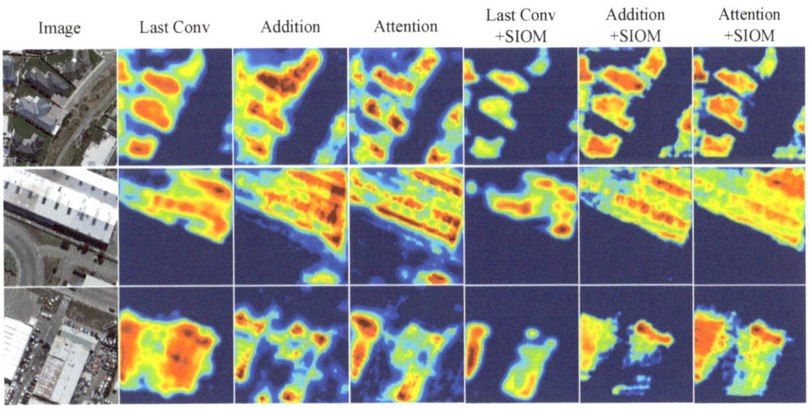

Figure 10. Visulizations of CAMs of different fusion strategies.

5.3. Performance of Hierarchical Feature Optimization in SIOM

The role of SIOM is to utilize the CAM with multi-scale information, which is effective in improving the integrality of the CAM. The core component of SIOM is hierarchical feature optimization, which is a learnable module that leverages the multi-level features to further improve the quality of CAM. It consists of three stages of optimization: the first two stages focus on mining the non-local relationships between the multi-level features to improve the integrity of the CAM, while the third stage leverages the local image information to further improve the fineness of the CAM. To evaluate the efficacy of hierarchical feature optimization, each stage of optimization is analyzed. The pixel correlation module (PCM) in SEAM is also included in the experiments for comparison, which also leverages feature relationships but only considers features from the last two convolutional layers. The experimental results are listed in Table 5, showing that the CAMs generated with hierarchical feature optimization in SIOM are better than those generated in PCM and GradCAM++. Specifically, the results of SIOM are better as the number of stages increases, especially when the third stage of local image information is included, which demonstrates the effectiveness of hierarchical optimization.

Table 5. The IoU (%), OA (%), and F1 score (%) of hierarchical feature optimization settings on WHU dataset.

	IoU	F1	OA
GradCAM++ (4 levels)	49.75	66.44	90.01
SEAM (SMCAM + PCM)	55.17	71.11	92.07
SIOM (Stage 1)	55.69	71.54	92.61
SIOM (Stage 1 + 2)	56.12	71.90	93.11
SIOM (Stage 1 + 2 + 3)	58.04	73.45	93.29

The CAMs produced by each stage of SIOM are displayed in Figure 11, and it is evident that the CAMs generated by SIOM are more comprehensive and clearer than those generated by GradCAM++ and PCM. As the number of stages increases, the clarity and fineness of the CAMs also improve. It is noteworthy that after including the image features in the third stage, the fineness is further enhanced, but the details are magnified to an excessive extent, which may not align with expectations (e.g., the texture of a roof). Despite this, the difference in semantic representation between the layers is still substantial, and, in practical applications, the direct use of image features may require additional processing.

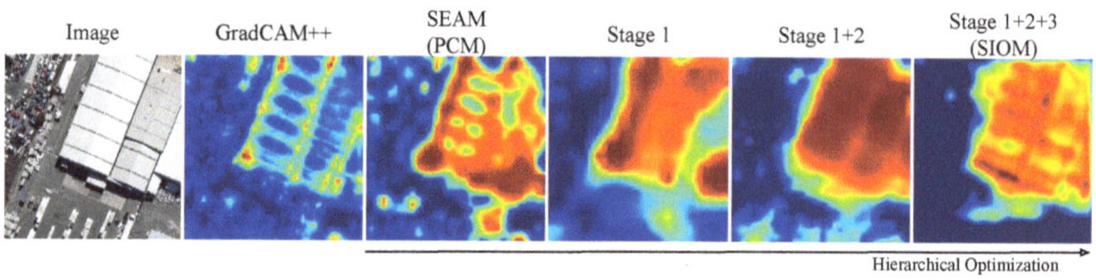

Figure 11. Visulizations of CAMs of different stages settings in hierarchical feature optimization.

5.4. *Effect of Scale Setting*

The proposed method employs a Siamese network that is trained using both the original image and a rescaled image from a multi-scale representation. It is crucial to discuss the potential impact of different scale settings on the method's performance. To achieve this, experiments are conducted on two datasets by varying the scale of the rescaled image while keeping the original image as the input. The scale of the rescaled image represents the ratio size of the original image. Since the task only involves building extraction, the F1 score is selected as the evaluation metric for these experiments.

The results are depicted in Figure 12. The Siamese network struggles to train effectively when the rescaled image and the original image are of the same size, as indicated by the dashed lines in the figure. From the results of the WHU dataset, the best accuracy is achieved when the scale of the rescaled image is within the range from [0.7, 0.8] to [1.2, 1.3]; secondly, the F1 score decreases and even fails when the scale is less than 0.6, while the F1 score also decreases slowly when the scale is larger than 1.5; thirdly, the F1 score decreases as the scale approaches one, as previously discussed. These observations suggest that the scale difference between the rescaled image and the original image must not be too large or too small. The results from the Chicago dataset are similar, except that the points with the highest scores are 0.8 for the WHU dataset and 0.7 and 1.4 for the Chicago dataset, which highlights the importance of considering dataset bias in the scale setting. In our initial experiments with the same scale setting as SEAM (0.3), the network fails to complete the training and even experiences a gradient explosion, further emphasizing the importance of finding an appropriate scale setting. For this reason, we set the scale ratio of the rescaled image to 0.8 in our experiments.

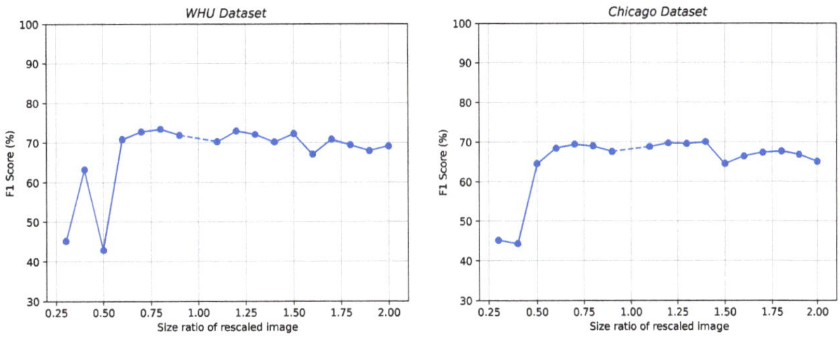

(a) WHU Dataset (b) Chicago Dataset

Figure 12. F1 score of CAM results under different scale settings on two datasets.

5.5. Limitations and Future Work

From the experimental results and the above analysis, the proposed method stands out among other WSSS methods by not only achieving highly accurate buildings but also ensuring completeness. However, the accuracy of building extraction using the proposed method is still considerably distant from that of the fully supervised segmentation approach. As shown in Figures 6 and 7, the CAM generated by the proposed method can sometimes blur the distinction between buildings and shadows, leading to inaccurate building edges. We hypothesize that this could be attributed to insufficient availability of supervisory information for differentiating buildings from shadows. Additionally, we have observed that our method performs better on the WHU dataset than on the Chicago dataset (Figure 5b), indicating that our approach is sensitive to dataset bias, which can hinder the generalization capability.

To further improve the performance of building extraction, incorporating more supervised information derived from underlying image features or semantic supervised information, such as additional category labels and image restoration information, may be a potential solution. Utilizing these extra supervisions is crucial for effectively and accurately distinguishing between buildings and non-buildings. Additionally, to overcome dataset bias, an unsupervised domain adaption approach using generative adversarial networks could be incorporated, thereby enabling the weakly supervised approach to be applied to a broader range of domains. However, developing specific models and conducting corresponding experiments is left for future work.

6. Conclusions

In this work, we introduce a novel scale-invariant multi-level context aggregation network for the task of weakly supervised building extraction from high-resolution imagery. Our approach focuses on two main aspects: (1) The self-attentive multi-level context aggregation module (SMCAM) is proposed to generate noise-free, fine-structured CAMs by shortening the back-gradient path and utilizing a self-attention mechanism to aggregate multi-level CAMs, allowing the model to learn the contributions of different CAMs for each pixel. (2) The scale-invariant optimization module (SIOM) is designed to narrow the supervision gap between segmentation and classification by leveraging multi-scale representations. This module includes a mutual learning mechanism for pixel-level auxiliary supervision through optimization with a multi-level constraint loss and a hierarchical optimization module to improve the CAM's completeness in a coarse-to-fine manner. The Siamese network integrates these two modules, leading to improved CAM quality in terms of both completeness and fineness, allowing for accurate building extraction results with only image-level labels.

The proposed method is evaluated through two experiments: CAM generation and building extraction. The results of the CAM generation experiment show that the proposed method outperforms state-of-the-art methods in the field of remote sensing and computer vision, as evidenced by its improved performance on both the WHU and Chicago datasets. The building extraction experiment results are similarly impressive, with the proposed method achieving an F1 score of 0.7759, an OA score of 0.9457, and an IoU score of 0.6339 for the WHU dataset, and an F1 score of 0.7411, an OA score of 0.8436, and an IoU score of 0.5887 for the Chicago dataset. These results demonstrate the effectiveness of the proposed method in accurately extracting buildings using only image-level labels.

Additionally, we present an analysis of the proposed modules. Our results indicate that the self-attentive aggregation component in SMCAM and the hierarchical optimization module in SIOM are both reasonable and effective. Furthermore, we examine the influence of the auxiliary classification branches on the network and the scale setting.

In our study, we have observed limitations in the ability of the proposed method to distinguish between shadows and buildings, as well as its sensitivity to dataset bias. In future work, we plan to incorporate additional supervision information to improve shadow differentiation and investigate domain adaptation techniques to mitigate the impact of dataset bias with the aim of achieving enhanced performance.

Author Contributions: Conceptualization, J.W. and X.Y.; methodology, J.W.; validation, X.Y., T.L. and X.G.; formal analysis, J.W. and L.S.; writing—original draft preparation, J.W. and X.Y.; writing—review and editing, L.S., T.L. and X.Y.; visualization, J.W. and X.Y.; funding acquisition, Z.L. All authors have read and agreed to the published version of the manuscript.

Funding: This work was funded by the Open Fund of Key Laboratory of Mine Environmental Monitoring and Improving around Poyang Lake, Ministry of Natural Resources (Grant No. MEMI-2021-2022-09) and partially by the Interdisciplinary Cultivation Fund under Project SWJTU (Southwest Jiaotong University).

Data Availability Statement: The data presented in this study are openly available in the WHU Building Dataset at gpcv.whu.edu.cn/data/building_dataset, accessed on 3 July 2018, and the Inria Aerial Image Labeling Dataset at project.inria.fr/aerialimagelabeling, accessed on 1 January 2017.

Conflicts of Interest: The authors declare no conflict of interest.

References

1. Maktav, D.; Erbek, F.; Jürgens, C. Remote Sensing of Urban Areas. *Int. J. Remote Sens.* **2005**, *26*, 655–659. [CrossRef]
2. Tomás, L.; Fonseca, L.; Almeida, C.; Leonardi, F.; Pereira, M. Urban Population Estimation based on Residential Buildings Volume Using IKONOS-2 Images and Lidar Data. *Int. J. Remote Sens.* **2016**, *37*, 1–28. [CrossRef]
3. Li, J.; Huang, X.; Tu, L.; Zhang, T.; Wang, L. A Review of Building Detection from Very High Resolution Optical Remote Sensing Images. *GISci. Remote Sens.* **2022**, *59*, 1199–1225. [CrossRef]
4. Guo, H.; Shi, Q.; Marinoni, A.; Du, B.; Zhang, L. Deep Building Footprint Update Network: A Semi-supervised Method for Updating Existing Building Footprint from Bi-temporal Remote Sensing Images. *Remote Sens. Environ.* **2021**, *264*, 112589. [CrossRef]
5. Haq, M.A. CNN Based Automated Weed Detection System Using UAV Imagery. *Comput. Syst. Sci. Eng.* **2022**, *42*, 837–849.
6. Kim, J.; Lee, M.; Han, H.; Kim, D.; Bae, Y.; Kim, H.S. Case Study: Development of the CNN Model Considering Teleconnection for Spatial Downscaling of Precipitation in a Climate Change Scenario. *Sustainability* **2022**, *14*, 4719. [CrossRef]
7. Haq, M.A.; Khan, M.A.R.; Talal, A. Development of PCCNN-based network intrusion detection system for EDGE computing. *Comput. Mater. Contin.* **2022**, *71*, 1769–1788.
8. Haq, M.A.; Khan, M.A.R. DNNBoT: Deep neural network-based botnet detection and classification. *Cmc-Comput. Mater. Continua* **2022**, *71*, 1729–1750.
9. Ronneberger, O.; Fischer, P.; Brox, T. U-Net: Convolutional Networks for Biomedical Image Segmentation. In Proceedings of the Medical Image Computing and Computer-Assisted Intervention, Munich, German, 21–26 July 2017; pp. 234–241.
10. Chen, L.C.; Papandreou, G.; Schroff, F.; Adam, H. Rethinking Atrous Convolution for Semantic Image Segmentation. *arXiv* **2017**, arXiv:1706.05587.
11. Zhao, H.; Shi, J.; Qi, X.; Wang, X.; Jia, J. Pyramid Scene Parsing Network. In Proceedings of the IEEE Conference on Computer Vision and Pattern Recognition, Honolulu, HI, USA, 21–26 July 2017; pp. 2881–2890.
12. Schuegraf, P.; Bittner, K. Automatic Building Footprint Extraction from Multi-resolution Remote Sensing Images Using a Hybrid FCN. *ISPRS Int. J. Geo-Inf.* **2019**, *8*, 191. [CrossRef]

13. Feng, W.; Sui, H.; Hua, L.; Xu, C.; Ma, G.; Huang, W. Building Extraction from VHR Remote Sensing Imagery by Combining an Improved Deep Convolutional Encoder-decoder Architecture and Historical Land Use Vector Map. *Int. J. Remote Sens.* **2020**, *41*, 6595–6617. [CrossRef]
14. Shao, Z.; Tang, P.; Wang, Z.; Saleem, N.; Yam, S.; Sommai, C. BRRNet: A Fully Convolutional Neural Network for Automatic Building Extraction from High-resolution Remote Sensing Images. *Remote Sens.* **2020**, *12*, 1050. [CrossRef]
15. Ma, J.; Wu, L.; Tang, X.; Liu, F.; Zhang, X.; Jiao, L. Building Extraction of Aerial Images by a Global and Multi-scale Encoder-decoder Network. *Remote Sens.* **2020**, *12*, 2350. [CrossRef]
16. Zhou, B.; Khosla, A.; Lapedriza, A.; Oliva, A.; Torralba, A. Learning Deep Features for Discriminative Localization. In Proceedings of the IEEE Conference on Computer Vision and Pattern Recognition, Las Vegas, NV, USA, 27–30 June 2016; pp. 2921–2929.
17. Saleh, F.S.; Aliakbarian, M.S.; Salzmann, M.; Petersson, L.; Álvarez, J.M.; Gould, S. Incorporating Network Built-in Priors in Weakly-Supervised Semantic Segmentation. *IEEE Trans. Pattern Anal. Mach. Intell.* **2017**, *40*, 1382–1396. [CrossRef] [PubMed]
18. Wei, Y.; Ji, S. Scribble-based Weakly Supervised Deep Learning for Road Surface Extraction from Remote Sensing Images. *IEEE Trans. Geosci. Remote Sens.* **2021**, *60*, 1–12. [CrossRef]
19. Ahn, J.; Kwak, S. Learning Pixel-Level Semantic Affinity with Image-Level Supervision for Weakly Supervised Semantic Segmentation. In Proceedings of the IEEE Conference on Computer Vision and Pattern Recognition, Salt Lake City, UT, USA, 18–22 June 2018.
20. Cao, Y.; Huang, X. A Coarse-to-fine Weakly Supervised Learning Method for Green Plastic Cover Segmentation Using High-resolution Remote Sensing Images. *ISPRS J. Photogramm. Remote Sens.* **2022**, *188*, 157–176. [CrossRef]
21. Zeng, Y.; Zhuge, Y.; Lu, H.; Zhang, L. Joint Learning of Saliency Detection and Weakly Supervised Semantic Segmentation. In Proceedings of the IEEE International Conference on Computer Vision, Seoul, Republic of Korea, 27 Octorber–2 November 2019.
22. Du, Y.; Fu, Z.; Liu, Q.; Wang, Y. Weakly Supervised Semantic Segmentation by Pixel-to-Prototype Contrast. In Proceedings of the IEEE Conference on Computer Vision and Pattern Recognition, New Orleans, LA, USA, 19–24 June 2022; pp. 4320–4329.
23. Chattopadhay, A.; Sarkar, A.; Howlader, P.; Balasubramanian, V.N. Grad-CAM++: Generalized Gradient-based Visual Explanations for Deep Convolutional Networks. In Proceedings of the IEEE Winter Conference on Applications of Computer Vision, Lake Tahoe, NV, USA, 12–15 March 2018; pp. 839–847.
24. Wang, Y.; Zhang, J.; Kan, M.; Shan, S.; Chen, X. Self-supervised Equivariant Attention Mechanism for Weakly Supervised Semantic Segmentation. In Proceedings of the IEEE Conference on Computer Vision and Pattern Recognition, Seattle, WA, USA, 14–19 June 2020.
25. Kwak, S.; Hong, S.; Han, B. Weakly Supervised Semantic Segmentation Using Superpixel Pooling Network. In Proceedings of the AAAI Conference on Artificial Intelligence, San Francisco, CA, USA, 4–9 February 2017.
26. Durand, T.; Mordan, T.; Thome, N.; Cord, M. WILDCAT: Weakly Supervised Learning of Deep Convnets for Image Classification, Pointwise Localization and Segmentation. In Proceedings of the IEEE Conference on Computer Vision and Pattern Recognition, Honolulu, HI, USA, 21–26 July 2017; pp. 642–651.
27. Chan, L.; Hosseini, M.S.; Plataniotis, K.N. A Comprehensive Analysis of Weakly-supervised Semantic Segmentation in Different Image Domains. *Int. J. Comput. Vis.* **2021**, *129*, 361–384. [CrossRef]
28. Chen, J.; He, F.; Zhang, Y.; Sun, G.; Deng, M. SPMF-Net: Weakly Supervised Building Segmentation by Combining Superpixel Pooling and Multi-scale Feature Fusion. *Remote Sens.* **2020**, *12*, 1049. [CrossRef]
29. Yan, X.; Shen, L.; Wang, J.; Deng, X.; Li, Z. MSG-SR-Net: A Weakly Supervised Network Integrating Multiscale Generation and Superpixel Refinement for Building Extraction from High-Resolution Remotely Sensed Imageries. *IEEE J. Sel. Top. Appl. Earth Obs. Remote Sens.* **2021**, *15*, 1012–1023. [CrossRef]
30. Li, Z.; Zhang, X.; Xiao, P.; Zheng, Z. On the Effectiveness of Weakly Supervised Semantic Segmentation for Building Extraction from High-resolution Remote Sensing Imagery. *IEEE J. Sel. Top. Appl. Earth Obs. Remote Sens.* **2021**, *14*, 3266–3281. [CrossRef]
31. Yan, X.; Shen, L.; Wang, J.; Wang, Y.; Li, Z.; Xu, Z. PANet: Pixelwise Affinity Network for Weakly Supervised Building Extraction From High-Resolution Remote Sensing Images. *IEEE Geosci. Remote Sens. Lett.* **2022**, *19*, 1–5. [CrossRef]
32. Fang, F.; Zheng, D.; Li, S.; Liu, Y.; Zeng, L.; Zhang, J.; Wan, B. Improved Pseudomasks Generation for Weakly Supervised Building Extraction from High-Resolution Remote Sensing Imagery. *IEEE J. Sel. Top. Appl. Earth Obs. Remote Sens.* **2022**, *15*, 1629–1642. [CrossRef]
33. Wang, J.; Shen, L.; Qiao, W.; Dai, Y.; Li, Z. Deep Feature Fusion with Integration of Residual Connection and Attention Model for Classification of VHR Remote Sensing Images. *Remote Sens.* **2019**, *11*, 1617. [CrossRef]
34. Ji, S.; Wei, S.; Lu, M. Fully Convolutional Networks for Multisource Building Extraction from an Open Aerial and Satellite Imagery Data Set. *IEEE Trans. Geosci. Remote Sens.* **2019**, *57*, 574–586. [CrossRef]
35. Maggiori, E.; Tarabalka, Y.; Charpiat, G.; Alliez, P. Convolutional Neural Networks for Large-scale Remote-sensing Image Classification. *IEEE Trans. Geosci. Remote Sens.* **2016**, *55*, 645–657. [CrossRef]
36. Maltezos, E.; Doulamis, A.; Doulamis, N.; Ioannidis, C. Building Extraction from LiDAR Data Applying Deep Convolutional Neural Networks. *IEEE Geosci. Remote Sens. Lett.* **2018**, *16*, 155–159. [CrossRef]

37. Zhang, Z.; Wang, Y. JointNet: A Common Neural Network for Road and Building Extraction. *Remote Sens.* **2019**, *11*, 696. [CrossRef]
38. Hosseinpour, H.; Samadzadegan, F.; Javan, F.D. CMGFNet: A Deep Cross-modal Gated Fusion Network for Building Extraction from Very High-resolution Remote Sensing Images. *ISPRS J. Photogramm. Remote Sens.* **2022**, *184*, 96–115. [CrossRef]
39. Li, Q.; Mou, L.; Hua, Y.; Shi, Y.; Zhu, X.X. CrossGeoNet: A Framework for Building Footprint Generation of Label-Scarce Geographical Regions. *Int. J. Appl. Earth Obs. Geoinf.* **2022**, *111*, 102824. [CrossRef]
40. Yuan, W.; Wang, J.; Xu, W. Shift Pooling PSPNet: Rethinking PSPNet for Building Extraction in Remote Sensing Images from Entire Local Feature Pooling. *Remote Sens.* **2022**, *14*, 4889. [CrossRef]
41. Chen, M.; Wu, J.; Liu, L.; Zhao, W.; Tian, F.; Shen, Q.; Zhao, B.; Du, R. DR-Net: An Improved Network for Building Extraction from High Resolution Remote Sensing Image. *Remote Sens.* **2021**, *13*, 294. [CrossRef]
42. Guo, H.; Du, B.; Zhang, L.; Su, X. A Coarse-to-fine Boundary Refinement Network for Building Footprint Extraction from Remote Sensing Imagery. *ISPRS J. Photogramm. Remote Sens.* **2022**, *183*, 240–252. [CrossRef]
43. Shu, Q.; Pan, J.; Zhang, Z.; Wang, M. MTCNet: Multitask Consistency Network with Single Temporal Supervision for Semi-supervised Building Change Detection. *Int. J. Appl. Earth Obs. Geoinf.* **2022**, *115*, 103110. [CrossRef]
44. Xia, L.; Zhang, X.; Zhang, J.; Yang, H.; Chen, T. Building Extraction from Very-high-resolution Remote Sensing Images Using Semi-supervised Semantic Edge Detection. *Remote Sens.* **2021**, *13*, 2187. [CrossRef]
45. Zheng, Y.; Yang, M.; Wang, M.; Qian, X.; Yang, R.; Zhang, X.; Dong, W. Semi-supervised Adversarial Semantic Segmentation Network Using Transformer and Multiscale Convolution for High-resolution Remote Sensing Imagery. *Remote Sens.* **2022**, *14*, 1786. [CrossRef]
46. Li, Q.; Shi, Y.; Zhu, X.X. Semi-supervised Building Footprint Generation with Feature and Output Consistency Training. *IEEE Trans. Geosci. Remote Sens.* **2022**, *60*, 1–17.
47. Meng, Y.; Chen, S.; Liu, Y.; Li, L.; Zhang, Z.; Ke, T.; Hu, X. Unsupervised Building Extraction from Multimodal Aerial Data Based on Accurate Vegetation Removal and Image Feature Consistency Constraint. *Remote Sens.* **2022**, *14*, 1912. [CrossRef]
48. Chen, H.; Cheng, L.; Zhuang, Q.; Zhang, K.; Li, N.; Liu, L.; Duan, Z. Structure-aware Weakly Supervised Network for Building Extraction from Remote Sensing Images. *IEEE Trans. Geosci. Remote Sens.* **2022**, *60*, 1–12. [CrossRef]
49. Wei, Y.; Feng, J.; Liang, X.; Cheng, M.M.; Zhao, Y.; Yan, S. Object Region Mining with Adversarial Erasing: A Simple Classification to Semantic Segmentation Approach. In Proceedings of the IEEE Conference on Computer Vision and Pattern Recognition, Honolulu, HI, USA, 21–26 July 2017; pp. 1568–1576.
50. Lee, J.; Kim, E.; Lee, S.; Lee, J.; Yoon, S. FickleNet: Weakly and Semi-supervised Semantic Image Segmentation Using Stochastic Inference. In Proceedings of the IEEE Conference on Computer Vision and Pattern Recognition, Long Beach, CA, USA, 15–20 June 2019; pp. 5267–5276.
51. Jiang, P.T.; Yang, Y.; Hou, Q.; Wei, Y. L2G: A Simple Local-to-global Knowledge Transfer Framework for Weakly Supervised Semantic Segmentation. In Proceedings of the IEEE Conference on Computer Vision and Pattern Recognition, New Orleans, LA, USA, 19–24 June 2022; pp. 16886–16896.
52. Kolesnikov, A.; Lampert, C.H. Seed, Expand and Constrain: Three Principles for Weakly-supervised Image Segmentation. In Proceedings of the European Conference on Computer Vision, Amsterdam, The Netherlands, 11–14 October 2016; pp. 695–711.
53. Huang, Z.; Wang, X.; Wang, J.; Liu, W.; Wang, J. Weakly-supervised Semantic Segmentation Network with Deep Seeded Region Growing. In Proceedings of the IEEE Conference on Computer Vision and Pattern Recognition, Salt Lake City, UT, USA, 18–22 June 2018; pp. 7014–7023.
54. Ahn, J.; Cho, S.; Kwak, S. Weakly Supervised Learning of Instance Segmentation with Inter-pixel Relations. In Proceedings of the IEEE Conference on Computer Vision and Pattern Recognition, Long Beach, CA, USA, 15–20 June 2019; pp. 2209–2218.
55. Chang, Y.T.; Wang, Q.; Hung, W.C.; Piramuthu, R.; Tsai, Y.H.; Yang, M.H. Weakly-supervised Semantic Segmentation via Sub-category Exploration. In Proceedings of the IEEE Conference on Computer Vision and Pattern Recognition, Seattle, WA, USA, 14–19 June 2020; pp. 8991–9000.
56. Fu, K.; Lu, W.; Diao, W.; Yan, M.; Sun, H.; Zhang, Y.; Sun, X. WSF-NET: Weakly Supervised Feature-fusion Network for Binary Segmentation in Remote Sensing Image. *Remote Sens.* **2018**, *10*, 1970. [CrossRef]
57. He, K.; Zhang, X.; Ren, S.; Sun, J. Deep Residual Learning for Image Recognition. In Proceedings of the IEEE Conference on Computer Vision and Pattern Recognition, Las Vegas, NV, USA, 27–30 June 2016; pp. 770–778.
58. Simonyan, K.; Zisserman, A. Very Deep Convolutional Networks for Large-Scale Image Recognition. In Proceedings of the International Conference on Learning Representations, San Diego, CA, USA, 7–9 May 2015.
59. Wang, X.; Girshick, R.; Gupta, A.; He, K. Non-local Neural Networks. In Proceedings of the IEEE Conference on Computer Vision and Pattern Recognition, Salt Lake City, UT, USA, 18–22 June 2018; pp. 7794–7803.
60. Su, H.; Jampani, V.; Sun, D.; Gallo, O.; Learned-Miller, E.; Kautz, J. Pixel-adaptive Convolutional Neural Networks. In Proceedings of the IEEE Conference on Computer Vision and Pattern Recognition, Long Beach, CA, USA, 15–20 June 2019; pp. 11166–11175.
61. Maggiori, E.; Tarabalka, Y.; Charpiat, G.; Alliez, P. Can Semantic Labeling Methods Generalize to Any City? The Inria Aerial Image Labeling Benchmark. In Proceedings of the 2017 IEEE International Geoscience and Remote Sensing Symposium, Fort Worth, TX, USA, 23–28 July 2017; pp. 3226–3229.

62. Maaten, L.; Hinton, G. Visualizing Data using t-SNE. *J. Mach. Learn. Res.* **2008**, *9*, 2579–2605.
63. Achanta, R.; Shaji, A.; Smith, K.; Lucchi, A.; Fua, P.; Süsstrunk, S. SLIC Superpixels Compared to State-of-the-art Superpixel Methods. *IEEE Trans. Pattern Anal. Mach. Intell.* **2012**, *34*, 2274–2282. [CrossRef] [PubMed]

remote sensing

MDPI

Article

Semantic Segmentation of China's Coastal Wetlands Based on Sentinel-2 and Segformer

Xufeng Lin [1], Youwei Cheng [1], Gong Chen [1], Wenjing Chen [1], Rong Chen [2], Demin Gao [1], Yinlong Zhang [2,3,4] and Yongbo Wu [2,3,*]

[1] College of Information Science and Technology, Nanjing Forestry University, Nanjing 210037, China; linxufeng@njfu.edu.cn (X.L.); youwei@njfu.edu.cn (Y.C.); gongchen@njfu.edu.cn (G.C.); wjchen@njfu.edu.cn (W.C.); dmgao@njfu.edu.cn (D.G.)

[2] College of Ecology and Environment, Nanjing Forestry University, Nanjing 210037, China; rongchen@njfu.edu.cn (R.C.); ylzhang@njfu.edu.cn (Y.Z.)

[3] Co-Innovation Center for Sustainable Forestry in Southern China, Nanjing Forestry University, Nanjing 210037, China

[4] National Positioning Observation Station of Hung-Tse Lake Wetland Ecosystem in Jiangsu Province, Huaian 223100, China

* Correspondence: yongbowu@njfu.edu.cn; Tel.: +86-025-85428935

Abstract: Concerning the ever-changing wetland environment, the efficient extraction of wetland information holds great significance for the research and management of wetland ecosystems. China's vast coastal wetlands possess rich and diverse geographical features. This study employs the SegFormer model and Sentinel-2 data to conduct a wetland classification study for coastal wetlands in Yancheng, Jiangsu, China. After preprocessing the Sentinel data, nine classification objects (construction land, *Spartina alterniflora (S. alterniflora)*, *Suaeda salsa (S. salsa)*, *Phragmites australis (P. australis)*, farmland, river system, aquaculture and tidal falt) were identified based on the previous literature and remote sensing images. Moreover, mAcc, mIoU, aAcc, Precision, Recall and F-1 score were chosen as evaluation indicators. This study explores the potential and effectiveness of multiple methods, including data image processing, machine learning and deep learning. The results indicate that SegFormer is the best model for wetland classification, efficiently and accurately extracting small-scale features. With mIoU (0.81), mAcc (0.87), aAcc (0.94), mPrecision (0.901), mRecall (0.876) and mFscore (0.887) higher than other models. In the face of unbalanced wetland categories, combining CrossEntropyLoss and FocalLoss in the loss function can improve several indicators of difficult cases to be segmented, enhancing the classification accuracy and generalization ability of the model. Finally, the category scale pie chart of Yancheng Binhai wetlands was plotted. In conclusion, this study achieves an effective segmentation of Yancheng coastal wetlands based on the semantic segmentation method of deep learning, providing technical support and reference value for subsequent research on wetland values.

Keywords: deep learning; semantic segmentation; SegFormer; coastal wetland; remote sensing images; machine learning

Citation: Lin, X.; Cheng, Y.; Chen, G.; Chen, W.; Chen, R.; Gao, D.; Zhang, Y.; Wu, Y. Semantic Segmentation of China's Coastal Wetlands Based on Sentinel-2 and Segformer. *Remote Sens.* **2023**, *15*, 3714. https://doi.org/10.3390/rs15153714

Academic Editors: Sawaid Abbas, Janet E. Nichol, Faisal M. Qamer and Jianchu Xu

Received: 30 May 2023
Revised: 18 July 2023
Accepted: 20 July 2023
Published: 25 July 2023

1. Introduction

The coastal wetland serves as an intermediary zone connecting marine and terrestrial ecosystems. It encompasses a vast expanse of coastal regions, intertidal areas and various aquatic environments, such as river systems, estuaries, salt marshes and sandy beaches. As a region of exquisite intricacy and ecological sensitivity, it undergoes profound influences from both the vast sea and the adjoining land. When it comes to the assessment of ecological service value, wetland ecosystems triumph over other ecosystems, such as oceans and forests. The second national wetland resources survey reveals that wetlands encompass a notable 5.58% of China's territory, with coastal wetlands alone constituting 10.85%. While the coastal wetland occupies less than 1% of the entire country's land area, it plays a vital role in purifying polluted water, mitigating floods, minimizing the dangers of storm surges and hurricanes, and providing a favorable natural environment.

The assessment of ecosystem service value serves as a pivotal bridge connecting ecosystem research and management decision-making. Its primary objective is to comprehensively comprehend the current status and dynamic trends of ecosystem services. The creation of a wetland map plays a crucial role in establishing the indispensable spatial and geographical foundation for the quantitative evaluation and calculation of wetland ecological service value. It also facilitates a comprehensive understanding of the position and function of coastal wetlands within the broader ecosystem, encompassing factors, such as hydrological characteristics and species diversity. Furthermore, the generation of a wetland map provides fundamental data and information support for the delineation of wetland protection zones and the implementation of wetland protection and restoration measures. This invaluable resource enables decision-makers to holistically consider the ecological value of wetlands and averts the destruction and wastage of valuable wetland resources.

High-resolution remote sensing images encompass a diverse array of distinctive attributes, spanning spectral, size, structure, shape, geometric factors, layout and other relevant characteristics. The spectral characterization of these images results in heightened spectral variation among similar features, reduced inter-class variation and an amplification of homospectral and heterospectral phenomena. As a consequence, the classification of high-resolution remotely sensed imagery has become increasingly intricate. The advancements in remote sensing technology have made a substantial impact on wetland research, offering an effective means to assess, monitor and manage wetland ecosystems. This technology facilitates the acquisition of high-resolution, multi-temporal and multi-spectral remote sensing image data specifically tailored to wetlands. By combining digital image processing with machine learning techniques, the automatic extraction and analysis of wetland vegetation, hydrological patterns, soil characteristics and meteorological information can be accomplished. This integration significantly enhances the efficiency and precision of obtaining valuable wetland information.

Traditional wetland extraction methods primarily employ manual visual interpretation techniques. This approach involves professionals acquiring specific target feature information on remote-sensing images through direct judgment or with the assistance of auxiliary interpretation instruments. For example, grayscale co-occurrence matrix [1], normalized difference index and scale invariant feature transformation are used [2]. The task of visual interpretation can be both arduous and time-consuming for those who interpret images. In addition, the quality of interpretation is limited by the interpreters' experience and knowledge of the area while the time required for information acquisition can be lengthy. For these reasons, visual interpretation is not a suitable method for independent monitoring of wetlands on a larger spatial scale. To address this issue, some scholars have taken steps to improve upon this method. Aaron Judah et al. [3] proposed a method to improve wetland classification based on multi-source remote sensing data. Sam Jackson et al. [4] used hyperspectral and lidar data to analyze coastal wetland vegetation. M. Amani et al. [5–7] used Google satellites to classify the wetlands.

As computer technology continues to advance and the use of multiple sources of data becomes increasingly prevalent, machine learning algorithms based on probabilistic statistics are supplanting traditional supervised and unsupervised classification methods as a more effective means of wetland classification. U. Maulik et al. [8] used semi-supervised SVM to classify wetland remote sensing images. Kazi Rifat Ahmed et al. [9] proposed a new method of wetland classification using optical index and machine learning. B. Fu et al. [10] mapped wetland vegetation with object- and pixel-based random forest algorithms. J. Munizaga, Ali Gonzalez-Perez and Ricardo Martínez Prentice [11–13] classified wetlands using machine learning and high-resolution imagery. Zou and Li [14,15] used landsat8 satellite imagery to achieve wetland classification. Tian et al. [16] used random forest to classify the vegetation of wetlands in arid areas of Xinjiang. L. F. Ruiz et al. [17–21] used Sentinel-1 images to classify wetland types.

Traditional classification methods based on spectral features, such as the maximum likelihood method, the minimum distance method and the K-mean clustering method, have suffered from reduced accuracy. Shallow learning algorithms, such as support vector machines and neural networks, are limited in their ability to effectively express complex functions and have restricted computational units. As the number of training samples and sample types increases, shallow models become ineffective in achieving desired classification results.

In recent times, deep learning algorithms have emerged as a promising approach for wetland information extraction, owing to their ability to abstract high-level features of images and efficiently reduce dimensionality. This development has been made possible by the advancements in image pattern recognition and artificial intelligence technology. B. Hosseiny et al. [22] proposed a spatiotemporally integrated deep learning model for wetland classification. Ali Jamali et al. [23–25] used Sentinel-1 and Sentinel-2 data to classify wetlands through Swin Transformer. Xingwei He et al. used Swin Transformer Embedding UNet for semantic segmentation of remote sensing images [26]. Z. Lv et al. used simple reading UNet for remote sensing image change detection [27]. Atharva Sharma et al. [28] applied a block-based convolutional neural network to extract roads and buildings from high-resolution remote sensing images simultaneously. Xin Li et al. [29] proposed a semantic segmentation network with dual attention depth fusion for large-scale satellite remote sensing images. Asif Raza et al. [30] used UNet++ to realize the change detection of high-resolution remote sensing images. Emmanuel Maggiori et al. [31] classified large-scale remote sensing images by the convolutional neural network. Transformer is a deep learning architecture commonly used in the field of natural language processing. It has also achieved some success in computer vision, which also includes the classification of some remote-sensing images [32–35].

The Yancheng Binhai Wetland, located in Jiangsu Province, stands as the first Binhai Wetland World Natural Heritage Site in China. In this piece of literature, we designate this area as our study site and employ Sentinel-2 to obtain high-resolution remote sensing data. We also utilize the SegFormer model, which is based on transformer architecture, to facilitate our wetland classification study. The contributions of our research are as follows: (1) We identify a set of feature variables that are suitable for the classification of coastal wetlands within China. (2) We summarize existing research and propose a comparative experimental method that can be applied to the classification of coastal wetlands. (3) We accomplish the semantic segmentation of coastal wetlands situated in Yancheng. (4) We put forward a loss function that combines CE_Loss and FocalLoss to address the commonly encountered issue of class imbalance in wetlands. This approach enhances the classification accuracy and generalization ability of our model. (5) We create a pie chart, which depicts the class proportions of coastal wetlands in Yancheng, for future wetland value research calculation, facilitating it as a point of reference.

2. Study Area and Data Preprocessing

2.1. Study Area

Yancheng Wetland is situated in the eastern coastal region of Jiangsu Province, China, as depicted in Figure 1. This wetland is the largest coastal wetland on the west coast of the Pacific Ocean and the edge of the Asian continent, with a sprawling area of approximately 4533 km^2. It constitutes 7/10 of the total mudflat area in Jiangsu Province and 1/7 of the nation's total mudflat area. This wetland has been included in the World Key Wetlands List and is renowned as the "Wetland Capital of the East". Yancheng coastal wetlands are chiefly located in the middle of the coast of northern Jiangsu Province, spanning the coastal regions of Ringshui, Binhai, Sheyang, Dafeng and Dongtai County. These wetlands can be found between 32°32–34°25 north latitude and 119°55–121°50 east longitude. The coastline of Yancheng extends 580 km, accounting for 2/3 of the entire coastline of Jiangsu Province, and the coastal mudflats cover an area of 4500 km^2. Based on preliminary calculations, the wetland area of Yancheng covers about 20% of the entire area of Yancheng, and it features two national key nature reserves. Yancheng wetlands are classified under the subtropical monsoon climate zone, experiencing hot and rainy summers and cold and dry winters. The average annual temperature ranges from 15–17°, and the annual precipitation amounts to roughly 1000 mm. Owing to tidal action, the mudflats within the area are continuously being silted up, resulting in the formation of a coastal wetland plain that is home to a diverse range of wetland types. The primary types of wetland vegetation cover include reeds, sod and rice grass.

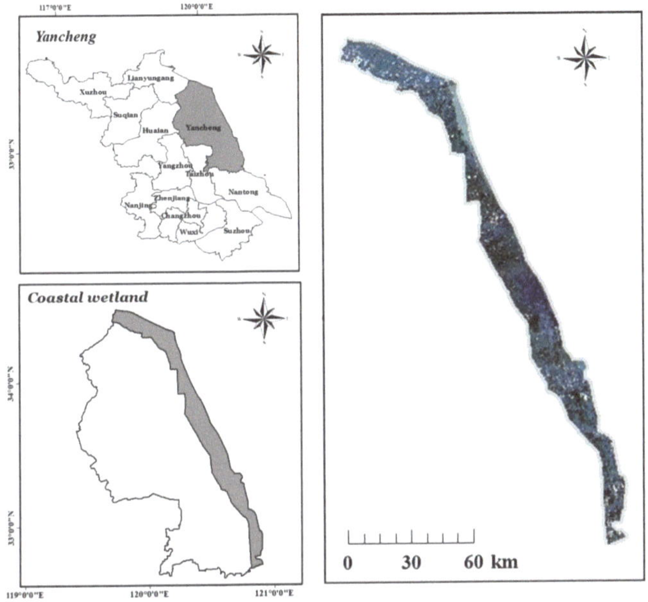

Figure 1. Yancheng Coastal Wetland, Jiangsu Province, China.

2.2. Data Collection

This study utilizes remote sensing images from the Sentinel-2 satellite of the Copernicus Data Centre of ESA, captured in August and September, as the data source for the Binhai wetland in Yancheng, Jiangsu Province. Sentinel-2 provides high spatial resolution (10 m) and multiple bands (13 bands) of data, encompassing vegetation, soil and water cover, inland waterways and coastal areas. For this research, remote sensing images captured during the summer season were selected. This choice was made because the hydrological characteristics of coastal wetlands are more distinct during this period, and the color variations of different wetland vegetation types in remote sensing images are more evident. These factors aid in manual visual labeling and other related tasks. Figure 2 presents an overview of the complete data processing procedure. We used software, such as SNAP and ENVI, to preprocess the source Sentinel data. This involved a sequential process of atmospheric correction, resampling, band synthesis and vector cropping to obtain the initial processed remote sensing images.

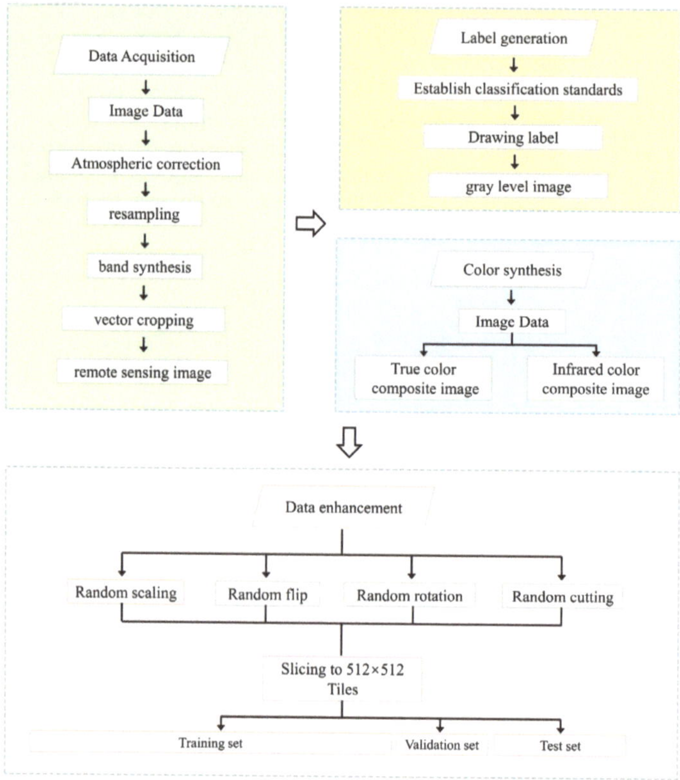

Figure 2. The flow chart of data processing.

As previously mentioned, we utilized snap and envi software to preprocess the source Sentinel data for this study. Atmospheric correction, resampling, band synthesis and vector cropping were sequentially employed to obtain the preliminary processed remote sensing images.

2.3. Wetland Classification

The concept of wetlands is comprehensive, encompassing natural or artificial, permanent or temporary marshes, peatlands or water areas, static or flowing, and can contain freshwater, brackish or saltwater bodies, including waters up to 6 m deep at low tide. Wetlands can be classified into various types, which can be broadly categorized into natural and artificial wetlands. For this study, we developed a wetland classification system by integrating previous research and utilizing existing satellite images. As presented in Table 1, the wetland classification system comprises eight types of wetlands, including Construction land, *S. alterniflora*, *S. salsa*, *P. australis*, Farmland, River system, Aquaculture and Tidal flat.

2.4. Production of Data Sets

For this study, we utilized Sentinel-2 satellite images with a resolution of 10 m. The Yancheng Binhai wetland is distributed along the coastline as a whole and has a strip shape with a size of 14,516 × 20,956. We initially cropped the large image into 10 images with a size of 2922 × 2247, based on the strip distribution characteristics of the wetland. Additionally, we manually labeled the images through visual inspection, in accordance with the wetland classification system mentioned previously. The specific categories corresponding to colors are presented in Table 2.

Table 1. Wetland classification criteria.

First Classification	Secondary Classification	Remote Sensing Image	Geometric Feature
Natural wetland	*S. alterniflora*	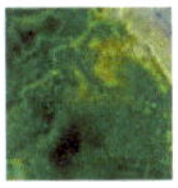	Green, single texture, redundancy and continuity
	S. salsa	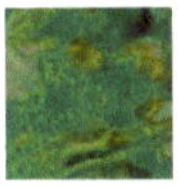	Pale yellow, flat, longitudinal stripes
	P. australis		Brown, green, monotonous texture
	River system		Light blue, striped, slender
	Tidal falt		Yellow, striped and distributed along the coast.
Artificial wetland	Farmland		Green, striped, distributed along the river.
	Aquaculture		Light color, regular grid, distributed along lakes.
Non-wetland	Construction land		Light gray, light yellow, textured and regular in shape.

Table 2. Wetland Category Colors.

Class Name	Color	Color Channel (R, G, B)	Color Channel (Gray)
Construction land	red	200, 0, 0	1
S. alterniflora	gold yellow	250, 200, 0	2
S. salsa	olive	200, 200, 0	3
P. australis	lime green	150, 250, 0	4
Farmland	green	0, 200, 0	5
River system	blue	0, 0, 200	6
Aquaculture	light blue	0, 150, 200	7
Tidal falt	deep pink	200, 150, 150	8
Background	white	255, 255, 255	0

2.5. Data Enhancement and Data Filtering

Our dataset comprises images in four bands, namely red (0.63–0.69 μm), green (0.52–0.59 μm), blue (0.45–0.52 μm) and near-infrared (0.77–0.89 μm). We generated true color synthetic images and NIR synthetic images by utilizing different band combinations. To enhance the training speed of our dataset, we converted RGB three-channel labels into single-channel labels and employed grayscale maps as the labels for training. Figure 3 displays our trained remote-sensing images and the corresponding labels.

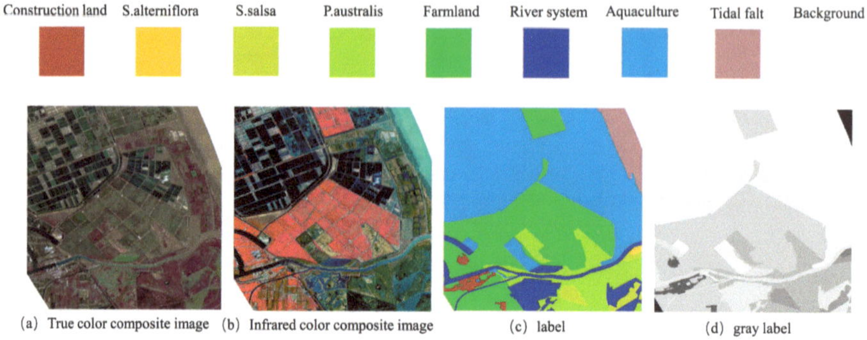

Figure 3. Remote sensing images and labels in the training set.

In the literature [36–39], the authors have shown that image cropping can improve the local effect of prediction. To augment the dataset, we initially apply random zooming and rotation/flip operations to the images in the dataset, followed by random cropping based on sizes of 128 × 128, 256 × 256, and 512 × 512. Additionally, we exclude images with excessive background and retain those containing some boundary information, ultimately accumulating 15,443 samples, which we subsequently adjust to a uniform size of 512 × 512. We partition the entire dataset into a training set, validation set, and test set in a 7:1:2 ratio, resulting in the final dataset.

3. Method

3.1. Model

Enze Xie et al. introduced the SegFormer model in 2021, which is an image segmentation model based on Transformers [40]. In contrast to conventional convolutional neural networks, SegFormer exhibits superior scalability and adaptability by employing multiple layers of Transformer encoders for feature learning. By conducting multi-layer feature extraction and employing context awareness of images, SegFormer can proficiently comprehend the semantic information in images, and thus, generate precise segmentation outcomes.

SegFormer comprises two primary modules, as illustrated in Figure 4. The first module is a hierarchical Transformer encoder that generates both high-resolution coarse features and low-resolution fine features. The second module is a lightweight ALL-MLP decoder that fuses these multi-level features to generate the final semantic segmentation masks.

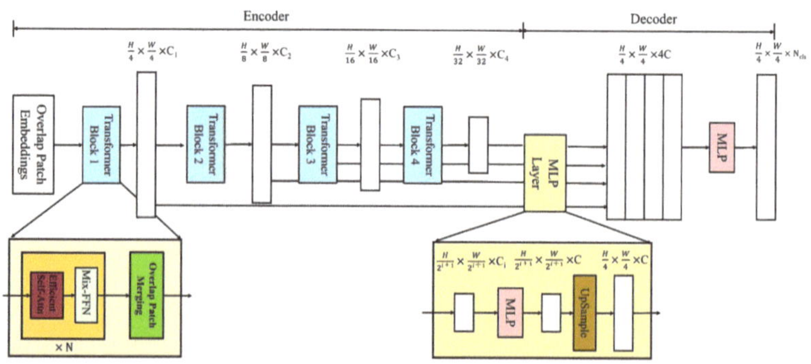

Figure 4. Model structure.

3.1.1. Encoder

The primary component of the encoder is a multi-layer transformer block. The Transformer module within the encoder, known as MiT, utilizes the overlap patch embeddings (OPE) structure for feature extraction and downsampling of the input image. The resulting features are then fed into the Multi-headed Self-attention layer and the Mix-FFN layer. Let us elaborate on these aforementioned structures.

3.1.2. Overlapped Patch Embeddings

In the SegFormer architecture, the overlap patch embedding module is utilized to extract image features. This module partitions the input image into multiple overlapping patches and generates an embedding vector for each of them. The overlap region between adjacent patches provides a broader perceptual field for capturing more contextual information while preserving the local information of each patch. This approach reduces information loss and mitigates the issue of boundary artifacts, thereby enhancing the robustness and generalization capability of the model.

3.1.3. Overlapped Patch Merging

In SegFormer, the Overlapped Patch Merging (OPM) technique is utilized to merge overlapping patches into a complete image. This module is inspired by ViT and is designed to combine non-overlapping images or feature blocks in ViT. Chunking may produce boundary effects, resulting in segmentation outcomes with discontinuous regions due to the absence of shared information between cropped blocks. Therefore, OPM leverages an overlapping patch merging process to eliminate these discontinuous edges by fusing adjacent blocks with similar or identical category labels to ensure spatial semantic coherence. The module can be implemented using a convolutional layer with Kernel = 7, Stride = 4 and Padding = 3.

3.1.4. Efficient Self-Attention

Efficient Self-Attention is a lightweight self-attention mechanism utilized for pixel-level feature extraction. In comparison to the conventional Self-Attention approach, it minimizes the computational load and number of parameters, thereby improving the model's efficiency during runtime and mitigating the issue of overfitting.

In the original multi-head self-attention process, each of the heads Q, K, V has the same dimensions $N \times C$, where $N = H \times W$ is the length of the sequence, the self-attention is estimated as:

$$Attention(Q, K, V) = Soft\,max(\frac{QK^T}{\sqrt{d_{head}}})V \tag{1}$$

where Q is the query vector, K is the key vector and V is the value vector. d_{head} is the latitude size of each vector, and Softmax is a normalization function.

The traditional Self-Attention needs to calculate the similarity matrix between any two positions during the computation, and the computational complexity of this process is $O\ (N^2)$, which generates a large computation and memory occupation and cannot satisfy the input of large-resolution images. To reduce the complexity, ESA adopts an idea similar to local attention and introduces the approximate ratio R to approximate the length of the input sequence.

$$\hat{K} = Reshape(\frac{N}{R}, C \cdot R)(K) \tag{2}$$

$$K = Linear(C \cdot R, C)(\hat{K}) \tag{3}$$

where Reshape function converts K into a sequence of $\frac{N}{R} * (C - R)$ shapes, Linear transforms $\hat{K}$ linearly and finally gets K of dimension $\frac{N}{R} * C$. The complexity of the ESA mechanism is reduced from $O(N^2)$ to $O\left(\frac{N^2}{R}\right)$. In the 4 stages of the model, R is set as (64,16,4,1).

3.1.5. Mix-FFN

Conventional Vision Transformer (ViT) uses Position Encoding (PE) to introduce position information. However, the resolution of PE is fixed. Therefore, when the test resolution and the training resolution are different, the position encoding needs to be interpolated, but it often leads to a decrease in accuracy. The Mix-FFN, the structure used in SegFormer, is shown in Figure 5. It takes into account the effect of zero padding on the leaked position information and uses 3 × 3 Conv directly in the feedforward network (FFN), whose equation is shown below.

$$X_{out} = MLP(GELU(Conv_{3\times3}(MLP(X_{in})))) + X_{in} \tag{4}$$

where X_{in} is the feature from the self-attention module. Mix-FFN mixes a 3 × 3 convolution and an MLP into each FFN.

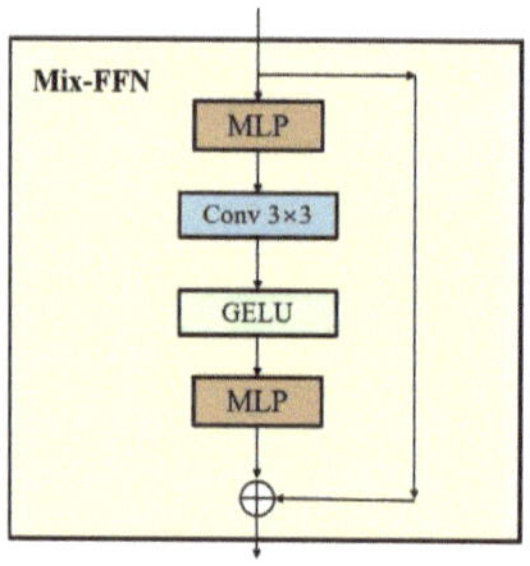

Figure 5. Mix-FFN.

3.2. Lightweight All-MLP Decoder

The Lightweight All-MLP Decoder is a compact decoder that processes the feature maps generated by the encoder output to obtain the final semantic segmentation outcome. This decoder has a reduced number of parameters, and the multi-level Transformer Encoder can obtain a broader perceptual field than the conventional CNN Encoder.

The proposed All-MLP decoder consists of four main steps. First, multi-level features F_i from the MiT encoder go through an MLP layer to unify the channel dimension. Then, in a second step, features are up-sampled to 1/4th and concatenated together. Third, an MLP layer is adopted to fuse the concatenated features F. Finally, another MLP layer takes the fused feature to predict the segmentation mask M with a $\frac{H}{4} \times \frac{W}{4} \times N_{cls}$ resolution, where N_{cls} is the number of categories. The formula is shown below.

$$\hat{F}_i = Linear(C_i, C)(F_i), \forall i \tag{5}$$

$$\hat{F}_i = Upsample(\frac{W}{4} \times \frac{W}{4})(\hat{F}_i), \forall i \tag{6}$$

$$F = Linear(4C, C)(Concat(\hat{F}_i)), \forall i \tag{7}$$

$$M = Linear(C, N_{cls})(F) \tag{8}$$

where M refers to the predicted mask and $Linear(C_{in}, C_{out})(\cdot)$ refers to a linear layer with C_{in} and C_{out} as input and output vector dimensions, respectively.

3.3. Model Training

In this experience, the gradient descent algorithm with momentum (Adam with weight decay, adamw) is chosen as the optimization algorithm for model training, and the model parameters are updated by minimizing the difference between the predicted and real features until the loss function reaches the minimum value. The momentum gradient descent method borrows the concept of momentum in physics when updating the parameters so that the update of parameters not only depends on the currently calculated gradient information but also takes into account the gradient information of the previous update steps. By using the exponentially weighted momentum m_t to replace the original gradient g_t for updating the model parameters, the model can avoid the oscillation problem during the training process and speed up the training of the model. m_t and the parameter update are calculated as shown in Equations.

$$m_t = \beta_1 m_{t-1} + (1 - \beta_1) g_t \tag{9}$$

$$v_t = \beta_2 v_{t-1} + (1 - \beta_2) {g_t}^2 \tag{10}$$

where g_t is the gradient of the current step. m_t is the first-order moment estimate of the gradient. v_t is the second-order moment estimate of the gradient. β_1 and β_2 are the corresponding exponential decay rates, respectively.

$$\hat{m}_t = \frac{m_t}{1 - (\beta_1)^t} \tag{11}$$

$$\hat{v}_t = \frac{v_t}{1 - (\beta_2)^t} \tag{12}$$

$$\alpha_t = \frac{\eta}{\sqrt{\hat{v}_t} + \varepsilon} \tag{13}$$

where η is the initial learning rate. ε is a very small constant to avoid the error of dividing by zero.

$$w_{t+1} = w_t - \alpha_t (\hat{m}_t + \lambda w_t) \tag{14}$$

where λ is the weight decay factor.

The deep learning server used in this study had a specific hardware and software configuration, which is detailed in Table 3. To ensure consistency, other experimental parameters were standardized, as shown in Table 4, to evaluate the models' performance on the dataset. The momentum descent algorithm with gradient was utilized, and the adamw optimizer, which incorporates the weight decay mechanism, was found to be faster and more accurate than other gradient descent algorithms with minimal human intervention. The CrossEntropyLoss function was used as the loss function for the experiments. Additionally, a preheating mechanism was implemented to compare the strengths and limitations of each model by setting the maximum Epoch to 70.

Table 3. The environment of the experiment.

GPU	Video Memory	Operating System	Language
NVIDIA GeForce RTX 3090	24 G	Linux	Python 3.9
CPU	**Random Access Memory**	**Deep Learning Framework**	**Cuda**
Random Access Memory Intel (R) Xeon (R) Gold 6330	60	Pytorch 1.7.0	Cuda 11.0

Table 4. Network parameters of the considered architectures.

Optimizer	Momentum	Learning Rate (LR)	LR Scheduler	Weight Decay	Loss Function	Warmup
AdamW	0.9	0.0001	Polynomial decay	0.01	CrossEntropyLoss	linear
Warmup Iters	**Warmup Ratio**	**Iters**	**Batch Size**	**Max Epoch**	**Validation Frequency**	
1500	1×10^{-6}	29,190	8	70	Each epoch	

3.4. Evaluation Metrics

To assess the semantic segmentation performance, we employed several widely-used metrics, including Mean Accuracy (*mAcc*), Overall Accuracy (*Acc*) and Mean Intersection over Union (*mIoU*).

$$mAcc = \frac{1}{n}\sum_{i=1}^{n}\frac{TP_i}{TP_i + FP_i} \quad (15)$$

$$aAcc = \frac{\sum_{i=1}^{n} TP_i}{\sum_{i=1}^{n}(TP_i + FN_i)} \quad (16)$$

where n is the total number of categories, TP_i is the number of samples that the model predicts as class i and are truly labeled as class i, FN_i is the number of samples that are class i but are wrongly predicted as other classes and FP_i is the number of samples that the model predicts as class i for all other classes.

$$mIoU = \frac{1}{k+1}\sum_{i=0}^{k}\frac{p_{ii}}{\sum_{j=0}^{k} p_{ij} + \sum_{j=0}^{k} p_{ij} - p_{ii}} \quad (17)$$

where p_{ii} represents the number of pixels predicted by category i to category i. $k + 1$ is the total number of categories, and p_{ij} is the number of pixels predicted by category i to category j.

The precision, recall, F_1-score and intersection over union (*IoU*) metrics are used to further evaluate the performance of various models on the test data set, which is calculated as follows.

$$Precision = \frac{TP}{TP + FP} \quad (18)$$

$$Recall = \frac{TP}{TP + FN} \quad (19)$$

$$F_1\text{-}score = 2 \times \frac{Precision \times Recall}{Precision + Recall} \quad (20)$$

where *TP* denotes the true positives, *FP* denotes the false positives and *FN* denotes false negatives. Based on these three metrics, we applied independent test set data to evaluate the performances with different models and different loss functions.

4. Experiments and Results

Results with Different Models

The same set of parameters was utilized on the test set to evaluate the performance of the models, and the results are depicted in Figure 6. SegFormer exhibited superior accuracy and robustness in multi-categorical semantic segmentation, as evidenced by its curve (yellow) outperforming other models across all metrics. At specific points, the SegFormer model achieved a mean Intersection over Union (mIoU) of 0.81, mean Accuracy (mAcc) of 0.87, and average Accuracy (aAcc) of 0.94. In contrast, the Unet model (orange) demonstrated poorer performance across various metrics.

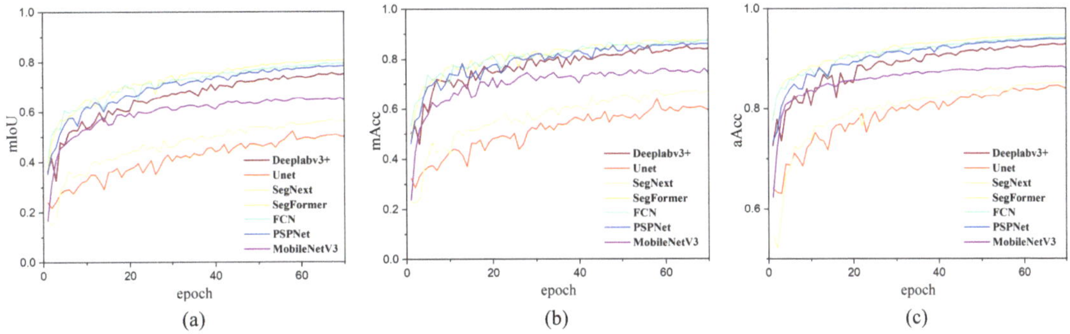

Figure 6. Comparison of the metrics of various models in model training (one validation per epoch). (**a**) The graph of mIoU. (**b**) The graph of mAcc. (**c**) The graph of aAcc.

Figure 7 shows the refined metrics of SegFormer for each class. As illustrated in Figure 7a, the model converges at approximately epoch 80. Figure 7b showcases some variations in the prediction outcomes for different categories, with construction land, s.alterniflora and farmland exhibiting better prediction results, with an IoU close to 90%. The river system's prediction effect is not satisfactory, with an IoU of about 40%. Figure 7c indicates that the overall prediction performance is good, with minor fluctuations in the Acc curve but in an upward trend. The accuracy of most categories can reach 90%, with the river system being the least accurate, with an accuracy of 50%.

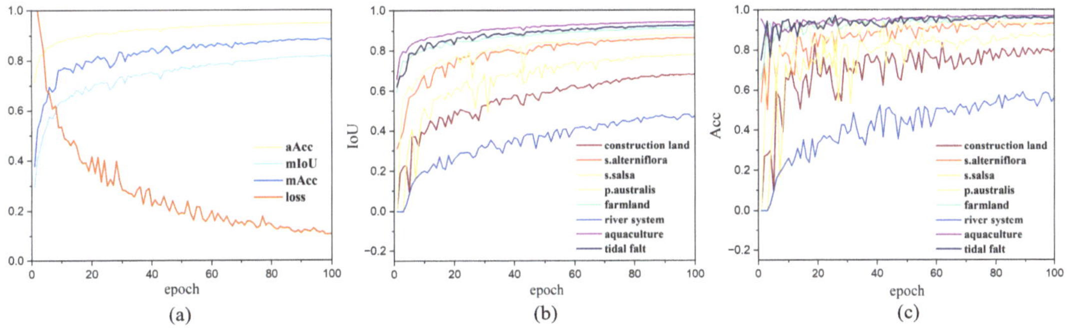

Figure 7. The evaluation results of segformer. (**a**) The iterative graph of aAcc, mIoU, mAcc and loss. (**b**) IoU curve for different classes. (**c**) Acc curve for different classes.

To further assess the model's classification effectiveness, Precision, Recall and F_1-score were selected, and the final test results on the test dataset are presented in Table 5. The table illustrates different prediction effects for each class, with significant differences in the indices for each model. SegFormer demonstrates the highest mPrecision (0.901) and the lowest (0.754), the highest mRecall (0.876) and the lowest (0.642), and the highest mF_1-score (0.887) and the lowest (0.632). These results indicate that SegFormer is the optimal model for remote sensing wetland classification, accurately identifying various categories of pixels and exhibiting better recognition of the shape and contour of the target object with high accuracy and consistency. Most categories achieved good results across all metrics in each model, but a few classes exhibited poor prediction results. Among the nine categories, the river system consistently had low indices, with the highest Recall being only 0.549. Therefore, the river system proves to be a challenging category in wetland classification.

Table 5. The comparison between various models in terms of Precision, Recall and F-1 score (CL = Construction land, *S. alterniflora* = S.a, *S. salsa* = S.s, *P. australis* = P.a, Farmland = F, River system = RS, Aquaculture = A, Tidal falt = T, Background = B, mRecall = MR, mPrecision = MP and mF-score = MF).

Model	CL	S.a	S.s	P.a	F	RS	A	T	B	MP	MR	MF
Deeplabv3+										0.859	0.843	0.850
Precision	0.716	0.881	0.848	0.913	0.927	0.626	0.962	0.937	0.919			
Recall	0.727	0.897	0.762	0.910	0.946	0.507	0.954	0.945	0.935			
F-1 score	0.722	0.889	0.803	0.911	0.937	0.560	0.958	0.941	0.927			
Unet										0.837	0.798	0.826
Precision	0.698	0.875	0.805	0.870	0.866	0.744	0.875	0.926	0.928			
Recall	0.746	0.840	0.745	0.817	0.869	0.524	0.942	0.824	0.933			
F-1 score	0.721	0.857	0.774	0.843	0.867	0.618	0.908	0.872	0.930			
SegNext										0.754	0.672	0.695
Precision	0.574	0.764	0.552	0.799	0.834	0.616	0.890	0.885	0.872			
Recall	0.342	0.730	0.555	0.679	0.903	0.164	0.904	0.873	0.901			
F-1 score	0.429	0.746	0.554	0.734	0.867	0.259	0.897	0.879	0.886			
SegFormer										**0.901**	**0.876**	**0.887**
Precision	0.791	0.924	0.917	0.923	0.948	0.745	0.968	0.957	0.933			
Recall	0.781	0.929	0.841	0.947	0.956	0.541	0.970	0.960	0.965			
F-1 score	0.786	0.926	0.877	0.935	0.952	0.627	0.969	0.959	0.948			
FCN										0.892	0.870	0.879
Precision	0.798	0.910	0.904	0.921	0.937	0.729	0.971	0.946	0.908			
Recall	0.751	0.926	0.826	0.940	0.957	0.549	0.961	0.958	0.960			
F-1 score	0.774	0.918	0.863	0.930	0.947	0.626	0.966	0.952	0.933			
PSPNet										0.879	0.864	0.869
Precision	0.721	0.893	0.898	0.922	0.938	0.702	0.965	0.946	0.923			
Recall	0.769	0.927	0.810	0.927	0.947	0.531	0.961	0.952	0.951			
F-1 score	0.744	0.910	0.852	0.924	0.943	0.604	0.963	0.949	0.937			
MobileNetV3										0.792	0.761	0.771
Precision	0.601	0.812	0.696	0.848	0.867	0.563	0.924	0.902	0.915			
Recall	0.566	0.800	0.745	0.845	0.912	0.277	0.917	0.901	0.882			
F-1 score	0.583	0.806	0.720	0.846	0.889	0.372	0.921	0.901	0.898			

In order to compare the accuracy and stability of different models more intuitively, we selected test set images for prediction. Figure 8 showcases a remote sensing image of the Yancheng Wetland Rare Birds National Nature Reserve in Jiangsu Province, which represents a typical coastal wetland. This area is characterized by a distinct estuarine delta landscape and consists of various types of wetland environments, including freshwater marshes, lakes, beaches and coastal wetlands of the Yangtze River Delta. It is home to a diverse range of species and holds high ecological conservation value.

The predicted results are presented in Figure 8. Combining the images and labels, the distribution of *S. alterniflora*, *S. salsa*, *P. australis* and Tidal falt in this reserve is relatively concentrated, accounting for most of the reserve. The identification of these four categories was generally consistent among the models, and all models except Unet were able to extract clear boundary information. Deeplabv3+, SegFormer, FCN, Unet and PSPNet were able to identify the river system at the edge of the protected area while the other models, SegNext and MobileNetV3, were able to extract the river system relatively accurately. However, the MobileNetV3 extraction results are finer than the actual ones. It is observed that the SegNext model exhibits inaccuracies in differentiating between paddy fields, ponds and rivers in arable land, leading to misclassifications of the River system and Aquaculture categories. Notably, the part marked by the larger red box in the labeled image is the scattered River system. In the prediction of this part, most models' segmentation results produce too many isolated points, with relatively smooth and lacking contour features. In contrast, SegFormer significantly outperforms other methods in terms of result accuracy and edge clarity. Additionally, in the other red box, most models ignore the river due to its small and slender area, with only SegFormer extracting a portion of the water body.

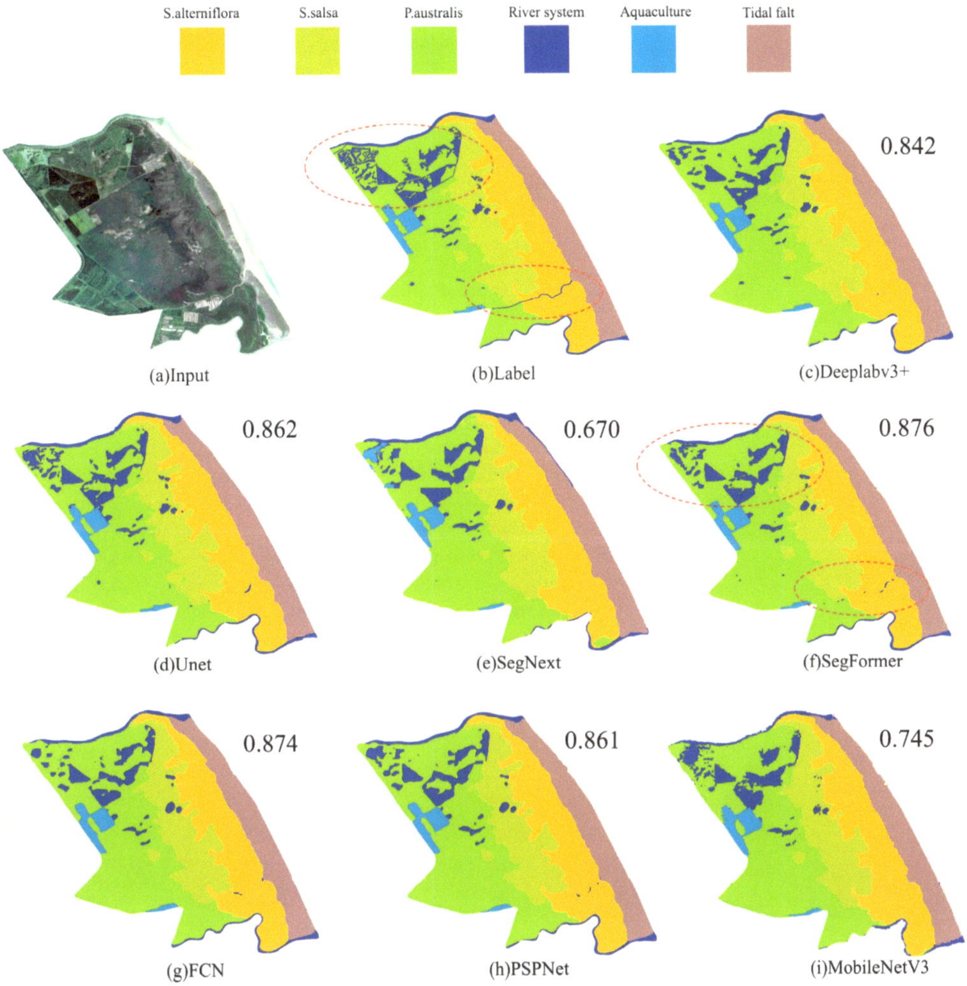

Figure 8. A comparison image about the predicted images of different deep learning models.

Figure 9 presents a comparison between the prediction results of deep learning methods and traditional methods, including supervised classification based on the statistical maximum likelihood method, machine learning-based neural networks and supervised classification based on support vector machines (SVM). While the traditional methods mentioned above may outperform deep learning in extracting boundary information, they suffer from misclassification issues. For example, the maximum likelihood method misclassifies *P. australis* as *S. salsa* while the neural network exhibits the opposite misclassification. SVM, on the other hand, struggles to accurately differentiate the river system from Aquaculture. These findings highlight the limitations of traditional methods when it comes to precise classification in complex wetland environments.

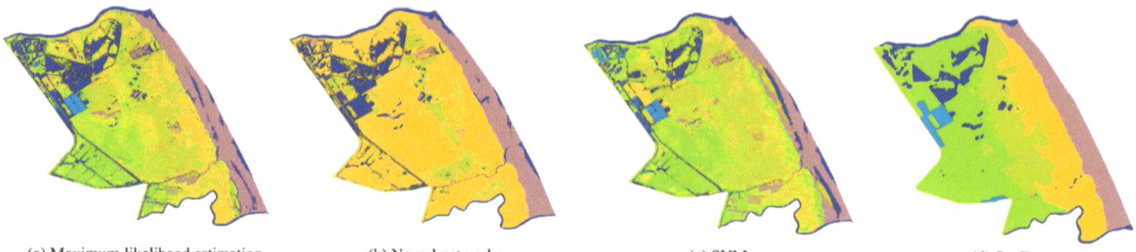

(a) Maximum likelihood estimation (b) Neural networks (c) SVM (d) SegFormer

Figure 9. Prediction graphs compared to traditional methods.

In order to further investigate the classification capability of the deep learning model mentioned above for small-scale features, we conducted additional comparisons using various sets of images, as illustrated in Figure 10. These images contained a diverse range of feature characteristics. From the prediction results, it can be inferred that the River system remains a challenging aspect of the prediction process. The SegFormer-based prediction method outperforms the others, as it can effectively extract most of the features in the water body. PSPNet and FCN methods are also effective but not as much as the SegFormer approach. In summary, the SegFormer model, based on deep learning, exhibits an impressive ability to perform semantic segmentation of wetland remote sensing images and is capable of learning and extracting the features inherent in wetlands. It can effectively segment boundaries in complex ground object environments and extract small-scale ground objects that are affected. As one of the deep learning methods, with the emergence of higher resolution images and more powerful computing devices, it is expected to exhibit even faster segmentation efficiency and higher accuracy than traditional methods.

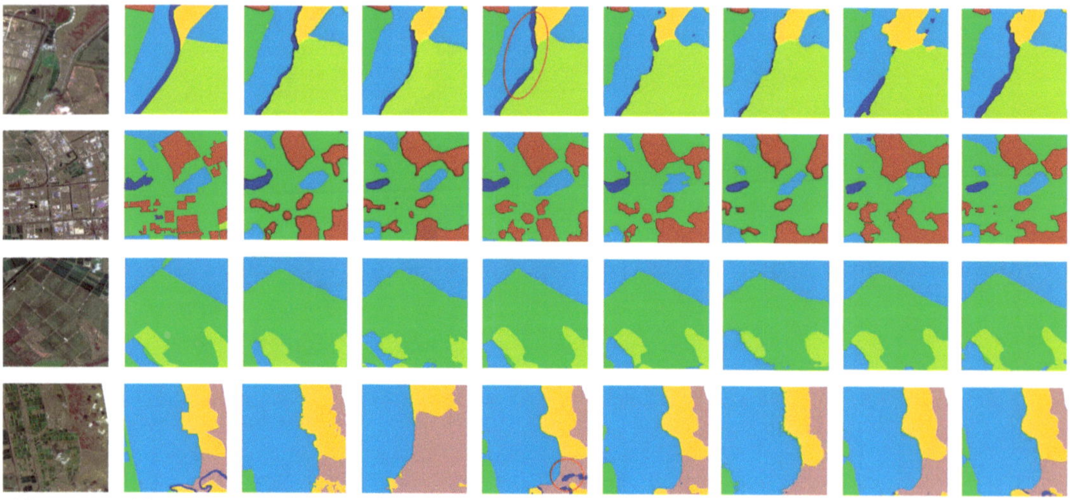

(a) Input (b) Label (c) Unet (d) SegNext (e) SegFormer (f) PSPNet (g) MobileNetV3 (h) Deeplabv3+ (i) FCN

Figure 10. Predictive graphs of different models.

5. Discussion

5.1. Results with Different Loss Functions

As illustrated in Figure 11, there exists a specific class imbalance in the dataset. When considering the relationship between the proportion of features in the training samples and the accuracy of the model's extraction results, it was observed that the extraction accuracy was higher for the feature classes that were more adequately represented in the dataset, such as Tidal falt, Aquaculture, Farmland, *P. australis* and *S. alterniflora*, for each model. Conversely, the extraction accuracy was relatively poor for categories that accounted for a smaller proportion of the river system, such as *S.salsa* and Construction land dataset. Due to the limited information contained in the samples for these categories, it becomes difficult for the network to learn their distribution pattern, and it is challenging to effectively identify them.

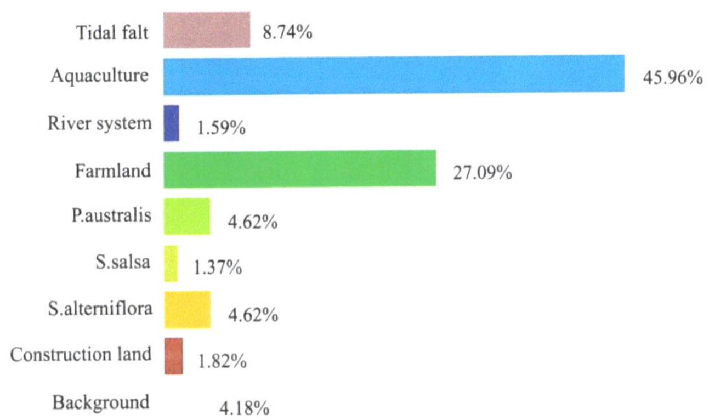

Figure 11. The proportion of the Number of Pixels (%).

In order to improve the extraction accuracy of the model, we explored multiple loss functions. Specifically, we chose the best model, SegFormer, and modified only the loss function while keeping the other parameters unchanged. For this study, we considered CrossEntropyLoss (CE_Loss), LovaszLoss, Focal Loss and CrossEntropyLoss + FocalLoss as loss functions, and the results are presented in Table 6. The findings in Table 6 indicate that, without any changes to the deep learning algorithm framework and optimization, LovaszLoss performs the worst, with its various metrics being lower than the other loss functions. FocalLoss is slightly inferior to CE_Loss, but it improves the performance of ConStruction land classification in terms of IoU, Acc and F-score metrics. The combination of CE_Loss and FocalLoss enhances the classification performance of the model under this problem. Specifically, the IoU and F-score for River system classification improved by 0.28% and 0.27%, respectively, compared to CE_Loss. Additionally, Acc for *S. salsa* classification improved by 0.56% while Acc and F-score for Construction land classification improved by 1.41% and 1.04%, respectively. Overall indexes, such as mIoU, mAcc, aAcc and mF-score, also improved by about 0.1% compared to CE_Loss. It can be concluded that, under the condition of unbalanced wetland classification, the combined loss function of CE_Loss and FocalLoss can enhance the classification accuracy of difficult classes, alleviate overfitting and misclassification of the model caused by class imbalance and improve the performance and generalization ability of the model.

5.2. Coastal Wetland Detection

Changes in precipitation and temperature can indeed have an impact on the landscape pattern of coastal wetlands. The gradual increase in temperature can lead to increased evaporation of surface water, reduced precipitation and limited availability of water resources. This puts additional strain on coastal wetlands and disrupts their ecological balance to some extent. Additionally, the growing population in Yancheng has contributed to the degradation of coastal wetlands. The demand for ecological resources provided by wetlands has risen, putting tremendous pressure on these fragile ecosystems. Excessive development and utilization have further worsened the situation, resulting in a significant decrease in the coastal wetland area. Considering the rapidly changing wetland environment, it is crucial to implement efficient and timely detection methods. These methods will help monitor and assess the status of coastal wetlands effectively. By detecting changes in land cover, vegetation and other relevant parameters, we can gain valuable insights into the health and conservation needs of these ecosystems. Timely detection and monitoring enable us to take prompt action and implement appropriate management strategies to mitigate the negative impacts and ensure the long-term sustainability of coastal wetlands.

Table 6. Comparison of different loss functions.

Loss Function	IoU	Acc	F-Score	mIoU	mAcc	aAcc	MR	MP	MF
CE_Loss				81.05	87.64	94.84	87.64	90.06	88.65
River system	45.62	54.07	62.65						
S. salsa	78.16	84.08	87.74						
Construction land	64.71	78.07	78.57						
LovaszLoss				77.72	87.62	92.84	87.62	85.31	86.31
River system	39.08	55.99	56.2						
S. salsa	76.1	86.64	86.43						
Construction land	58.23	85.24	73.6						
FocalLoss				80.48	87.46	94.78	87.46	89.34	88.24
River system	43.91	53.44	61.02						
S. salsa	75.63	82.97	86.12						
Construction land	65.48	79.27	79.14						
CE_Loss + FocalLoss				81.15	87.73	94.91	87.73	89.99	88.73
River system	45.9	56.02	62.92						
S. salsa	77.25	84.64	87.16						
Construction land	66.12	77.16	79.61						

Traditional wetland inspection mainly relies on manual surveys and visual inspections of remote sensing images [41–43], which consume significant human and material resources. While various digital image processing methods have been proposed to improve the classification accuracy of detailed information in remote sensing images, they are often limited in their ability to meet the requirements of complex image classification [44–46]. Although machine learning methods have been shown to enhance classification accuracy [47–50], they require image preprocessing and feature enhancement to achieve better classification results [51]. Moreover, the degree of automation is not high, and significant improvements in the correct rate are already difficult to achieve. In this paper, we propose more specific methods and ideas for wetland classification, building upon existing deep learning methods and demonstrating their feasibility and research value.

In this study, we utilized high-resolution images provided by Sentinel-2 to count the percentage distribution of feature classes in the study area, enabling subsequent calculation of ecological values. The specific proportional information is illustrated in Figure 12.

Figure 12 reveals that Aquaculture, Farmland and Tidal flat constitute the predominant components of Yancheng's coastal wetland, accounting for 42.2%, 29.7% and 14.7% respectively. Notably, *S. salsa* and *P. australis*, which represent unique vegetation within the wetland, encompass just 1% and 3.6%, respectively, totaling an area of 44.79 km^2 and 100.84 km^2. The utilization of deep learning techniques enables the swift detection of changes in Yancheng's coastal wetland, as well as variations in the proportion of different species. The value of wetland ecosystem services serves as a monetized representation illustrating the collective contribution of wetland ecosystems to human well-being. It encompasses an array of services, such as product supply, ecological regulation, and spiritual and cultural benefits. When evaluating the wetland ecosystem service value of Yancheng, it is crucial to select appropriate evaluation indices and systems aligned with the region's available resources and market conditions. The wetland mapping provides valuable spatial and geographic foundations that facilitate effective assessment of the wetland ecosystem service value.

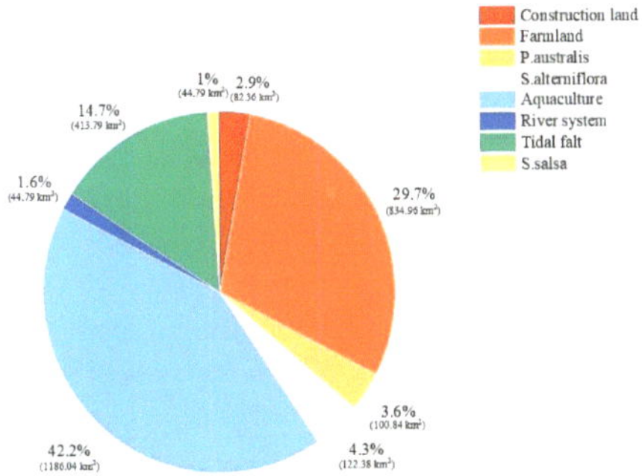

Figure 12. Pie chart of feature class scales for the study area.

Regarding the significant components of Yancheng's coastal wetland, namely Aquaculture, farmland and mudflat, their construction should prioritize enhancing the ecosystem service value concerning product supply, water flow regulation and water purification. As for the special vegetation occupying a comparatively smaller proportion, a comprehensive assessment of their ecosystem service value should concentrate on aspects related to species conservation, tourism and recreation, as well as scientific and educational services.

6. Conclusions

In this study, we utilized high-resolution images provided by Sentinel-2 and constructed an information extraction model of the Yancheng Binhai wetland based on SegFormer. Based on the relevant literature and remote sensing images, we classified the wetlands in the area into nine species, namely, Construction land, *S. alterniflora*, *S. salsa*, *P. australis*, Farmland, River system, Aquaculture and Tidal falt. Remote sensing images were preprocessed using atmospheric correction, resampling, band synthesis and vector cropping. The labeled images were drawn, and data enhancement methods, such as random deflation, random flip and random crop, were applied. Finally, the training images were standardized to 512 × 512 size and divided into a training set, validation set and test set in a ratio of 7:1:2. Evaluation metrics, such as mIoU, mAcc, aAcc, mRecall, mPrecision and mF-score were used. While ensuring the consistency of other parameters, we conducted comparison experiments and found that the SegFormer model exhibited better learning and extraction ability for features contained in wetlands, with various metrics, such as mIoU (0.81), mAcc (0.87), aAcc (0.94), mPrecision (0.901), mRecall (0.876) and mF-score (0.887), being higher than those of other models. We compared the prediction images of various deep learning methods for the key protected area of Yancheng Binhai Wetland and found that SegFormer's prediction results were superior to those of other methods in terms of accuracy and edge clarity, effectively segmenting the boundary in the complex feature environment and extracting small-scale features under influence more accurately. We also studied the common class imbalance in wetlands and concluded that the loss function using the combination of CE_Loss and FocalLoss can improve the classification accuracy of difficult classes and enhance the performance and generalization ability of the model. Finally, we extracted the class proportion information of the Yancheng coastal wetland, which provides technical support and reference value for subsequent wetland value research and research ideas for related coastal wetland information extraction research in China.

Author Contributions: X.L. was accountable for formulating the programs and composing the primary manuscript. Y.C. and G.C. aided in the gathering and analysis of data. W.C. assisted in data collection and produced the figures. R.C. supported in processing remote sensing imagery and generating labels. D.G. played a crucial role in designing and drafting the project, lending vital assistance in manuscript revising. Y.Z. furnished guidance and priceless perspectives on data sources and classification methods. Y.W. charted the project, edited the manuscript and provided constructive feedback throughout the process. All authors have read and agreed to the published version of the manuscript.

Funding: This project is sponsored by the the Jiangsu Forestry Science & Technology Innovation and Extension Project (Project No: LYKJ[2022]02) and Provincial Student Innovation Training Program (Project No: 202310298113Y).

Data Availability Statement: The data are not publicly available due to the confidentiality of the research projects.

Conflicts of Interest: The authors declare no conflict of interest.

References

1. Bansal, S.; Katyal, D.; Garg, J.K. A novel strategy for wetland area extraction using multispectral MODIS data. *Remote Sens. Environ.* **2017**, *200*, 183–205. [CrossRef]
2. LoweDavid, G. Distinctive Image Features from Scale-Invariant Keypoints. *Int. J. Comput. Vis.* **2004**, *60*, 91–110.
3. Judah, A.; Hu, B. An Advanced Data Fusion Method to Improve Wetland Classification Using Multi-Source Remotely Sensed Data. *Sensors* **2022**, *22*, 8942. [CrossRef] [PubMed]
4. Suir, G.M.; Jackson, S.; Saltus, C.; Reif, M.K. Multi-Temporal Trend Analysis of Coastal Vegetation Using Metrics Derived from Hyperspectral and LiDAR Data. *Remote Sens.* **2023**, *15*, 2098. [CrossRef]
5. Amani, M.; Brisco, B.; Mahdavi, S.; Ghorbanian, A.; Moghimi, A.; DeLancey, E.R.; Merchant, M.A.; Jahncke, R.; Fedorchuk, L.; Mui, A.; et al. Evaluation of the Landsat-Based Canadian Wetland Inventory Map Using Multiple Sources: Challenges of Large-Scale Wetland Classification Using Remote Sensing. *IEEE J. Sel. Top. Appl. Earth Obs. Remote Sens.* **2021**, *14*, 32–52. [CrossRef]
6. Li, A.; Song, K.; Chen, S.; Mu, Y.; Xu, Z.; Zeng, Q. Mapping African wetlands for 2020 using multiple spectral, geo-ecological features and Google Earth Engine. *ISPRS J. Photogramm. Remote Sens.* **2022**, *193*, 252–268. [CrossRef]
7. Gxokwe, S.; Dube, T.; Mazvimavi, D. Leveraging Google Earth Engine platform to characterize and map small seasonal wetlands in the semi-arid environments of South Africa. *Sci. Total Environ.* **2021**, *803*, 150139. [CrossRef]
8. Maulik, U.; Chakraborty, D. Learning with transductive SVM for semisupervised pixel classification of remote sensing imagery. *Isprs J. Photogramm. Remote Sens.* **2013**, *77*, 66–78. [CrossRef]
9. Ahmed, K.R.; Akter, S.; Marandi, A.; Schüth, C. A simple and robust wetland classification approach by using optical indices, unsupervised and supervised machine learning algorithms. *Remote Sens. Appl. Soc. Environ.* **2021**, *23*, 100569. [CrossRef]
10. Fu, B.; Wang, Y.; Campbell, A.D.; Li, Y.; Zhang, B.; Yin, S.; Xing, Z.; Jin, X. Comparison of object-based and pixel-based Random Forest algorithm for wetland vegetation mapping using high spatial resolution GF-1 and SAR data. *Ecol. Indic.* **2017**, *73*, 105–117. [CrossRef]
11. Gonzalez-Perez, A.; Abd-Elrahman, A.H.; Wilkinson, B.E.; Johnson, D.J.; Carthy, R.R. Deep and Machine Learning Image Classification of Coastal Wetlands Using Unpiloted Aircraft System Multispectral Images and Lidar Datasets. *Remote Sens.* **2022**, *14*, 3937. [CrossRef]
12. Munizaga, J.; García, M.; Ureta, F.; Novoa, V.; Rojas, O.; Rojas, C. Mapping Coastal Wetlands Using Satellite Imagery and Machine Learning in a Highly Urbanized Landscape. *Sustainability* **2022**, *14*, 5700. [CrossRef]
13. Prentice, R.M.; Peciña, M.V.; Ward, R.D.; Bergamo, T.F.; Joyce, C.; Sepp, K. Machine Learning Classification and Accuracy Assessment from High-Resolution Images of Coastal Wetlands. *Remote Sens.* **2021**, *13*, 3669. [CrossRef]
14. Xing, L.; Wang, H.; Fan, W.; Chen, C.; Li, T.; Wang, G.; Zhai, H. Optimal Features Selection for Wetlands Classification Using Landsat Time Series. In Proceedings of the IGARSS 2018—2018 IEEE International Geoscience and Remote Sensing Symposium, Valencia, Spain, 22–27 July 2018; IEEE: Piscataway, NJ, USA, 2018; pp. 8385–8388.
15. Zou, Q.; Chen, C. Transferring Deep Belief Networks for the Classification of LANDSAT8 Remote Sensing Imagery. *J. Phys. Conf. Ser.* **2020**, *1544*, 012106. [CrossRef]
16. Tian, S.; Zhang, X.; Tian, J.; Sun, Q. Random Forest Classification of Wetland Landcovers from Multi-Sensor Data in the Arid Region of Xinjiang, China. *Remote Sens.* **2016**, *8*, 954. [CrossRef]
17. Ruiz, L.F.C.; Guasselli, L.A.; Simioni, J.P.D.; Belloli, T.F.; Fernandes, P.C.B. Object-based classification of vegetation species in a subtropical wetland using Sentinel-1 and Sentinel-2A images. *Sci. Remote Sens.* **2021**, *3*, 100017. [CrossRef]
18. Liu, H.; Jiang, Q.; Ma, Y.; Yang, Q.; Shi, P.; Zhang, S.; Tan, Y.; Xi, J.; Zhang, Y.; Liu, B.; et al. Object-Based Multigrained Cascade Forest Method for Wetland Classification Using Sentinel-2 and Radarsat-2 Imagery. *Water* **2022**, *14*, 82. [CrossRef]

19. Zhao, Y.; Mao, D.; Zhang, D.; Wang, Z.; Du, B.; Yan, H.; Qiu, Z.; Feng, K.; Wang, J.; Jia, M. Mapping Phragmites australis Aboveground Biomass in the Momoge Wetland Ramsar Site Based on Sentinel-1/2 Images. *Remote Sens.* **2022**, *14*, 694. [CrossRef]
20. Yang, H.; Liu, X.; Chen, Q.; Cao, Y.B. Mapping Dongting Lake Wetland Utilizing Time Series Similarity, Statistical Texture, and Superpixels With Sentinel-1 SAR Data. *IEEE J. Sel. Top. Appl. Earth Obs. Remote Sens.* **2022**, *15*, 8235–8244. [CrossRef]
21. Garba, S.I.; Ebmeier, S.K.; Bastin, J.F.; Mollicone, D.; Holden, J. The Detection of Wetlands And Wetland Fragmentation Using Sentinel 1 And 2 Imagery: The Example of Southern Nigeria. *Res. Sq.* **2021**, preprint.
22. Hosseiny, B.; Mahdianpari, M.; Brisco, B.; Mohammadimanesh, F.; Salehi, B. WetNet: A Spatial-Temporal Ensemble Deep Learning Model for Wetland Classification Using Sentinel-1 and Sentinel-2. *IEEE Trans. Geosci. Remote Sens.* **2021**, *60*, 4406014. [CrossRef]
23. Jamali, A.; Mahdianpari, M. Swin Transformer and Deep Convolutional Neural Networks for Coastal Wetland Classification Using Sentinel-1, Sentinel-2, and LiDAR Data. *Remote Sens.* **2022**, *14*, 359. [CrossRef]
24. Jamali, A.; Mohammadimanesh, F.; Mahdianpari, M. Wetland Classification with Swin Transformer Using Sentinel-1 and Sentinel-2 Data. In Proceedings of the IGARSS 2022—2022 IEEE International Geoscience and Remote Sensing Symposium, Kuala Lumpur, Malaysia, 17–22 July 2022; IEEE: Piscataway, NJ, USA, 2022; pp. 6213–6216.
25. Jamali, A.; Mahdianpari, M.; Brisco, B.; Mao, D.; Salehi, B.; Mohammadimanesh, F. 3DUNetGSFormer: A deep learning pipeline for complex wetland mapping using generative adversarial networks and Swin transformer. *Ecol. Inform.* **2022**, *72*, 101904. [CrossRef]
26. He, X.; Zhou, Y.; Zhao, J.; Zhang, D.; Yao, R.; Xue, Y. Swin Transformer Embedding UNet for Remote Sensing Image Semantic Segmentation. *IEEE Trans. Geosci. Remote Sens.* **2022**, *60*, 4408715. [CrossRef]
27. Lv, Z.; Huang, H.; Gao, L.; Benediktsson, J.A.; Zhao, M.; Shi, C. Simple Multiscale UNet for Change Detection With Heterogeneous Remote Sensing Images. *IEEE Geosci. Remote Sens. Lett.* **2022**, *19*, 2504905. [CrossRef]
28. Sharma, A.; Liu, X.; Yang, X.; Shi, D. A patch-based convolutional neural network for remote sensing image classification. *Neural Netw.* **2017**, *95*, 19–28. [CrossRef] [PubMed]
29. Li, X.; Xu, F.; Lyu, X.; Gao, H.; Tong, Y.; Cai, S.; Li, S.; Liu, D. Dual attention deep fusion semantic segmentation networks of large-scale satellite remote-sensing images. *Int. J. Remote Sens.* **2021**, *42*, 3583–3610. [CrossRef]
30. Raza, A.; Huo, H.; Fang, T. EUNet-CD: Efficient UNet++ for Change Detection of Very High-Resolution Remote Sensing Images. *IEEE Geosci. Remote Sens. Lett.* **2022**, *19*, 3510805. [CrossRef]
31. Maggiori, E.; Tarabalka, Y.; Charpiat, G.; Alliez, P. Convolutional Neural Networks for Large-Scale Remote-Sensing Image Classification. *IEEE Trans. Geosci. Remote Sens.* **2017**, *55*, 645–657. [CrossRef]
32. Bazi, Y.; Bashmal, L.; Al Rahhal, M.M.; Dayil, R.A.; Ajlan, N.A. Vision Transformers for Remote Sensing Image Classification. *Remote Sens.* **2021**, *13*, 516. [CrossRef]
33. He, J.; Zhao, L.; Yang, H.; Zhang, M.; Li, W. HSI-BERT: Hyperspectral Image Classification Using the Bidirectional Encoder Representation From Transformers. *IEEE Trans. Geosci. Remote Sens.* **2020**, *58*, 165–178. [CrossRef]
34. Hong, D.; Han, Z.; Yao, J.; Gao, L.; Zhang, B.; Plaza, A.J.; Chanussot, J. SpectralFormer: Rethinking Hyperspectral Image Classification With Transformers. *IEEE Trans. Geosci. Remote Sens.* **2021**, *60*, 5518615. [CrossRef]
35. Mohammadimanesh, F.; Salehi, B.; Mahdianpari, M.; Brisco, B.; Motagh, M. Multi-temporal, multi-frequency, and multi-polarization coherence and SAR backscatter analysis of wetlands. *Isprs J. Photogramm. Remote Sens.* **2018**, *142*, 78–93. [CrossRef]
36. Brendel, W.; Bethge, M. Approximating CNNs with Bag-of-local-Features models works surprisingly well on ImageNet. *arXiv* **2019**, arXiv:1904.00760.
37. Azulay, A.; Weiss, Y. Why do deep convolutional networks generalize so poorly to small image transformations? *J. Mach. Learn. Res.* **2018**, *20*, 181–184.
38. Ilyas, A.; Santurkar, S.; Tsipras, D.; Engstrom, L.; Tran, B.; Madry, A. Adversarial Examples Are Not Bugs, They Are Features. *arXiv* **2019**, arXiv:1905.02175.
39. Geirhos, R.; Rubisch, P.; Michaelis, C.; Bethge, M.; Wichmann, F.; Brendel, W. ImageNet-trained CNNs are biased towards texture; increasing shape bias improves accuracy and robustness. *arXiv* **2018**, arXiv:1811.12231.
40. Xie, E.; Wang, W.; Yu, Z.; Anandkumar, A.; Álvarez, J.M.; Luo, P. SegFormer: Simple and Efficient Design for Semantic Segmentation with Transformers. *arXiv* **2021**, arXiv:2105.15203.
41. Lane, C.R.; Liu, H.; Autrey, B.C.; Anenkhonov, O.A.; Chepinoga, V.V.; Wu, Q. Improved Wetland Classification Using Eight-Band High Resolution Satellite Imagery and a Hybrid Approach. *Remote Sens.* **2014**, *6*, 12187–12216. [CrossRef]
42. Amani, M.; Mahdavi, S.; Bérard, O. Supervised wetland classification using high spatial resolution optical, SAR, and LiDAR imagery. *J. Appl. Remote Sens.* **2020**, *14*, 024502. [CrossRef]
43. Amani, M.; Foroughnia, F.; Moghimi, A.; Mahdavi, S. 3D Habitat Mapping Using High-Resolution Optical Satellite and Lidar Data. In Proceedings of the 2022 10th International Conference on Agro-geoinformatics (Agro-Geoinformatics), Quebec City, QC, Canada, 11–14 July 2022; pp. 1–5.
44. Kaplan, G.; Avdan, U. Evaluating the utilization of the red edge and radar bands from sentinel sensors for wetland classification. *Catena* **2019**, *178*, 109–119. [CrossRef]
45. Wu, N.; Shi, R.; Zhuo, W.; Zhang, C.; Zhou, B.; Xia, Z.; Tao, Z.; Gao, W.; Tian, B. A Classification of Tidal Flat Wetland Vegetation Combining Phenological Features with Google Earth Engine. *Remote Sens.* **2021**, *13*, 443. [CrossRef]

46. Li, X.; Xu, F.; Liu, F.; Xia, R.; Tong, Y.; Li, L.; Xu, Z.; Lyu, X. Hybridizing Euclidean and Hyperbolic Similarities for Attentively Refining Representations in Semantic Segmentation of Remote Sensing Images. *IEEE Geosci. Remote Sens. Lett.* **2022**, *19*, 5003605. [CrossRef]
47. Araya-López, R.A.; Lopatin, J.; Fassnacht, F.E.; Hernández, H.J. Monitoring Andean high altitude wetlands in central Chile with seasonal optical data: A comparison between Worldview-2 and Sentinel-2 imagery. *ISPRS J. Photogramm. Remote Sens.* **2018**, *145*, 213–224. [CrossRef]
48. Liu, J.; Feng, Q.; Gong, J.; Zhou, J.; Li, Y. Land-cover classification of the Yellow River Delta wetland based on multiple end-member spectral mixture analysis and a Random Forest classifier. *Int. J. Remote Sens.* **2016**, *37*, 1845–1867. [CrossRef]
49. Jiao, L.; Sun, W.; Yang, G.; Ren, G.; Liu, Y. A Hierarchical Classification Framework of Satellite Multispectral/Hyperspectral Images for Mapping Coastal Wetlands. *Remote Sens.* **2019**, *11*, 2238. [CrossRef]
50. Li, X.; Xu, F.; Liu, F.; Lyu, X.; Tong, Y.; Xu, Z.; Zhou, J. A Synergistical Attention Model for Semantic Segmentation of Remote Sensing Images. *IEEE Trans. Geosci. Remote Sens.* **2023**, *61*, 5400916. [CrossRef]
51. Li, X.; Xu, F.; Xia, R.; Li, T.; Chen, Z.; Wang, X.; Xu, Z.; Lyu, X. Encoding Contextual Information by Interlacing Transformer and Convolution for Remote Sensing Imagery Semantic Segmentation. *Remote Sens.* **2022**, *14*, 4065. [CrossRef]

remote sensing MDPI

Article

Active Pairwise Constraint Learning in Constrained Time-Series Clustering for Crop Mapping from Airborne SAR Imagery

Xingli Qin [1,2], Lingli Zhao [3,*], Jie Yang [2], Pingxiang Li [2], Bingfang Wu [1,4], Kaimin Sun [2] and Yubin Xu [5]

1 State Key Laboratory of Remote Sensing Science, Aerospace Information Research Institute, Chinese Academy of Sciences, Beijing 100101, China
2 State Key Laboratory of Information Engineering in Surveying, Mapping and Remote Sensing, Wuhan University, Wuhan 430079, China
3 School of Remote Sensing and Information Engineering, Wuhan University, Wuhan 430079, China
4 College of Resources and Environment, University of Chinese Academy of Sciences, Beijing 100049, China
5 China Academy of Civil Aviation Science and Technology, Beijing 100028, China
* Correspondence: zhaolingli@whu.edu.cn; Tel.: +86-159-7204-6289

Abstract: Airborne SAR is an important data source for crop mapping and has important applications in agricultural monitoring and food safety. However, the incidence-angle effects of airborne SAR imagery decrease the crop mapping accuracy. An active pairwise constraint learning method (APCL) is proposed for constrained time-series clustering to address this problem. APCL constructs two types of instance-level pairwise constraints based on the incidence angles of the samples and a non-iterative batch-mode active selection scheme: the must-link constraint, which links two objects of the same crop type with large differences in backscattering coefficients and the shapes of time-series curves; the cannot-link constraint, which links two objects of different crop types with only small differences in the values of backscattering coefficients. Experiments were conducted using 12 time-series images with incidence angles ranging from 21.2° to 64.3°, and the experimental results prove the effectiveness of APCL in improving crop mapping accuracy. More specifically, when using dynamic time warping (DTW) as the similarity measure, the kappa coefficient obtained by APCL was increased by 9.5%, 8.7%, and 5.2% compared to the results of the three other methods. It provides a new solution for reducing the incidence-angle effects in the crop mapping of airborne SAR time-series images.

Keywords: synthetic aperture radar (SAR); crop mapping; time-series images; constrained clustering; active constraint learning

check for updates

Citation: Qin, X.; Zhao, L.; Yang, J.; Li, P.; Wu, B.; Sun, K.; Xu, Y. Active Pairwise Constraint Learning in Constrained Time-Series Clustering for Crop Mapping from Airborne SAR Imagery. *Remote Sens.* **2022**, *14*, 6073. https://doi.org/10.3390/rs14236073

Academic Editors: Janet E. Nichol, Faisal M. Qamer, Jianchu Xu and Sawaid Abbas

Received: 21 October 2022
Accepted: 27 November 2022
Published: 30 November 2022

1. Introduction

Food security is the foundation and guarantee of human health and social stability. Timely and accurate crop mapping can assist with crop production management, crop yield prediction, crop disaster warning, and crop planting planning, which is of considerable importance to ensure food security [1–3]. Remote sensing technology has become an important means of crop mapping due to its wide coverage, low cost, near real-time performance, and the large amount of information [1,4–6].

Among the numerous earth observation sensors, synthetic aperture radar (SAR) systems on airborne platforms have been gaining popularity in crop mapping applications. On the one hand, the airborne platforms have the advantages of good flexibility and reusability and can guarantee regular and high-frequency observations throughout the growing season, which increases their suitability for agricultural monitoring. On the other hand, as an activate microwave sensor, SAR provides unique information for crop mapping: the microwave signal is sensitive to the dielectric and geometrical properties of crops [7], and the acquired backscattering coefficients are synthesized information of the land cover type, terrain slope, surface roughness, local incidence angle, and dielectric constant [8].

Thus, rich polarimetric features can be extracted by various polarimetric decomposition algorithms [9–13].

The accuracy of crop mapping in a single image is affected by the phenomenon of "same object with different spectra" and "different objects with the same spectra." Several researchers currently use time-series analysis technology to map crops from time-series images based on the differences in the growth trends of the crops over the whole growth period. These methods can be divided into two main categories: supervised and unsupervised methods. In the supervised methods, many researchers have focused on finding remarkably effective features and classifiers to improve the accuracy of specific crop mapping tasks [14–20]. This kind of method can achieve high accuracy but usually requires a large number of training samples, and the acquisition of high-quality training samples can be remarkably time-consuming. Considering the difficulty of the collection of high-quality training samples, the unsupervised crop mapping approaches have attracted increasing attention in recent years. In the unsupervised approaches, some of the methods use crop phenology to establish rules for distinguishing the different crops [21,22]. However, the establishment of rules is highly dependent on expert knowledge and specific crops, which leads to weak universality. Therefore, other methods have been developed using time-series clustering technology [23–28]. Time-series clustering is a process of combining homogeneous time-series data based on a certain similarity measure [29], and its accuracy is mainly affected by the similarity measure criterion, the time-series data representation method, and the clustering method [29]. Time-series clustering technology can be used to cluster the time-series curves of crops effectively due to the differences in the growth cycles and trends of different crops.

However, when applied to the airborne SAR imagery, the crop mapping accuracy may be reduced by the large range of incidence angles. More specifically, when the incidence angles of the images substantially vary, the backscattering coefficients of the same type of crops at the near and far ranges of the image might be quite different, complicating the clustering of these crops into a single group. For the incidence angle effects, some studies [24,30,31] removed the regions with excessively large incidence angles from the image, which cannot fully exploit the image data. Other studies use incidence-angle correction techniques to address these problems, such as normalizing the backscattering coefficient with respect to the incidence angle [32], model and correct for the effect of variation in incidence angle in azimuth line [33], a cosine correction estimated through linear regression [34], and a histogram matching procedure, where the adjustment was based on the lowest two central moments, the mean and variance [35]. However, these techniques are usually suitable for specific data with a small range of incidence angles or specific types of ground objects.

Constrained clustering methods are introduced in this paper to solve the incidence angle effects. Constrained clustering (which is also known as semi-supervised clustering) can introduce background knowledge (also known as side information) to guide a clustering algorithm [36]. The constrained clustering methods focus on the most commonly used K-means based algorithms, in which the clustering algorithm or the objective function is modified such that user constraints are used to guide the algorithm for remarkably appropriate data partitioning [36]. In addition, the current study focused on the most widely used instance-level pairwise constraints [37]: must-link constraints (MLCs) and cannot-link constraints (CLCs). An MLC ensures that two instances should remain in the same cluster, while a CLC ensures that two instances should be in two different clusters [38]. The quality of the constraints plays a critical part in constrained clustering. However, the commonly used constraint generation methods in many studies [36,39–43] are random sampling and manual selection. In random sampling, the constraints are generated by randomly taking pairs of points and generating an MLC or CLC depending on whether they belong to the same class or not. Consequently, the quality of the constraints cannot be guaranteed. In the manual selection, the speckle noise inherent in SAR images will aggravate the difficulty of the identification of constraints. Some studies [44–47] have attempted to introduce remarkably effective constraint information, but their methods are only applicable in specific domains.

Therefore, an active pairwise constraint learning method (APCL) is introduced to achieve reliable crop mapping from airborne SAR images via constrained time-series clustering. The proposed method integrates the batch-mode active selection, the information contained in the incidence angle, and the characteristics of the crop time-series curves to learn informative instance-level constraints. The method was designed for SAR images with a wide range of incidence angles and was verified in the experiments conducted in this study, where crops from the uninhabited aerial vehicle synthetic aperture radar (UAVSAR) images were mapped using several constrained clustering methods and time-series similarity measures.

2. Study Area and Experimental Data

The study area located in the southwest of Winnipeg, Manitoba, Canada, is covered by various crops, as shown in Figure 1a,b. Four types of crops were considered in the crop mapping experiments: oat, corn, canola, and soybean. The reference map for the crop distribution is shown in Figure 1c, which was established in accordance with the land-cover classification map produced by the NASA National Snow and Ice Data Center Distributed Active Archive Center (NSIDC DAAC) [48].

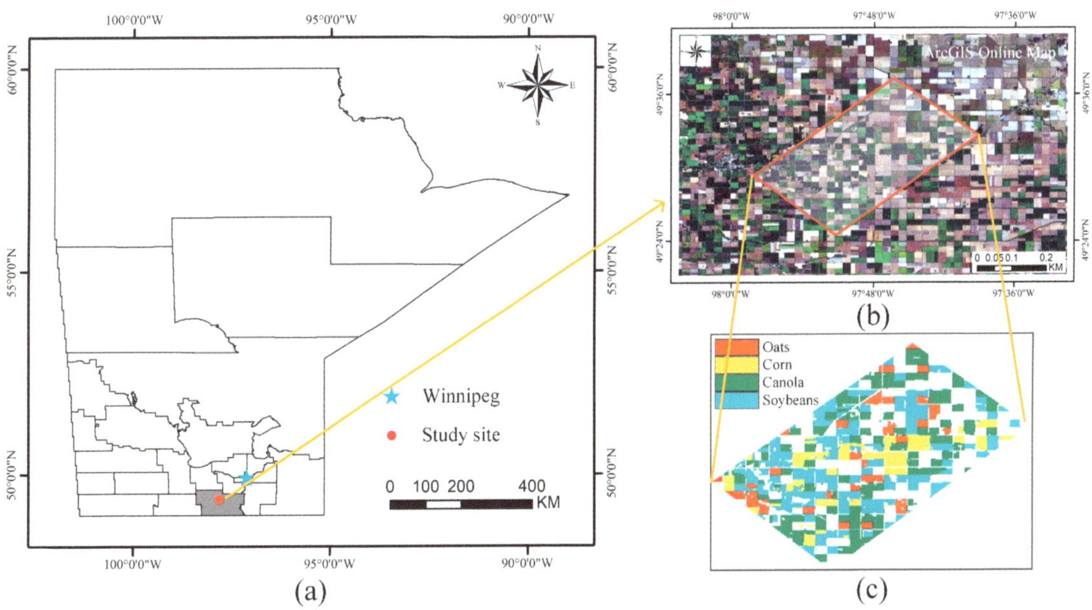

Figure 1. Overview of the study area: (**a**) location of the study area; (**b**) overview of the study area in ArcGIS Online Map; (**c**) reference map of crop distribution.

The data used in this study were the UAVSAR images obtained during the Soil Moisture Active Passive Validation Experiment 2012 (SMAPVEX2012). The SMAPVEX12 campaign lasted 43 days (7 June–19 July 2012), and the soil moisture and vegetation conditions changed significantly during the time period. A total of 13 days of UAVSAR data were acquired by the end of the campaign [30]. Twelve UAVSAR images of the same flight line were used for crop mapping. The images were ground-range projected and multi-looked data (GRD) whose preprocessing procedures include radiometric calibration, geometric correction (WGS-84), and multi-look processing (15 pixels in range and 60 pixels in azimuth directions). These images were resampled from 1134 × 795 (pixels) to

378 × 265 (pixels) to reduce the time spent in the experiment. The images are multi-looked; thus, any speckle-filtering process was not conducted on the images. A description of the data is provided in Table 1, and the corresponding Pauli RGB images are shown in Figure 2. The local incidence angles of the pixels in the images are shown in Figure 3, where the minimum value is 21.2° (near range) and the maximum value is 64.3° (far range). The local incidence angle is defined as the angle between the radar line-of-sight and the surface normal vector [33,49,50]. The terrain of the study area can be characterized as flat; thus, the incidence angle and local incidence angle are assumed to be the same in this paper. The numbers of pixels for each crop type are presented in Table 2.

Table 1. Description of the UAVSAR time-series images.

Sensor	Look Direction	Band (Frequency)	Incidence Angle (°)	Image Size (Pixels)	Observation Date (2012)
UAVSAR	Left	L (1.26 GHz)	21.2–64.3	378 × 265	17, 19, 22, 23, 25, 27, 29 June 5, 10, 13, 14, 17 July

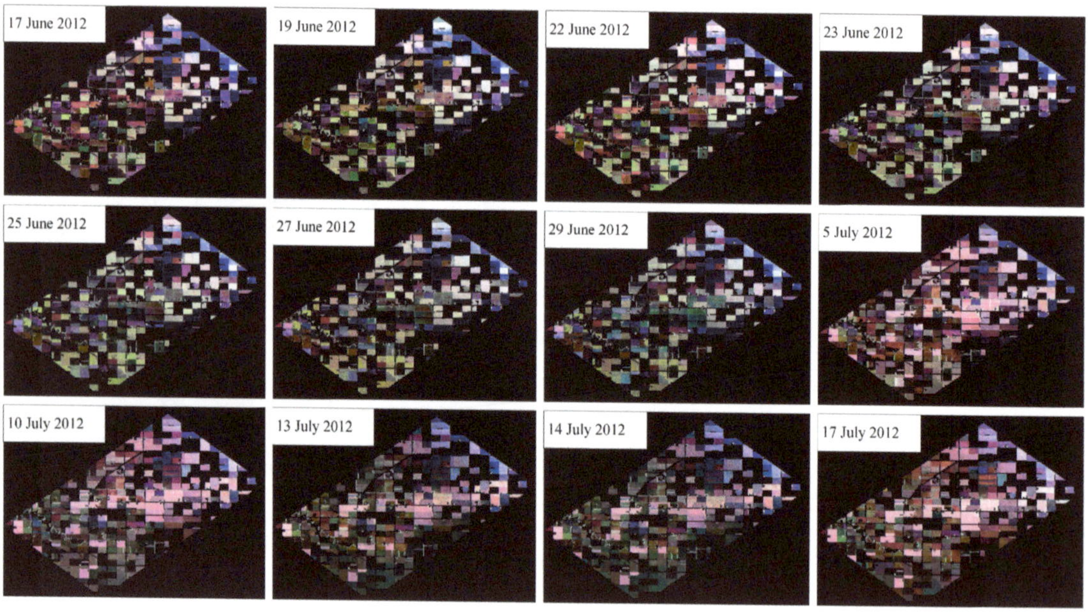

Figure 2. Pauli RGB images of the UAVSAR time-series images, with red $|S_{HH} - S_{VV}|^2$, green $4|S_{HV}|^2$, and blue $|S_{HH} + S_{VV}|^2$.

Table 2. Number of pixels of each crop.

Crop Type	Number of Pixels
Oats	3454
Corn	4395
Canola	10,426
Soybean	12,330
Total	30,605

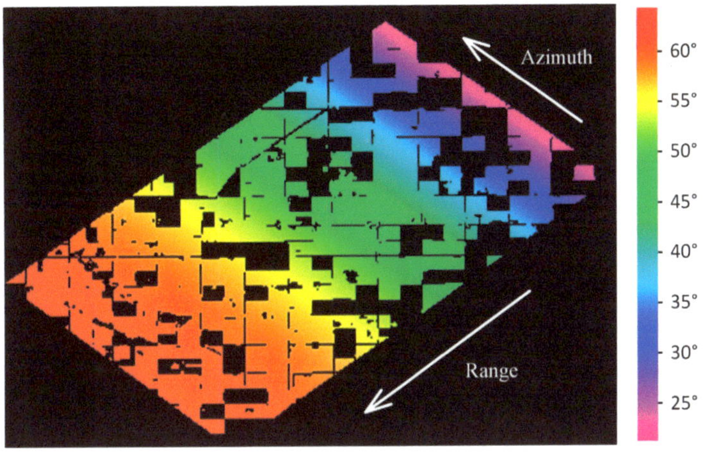

Figure 3. Local incidence angles of the samples.

The Pauli RGB images reveal that the backscattering characteristics of the same crop at the near and far ranges are quite different. The backscattering coefficients are mainly affected by the terrain slope, soil moisture, geometric structure, and incidence angle. Among these factors, the topography is flat to gently undulating, with slopes from 0% to 2%; the types of soil in this area are loam and clay, which have similar soil moisture contents [30,51]; and the geometric structures of the same crops in this area are similar. According to NOAA Global Surface Summary of Day [52], during the period of data acquisition, the average temperature in this area is 20.3 °C, and the accumulated precipitation is 4.5 cm. Therefore, the large variation of the incidence angle (21.2°–64.3°) is considered in this paper to be the main factor causing the large difference in backscattering between the near and far ranges of the same crop. In addition, the shape of cropland in this region is relatively regular, which makes it more conducive to study the impact of the incidence angle effect.

Previous works [20,53,54] have proven that among the polarization channels of HH, HV, and VV, the HV polarization is the most effective channel for discriminating different types of crops in the crop mapping of polarimetric SAR time-series images because the HV is remarkably sensitive to the volume scattering, which supports the value of structural features in distinguishing across crop types [20]. Therefore, only the backscattering coefficient of HV polarization (S_{HV}) was used for the crop mapping experiments in this paper; that is, the time-series curve for each sample S_i was constructed using the corresponding S_{HV} values in all 12 images (i.e., $S_i = \left\{S_{HV}^{t1}, S_{HV}^{t2}, \cdots, S_{HV}^{tn}\right\}$). In addition, using only S_{HV} rather than all the polarimetric features can simplify the analysis of the characteristics of time-series curves and provide guidance for crop mapping using single and dual-polarization data.

3. Methods

The constrained clustering methods using instance-level constraints were introduced to improve the reliability and stability of time-series clustering results. However, the effectiveness of pairwise constraints from the commonly used random sampling remains uncertain. Therefore, considering the characteristics of UAVSAR time-series images, a new method for generating informative pairwise constraints—namely, "active pairwise constraint learning (APCL)," is proposed. The flowchart of APCL is shown in Figure 4, which comprises the following two main steps: (1) extraction of the candidate sample set and (2) construction of the pairwise constraints.

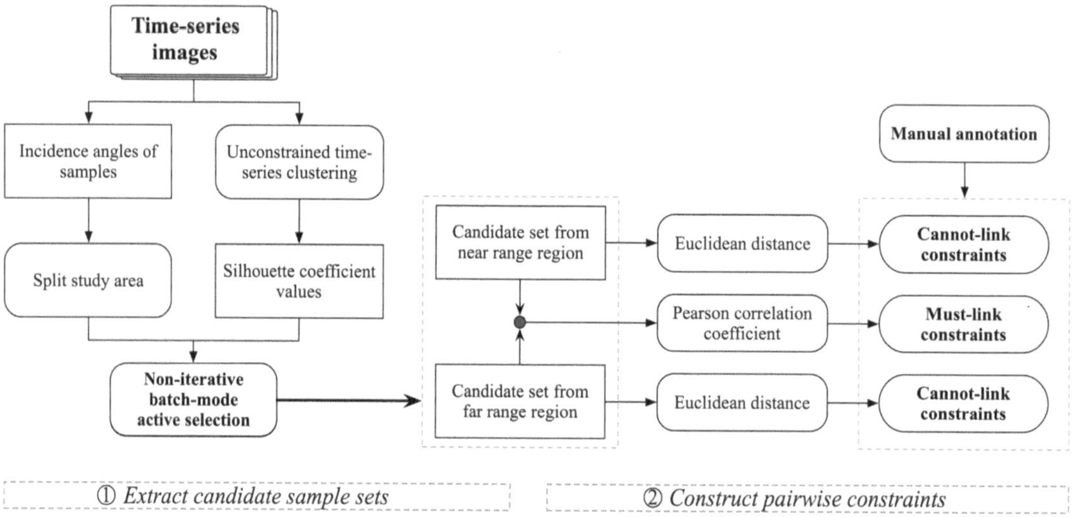

Figure 4. Flowchart of the proposed active pairwise constraint learning method.

3.1. Non-Iterative Batch-Mode Active Selection Theory

The construction of instance-level pairwise constraints is based on a group of samples with rich information (that are representative and diverse). Active learning is commonly used in classification tasks to select the informative samples, which are regarded as the most helpful to improve the classifier performance [55]. Two criteria are usually available for the measurement of the information of samples: (1) an uncertainty criterion and (2) a diversity criterion. The uncertainty criterion usually selects the samples near the class boundary because they are not easily categorized into a certain class by the classifier—that is, these samples have the highest uncertainties. Significant differences between a group of samples in the diversity criterion can effectively avoid knowledge redundancy; therefore, these samples have abundant information. The uncertainty and diversity criteria are considered in the proposed method to extract an informative candidate sample set.

However, conventional active learning techniques select only a single sample at each iteration for manual labeling and retrain the prediction model [56–59], which is inefficient and inapplicable to the clustering task of the paper. The batch-mode active learning (BMAL) technique is considered for selecting samples, in which a batch of samples are selected for manual annotation simultaneously at each round, to address the aforementioned limitations. The key issue of BMAL is to select the most informative batch of samples with as minimal redundancy to provide the most possible information to the prediction model [56,58,60,61]. Notably, differently from the conventional BMAL in classification tasks, a non-iterative batch-mode active selection scheme, in which the sample selection process was performed only once, is considered.

3.2. Extraction of the Candidate Sample Set

This step aims to extract a group of samples with representativeness and diversity based on the non-iterative batch-mode active selection scheme.

First, based on the study of Stanitsas et al. [62], the silhouette coefficient [63] is used for the measurement of the uncertainty of samples. The silhouette coefficient of a sample is calculated as follows:

$$S(i) = \frac{b(i) - a(i)}{max(a(i),\ b(i))} \tag{1}$$

where $a(i)$ is the average dissimilarity of sample I with all the other samples within the same cluster, and $b(i)$ is the lowest average dissimilarity of sample i to each of the clusters to which it is not assigned. The range of $S(i)$ is between -1 and 1.

The samples with low silhouette coefficients are likely to be assigned to the wrong clusters; therefore, these samples will also have the highest uncertainties. The samples were first clustered using unconstrained time-series clustering, and then their silhouette coefficients were calculated on the basis of the clustering results to calculate the silhouette coefficient of each sample.

Two regions with equal space spans were then separated from the study area along the range direction: (1) the near range region (with incidence angles ranging from 21.2° to 47.2°, denoted by NR) and (2) the far range region (with incidence angles ranging from 54.0° to 64.3°, denoted by FR). The variation in the incidence angle values along the range direction is nonlinear; therefore, the corresponding incidence angle ranges of the two regions are different. The separation of NR and FR aims to extract two candidate sample sets (denoted by U_{NR} and U_{FR}, respectively) independently from the two regions and then improves the informativeness of the generated constraints, which is comprehensively discussed in Section 4.4.

Finally, based on the silhouette coefficients of the samples, two candidate sample sets with high uncertainty and considerable diversity were extracted from the two regions using a greedy strategy. The main ideas of the strategy are as follows: (1) the samples with low silhouette coefficients are preferred, (2) the selected samples should be evenly distributed in different clusters, and (3) the geometric distance in the feature space between samples inner a cluster should be larger than a specified threshold T. The first idea is to ensure that the selected samples have high uncertainty, and the last two ideas aim to ensure that the selected samples are of considerable diversity. In real programming implementation, achieving the greedy sample extraction strategy is easy, and various implementation methods are available [37]; however, its implementation is not comprehensively discussed. In the proposed method, the sizes of U_{NR} and U_{FR} are both set to one-third of the number of constraints to be constructed.

K-means [64] is used as the unconstrained clustering algorithm, with the ED as the similarity criterion, to visualize the efficiency of the proposed method. The extracted candidate sample sets U_{NR} and U_{FR} are shown in Figure 5, where the number of samples in each candidate sample set is 150.

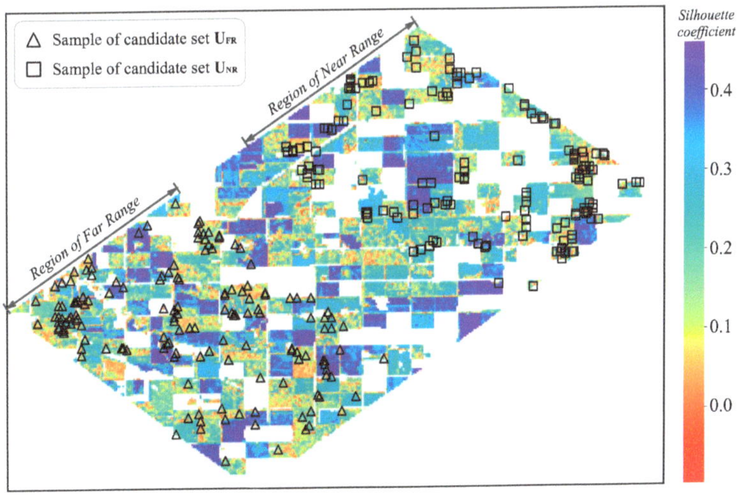

Figure 5. Visualization of the distribution of the candidate sample sets. The background image is the heatmap of the samples' silhouette coefficients.

3.3. *Construction of the Pairwise Constraints*

Two characteristics of the crops in the UAVSAR time-series images are considered when learning the pairwise constraints from U_{NR} and U_{FR}: (1) the time-series curves of the same type of crops from the near and far ranges may have high shape similarity but large spatial distances; and (2) the time-series curves of the different types of crops with small differences of incidence angle may be close to each other in the feature space, thereby leading to misclassification. Consequently, the MLCs and CLCs are constructed on the basis of the first and second characteristics, respectively. The detailed procedure of constructing the N_{MLC} MLCs and N_{CLC} CLCs is presented below.

3.3.1. Construction of the Must-Link Constraints (MLCs)

First, the Pearson correlation coefficient is used to construct a shape similarity matrix $M_{Pearson}$:

$$M_{Pearson}(i,j) = r\left(s^i_{NR}, s^j_{FR}\right) \tag{2}$$

where s^i_{NR} and s^j_{FR} are the ith sample of U_{NR} and the jth sample of U_{FR}, respectively. Thus, each element in $M_{Pearson}$ corresponds to two samples from U_{NR} and U_{FR}.

Second, the elements in $M_{Pearson}$ are sorted in accordance with their Pearson correlation coefficient values from small to large. Then, the pairs of samples, wherein the samples in each pair are of the same crop type according to the order are chosen until N_{MLC} pairs of samples, are selected. These selected sample pairs are the required MLCs. Furthermore, each sample can be selected at most two times to guarantee the diversity of the constraints.

The learned MLCs have the following features. (1) Each pair of constraints links the samples of the same class from different incidence angle regions (which means that their backscattering coefficients markedly differ). (2) The shape similarity between the time-series curves of linked samples is low (because their correlation coefficients are small). (3) The linked samples are from the candidate sample set (indicating their high clustering uncertainties). Therefore, the linked samples are the most difficult to cluster correctly.

3.3.2. Construction of the Cannot-Link Constraints (CLCs)

Two sets of CLCs are constructed from U_{NR} and U_{FR}. Taking the construction of the CLCs from U_{FR} as an example, the detailed process is as follows.

Firstly, the samples in candidate set U_{FR} are used to construct an ED matrix M_{ED}:

$$M_{ED}(i,j) = ED\left(s^i_{FR}, s^j_{FR}\right) \tag{3}$$

where s^i_{FR} is the ith sample of U_{FR}, s^j_{FR} is the jth sample of U_{FR}, and $M_{ED}(i,j) = M_{ED}(j,i)$. That is, M_{ED} is a symmetric matrix.

The elements in the upper triangular matrix are then sorted in accordance with their ED values from small to large, where each element corresponds to a pair of samples from U_{FR}.

Finally, the $N_{CLC}/2$ pairs of samples, which are different types of crops, were selected in accordance with the order and then used as a set of CLCs. In addition, each sample was selected at most two times to guarantee the diversity of the constraints. Another set of CLCs of quantity $N_{CLC}/2$ was constructed from U_{NR} using the above procedure, and a group of CLCs of quantity N_{CLC} can then be obtained.

The linked samples in the above flowchart are from the same region (region FR or NR) and are close in the feature space, allowing easy clustering into the same group. Therefore, using these samples as CLCs could help improve the clustering.

K-means was used as the unconstrained clustering algorithm and the ED as the similarity measure for the active constraint learning to visualize the distribution of the constraints constructed by the proposed method (APCL). The constraints of another two constraint learning methods were used for comparison. The first method was random sampling based on regions of incidence angle (denoted by RSRIA), which randomly generates CLCs within the same incidence angle region and generates MLCs between different incidence angle regions. The second method is the commonly used random sampling. The number of each kind of pairwise constraint was set to 200 for both constraint learning methods. The distributions of the obtained constraints are shown in Figure 6. Compared with the constraints of random sampling, the constraints of the proposed method link the samples with low silhouette coefficients, and all these samples are from either the near or the far range.

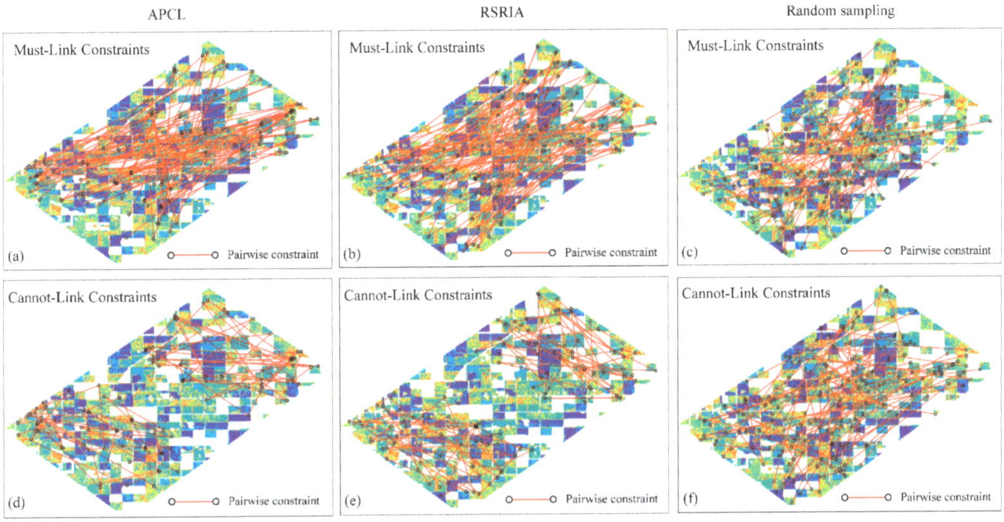

Figure 6. Distribution of the pairwise constraints, where the background is the heatmap of the samples' silhouette coefficients: (**a**) MLCs of APCL; (**b**) MLCs of RSRIA; (**c**) MLCs of random sampling; (**d**) CLCs of APCL; (**e**) CLCs of RSRIA; (**f**) CLCs of random sampling.

4. Results and Discussion

This section includes four parts. In the first part, a qualitative analysis of impact of the incidence angle effect on crop separability is presented via the time-series curves of various crop types with different incidence angles. In the second part, we present the results of numerous experiments that were conducted to illustrate the efficiency of the proposed method. Only 20% of the total number (30,605) of samples of the time-series images were used in each experiment due to the large number of experiments and the time-consuming time-series clustering. In the third part, all the samples in the study area were used for the clustering, and the corresponding crop classification maps were then generated. In the experiment, when generating MLCs and CLCs, the processes of identifying whether the two selected samples should be linked were based on the ground truth data (in practice, this method should be conducted manually). In the fourth part, a quantitative evaluation of the informativeness of the constraints was conducted. All the experiments were performed on a 64-bit Window 10 PC, and the programming language was Python 3.

4.1. Impact of the Incidence Angle Effect on Crop Separability

First, considering that the incidence angle has different impacts on different regions, the following three regions with equal space spans were separated from the study area along the range direction to further discuss the temporal characteristics of the crop under different incidence angles: a region with small incidence angles (incidence angles ranging from 21.2° to 32.0°, denoted by Inc_S); a region with medium incidence angles (incidence angles ranging from 50.0° to 56.0°, denoted by Inc_M); and a region with large incidence angles (incidence angles ranging from 60.0° to 64.3°, denoted by Inc_L). One hundred samples of each type of crop were then sampled from these three regions, as follows:

$$S_{HV} = \frac{1}{100}\sum_{i=1}^{100} S_{HV}^{i} \tag{4}$$

The averaged S_{HV} of the selected samples of each crop was used to construct the time-series curves. A total of 12 time-series curves were generated for the four crops from the three regions, as shown in Figure 7.

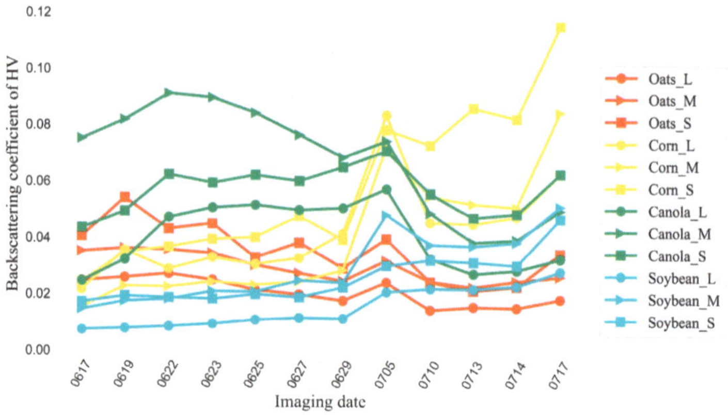

Figure 7. Time-series curves of the crops in different regions. "_L" denotes that the crop is from the region of Inc_L, "_M" denotes that the crop is from the region of Inc_M, and "_S" denotes that the crop is from the region of Inc_S.

These findings show that the S_{HV} values of all the crops substantially change over time. For oats, the shape of the time-series curve in the region of Inc_L is similar to that in the region of Inc_M, and that in the region of Inc_S markedly changes. For corn, the S_{HV} of the time-series curve in Inc_S is significantly higher than those of Inc_M and Inc_L after day 0705. For canola, the S_{HV} of the time-series curve in Inc_M is significantly higher than those of Inc_S and Inc_L before day 0629. For soybean, the shapes of the three time-series curves are similar. Thus, the different characteristics of the time-series curves of some crops are attributed to the different incidence angles.

Three similarity matrices based on the Euclidean distance (ED), dynamic time warping (DTW), and Pearson correlation coefficient were generated as shown in Figure 8 to further evaluate the distinguishability of these crop time-series curves. The ED is a lock-step measure (one-to-one) and sensitive to scaling, and the Pearson correlation coefficient is invariant to the scale and location of the data points [65]. DTW [66] uses a dynamic programming (DP) method to find an optimal alignment between the two time series via nonlinear warping, thus increasing its suitability for evaluating the similarity of time series of different lengths.

Figure 8. Similarity matrices for the crop time-series curves: (**a**) similarity matrix based on the ED, where 0 denotes the highest similarity; (**b**) similarity matrix based on the DTW, where 0 denotes the highest similarity; (**c**) similarity matrix based on the Pearson correlation coefficient, where 1 denotes the highest similarity.

These findings show that the similarities between some curves of different crops are relatively high. Therefore, many crop samples will be easily misclassified in the time-series clustering. For instance, in Figure 8a, the ED between Oats_L and Oats_S is larger than that between Oats_L and Soybean_L and Soybean_S. Thus, when using the ED as the similarity metric in time-series clustering, the samples of oats in Inc_L are likely to be clustered into one group with the soybean samples. The ED between the soybean samples and their time-series curves is relatively small, indicating that they are less likely to be misclassified in the process of time-series clustering. Figure 8c shows that the similarities between corn and soybean are relatively high, indicating that the samples of these crops would be easily misclassified when using Pearson correlation coefficient as the similarity metric.

The characteristics of the crops in the UAVSAR time-series images can be summarized as follows. (1) The time-series curves of crops change significantly with time. (2) The time-series curves of the same type of crops have similar shapes in all the regions of Inc_S, Inc_M, and Inc_L. (3) The spatial distance between the time-series curves of the same crops in the feature space may be substantially large due to their difference in incidence angle. (4) The time-series curves of different crops may have considerable shape similarities or small spatial distances. Among these characteristics, the first two characteristics are beneficial to the clustering of the same crops from different regions, and the latter two characteristics are not conducive to the clustering.

4.2. Constrained Clustering Accuracy Analysis

The experimental settings of each group of experiments in the first part are presented in Table 3. Each group of experiments was repeated five times, and the average accuracy was taken as the final result for analysis. In each group of experiments, 20% of the total samples were first randomly sampled, and then these samples were used in the experiments to make the results comparable. Finding the optimal number of initial clustering centroids is difficult for all scenarios because it may be different under various clustering methods, similarity criteria, and types of constraints. The optimal number was simply set as eight (two times the number of crop types), and the number of MLCs was set similarly to that of the CLCs.

Table 3. Experimental settings of each group of experiments.

Constrained Clustering Algorithm	1. COP-KMeans, 2. PC-KMeans, 3. MIP-KMeans
Time-series similarity criterion	1. Euclidean distance (ED) 2. Dynamic time warping (DTW) 3. Pearson correlation coefficient (Pearson)
Constraint generation method	1. Active pairwise constraint learning (APCL) 2. Random sampling based on regions of incidence angle (RSRIA) 3. Random sampling
Number of pairwise constraints	{0,150,300,450,600,750,900,1050,1200,1350,1500,1650,1800,1950,2100,2250,2400}
Values of sample's time series	Backscattering coefficients of HV polarization in the 12 images

The three constrained clustering algorithms used in the experiments are all K-means-based methods, which are suitable for time-series application because they iteratively calculate distances to cluster and update these centroids [42]. Their iterative processes are similar, but their usage of constraints is quite different. Therefore, the use of these methods is helpful to analyze how the constraints work and evaluate the effectiveness of constraints from different methods.

In COP-KMeans [39], before assigning a sample to a cluster, it first judges whether the sample is linked with clustered samples by the MLCs (if any) and assigns the sample to the same cluster; otherwise, it is assigned to the nearest cluster while not violating the CLC. However, if not all the constraints are satisfied, then the algorithm fails and no clustering result can be obtained. The following adjustments are performed in the experiments to ensure the COP-KMeans output clustering result: if a sample has constraint violations in each iteration, then the sample will not be assigned to any clusters; if samples that have constraint violations still exist at the end of the last iteration, then these samples will be assigned to the clusters with the fewest constraint violations. PC-KMeans [67] formulates the goal of clustering as minimizing a combined objective function: the sum of the total distance between the samples and their cluster centroids and the cost of violating the pairwise constraints. The constraint violation cost was a relative value in the experiments; that is, the cost of each constraint violation was set as 0.1 times the distance between the sample and the cluster centroid. MIP-KMeans [68] considers the sample constraints and objective function in its entire combinatorial configuration and uses a general optimization tool (the Gurobi optimizer) to solve it [42]. The maximum number of iterations of these algorithms was set as 200, and the convergence condition for these methods was that the change of position of any centroids is no more than a threshold (which was set as 0.0005).

Three time-series similarity measures were used in this study: the ED, DTW, and the Pearson correlation coefficient (denoted by Pearson). For each experiment, in the unconstrained clustering for calculating silhouette coefficient, the used time-series similarity measure was set to be the same as that used in the constrained clustering.

The normalized mutual information (NMI) [69] was used as the clustering evaluation index. The range of NMI is [0, 1], and a high value indicates an effective clustering performance. The mean and standard deviation of NMI for each group of experiments are shown in Figure 9.

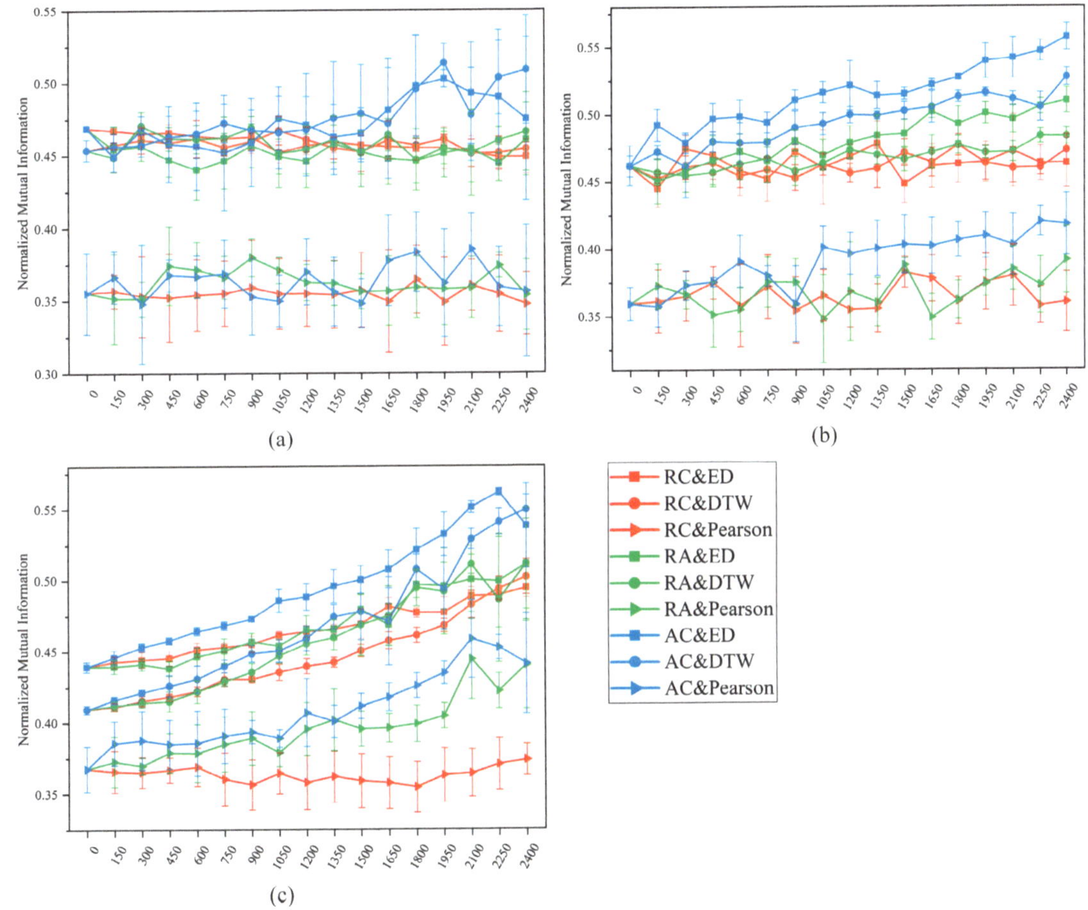

Figure 9. Performance plots show how the NMI increases with the number of pairwise constraints: (**a**) results of COP-KMeans; (**b**) results of PC-KMeans; (**c**) results of MIP-KMeans. The circular markers, the inverted triangular markers, and the square markers correspond to the results obtained using ED, DTW, and Pearson, respectively.

For convenience, in Figure 9, "RC" represents "constraints from random sampling," "RA" represents "constraints from RSRIA," and "AC" represents "constraints from APCL." For example, "RC&ED" represents the results obtained using constraints generated by random sampling and the ED as the similarity criterion.

As shown in Figure 9, the results are for unconstrained clustering when the number of the constraints is zero. The clustering processes of the three constrained clustering algorithms are consistent with that of the traditional K-means method when the number is zero and their NMI values are different. This phenomenon is due to the random selection of the initial clustering centroids of COP-KMeans and the generation of initial clustering centroids of PC-KMeans and MIP-KMeans in specific ways. For further details, please refer to [67,68].

These findings show that COP-KMeans is insensitive to the number of constraints, and only "AC&ED" and "AC&DTW" have better NMI than unconstrained clustering results when the number of pairwise constraints is larger than 1650. PC-KMeans can effectively utilize the constraint information; therefore, its accuracies are generally better than those of the two other algorithms. Among the results of PC-KMeans, the NMI curves of APCL are better than random sampling and RSRIA and all have an ascending trend, especially for the curve of "AC&ED." MIP-KMeans is remarkably sensitive to the number of constraints. Thus, all of the NMI curves show an ascending trend with the increase in the number of pairwise constraints, except for the curve of "RC&Pearson." In addition, when the number of constraints was 2250, the curve of "AC&ED" of MIP-KMeans achieved the best NMI value of 0.562 among all the results in Figure 9.

Overall, the experimental results demonstrate the superiority of the proposed APCL, which can improve the clustering accuracy effectively under various constrained clustering algorithms, time-series similarity criteria, and different numbers of constraints.

No obvious correlation exists between the running time of the clustering algorithms and the number of constraints in the above experiments. Therefore, the average running time of each algorithm was calculated under all numbers of constraints, as shown in Figure 10.

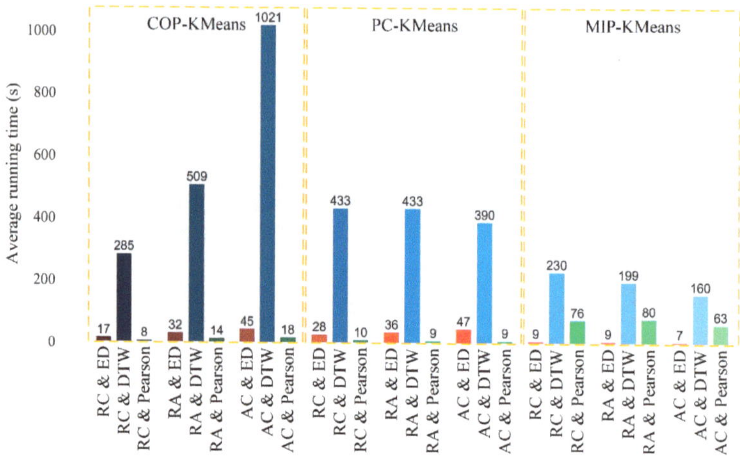

Figure 10. Average running time of each algorithm. The bars in the red range are the results using ED, the bars in the blue range are the results using DTW, and the bars in the green range are the results using Pearson.

The running times of different algorithms and time-series criteria are quite different. DTW is more time-consuming than ED and Pearson among all the methods due its high computational complexity. COP-KMeans with AC&DTW is the most time-consuming method because it requires additional iterative time to convergence. When using ED or DTW as the similarity measure, MIP-KMeans uses less running time than COP-KMeans and PC-KMeans. This finding is due to MIP-KMeans solving an overall optimal result in each iteration [68], which demonstrated a fast convergence speed, resulting in fewer iterations than the other two algorithms. However, in the results of using Pearson as the similarity measure, the running time of MIP-KMeans is significantly larger than COP those of KMeans and PC-KMeans. The reason is that the distinguishability of various crops was poor when using Pearson. Thus, MIP-KMeans had to spend additional time in performing global optimization in each iteration to find an overall solution in which the pairwise constraints were also satisfied. Therefore, the convergence speed of MIP-KMeans is slow, resulting in higher time consumption than the two other algorithms.

4.3. Crop Classification Accuracy Analysis

All the samples (the total number is 30,605) of the study area were used for time-series clustering, and PC-KMeans was used as the constrained clustering algorithm to further evaluate the effectiveness of APCL. The three time-series similarity criteria (ED, DTW, and Pearson) and the generation methods (RSRIA and random sampling) of the two constraints were also used for comparison. The total number of MLCs and CLCs was 12,000 (40% of the total samples), and the number of initial cluster centroids was set to eight. In addition, the results of PC-KMeans with zero constraints were used as the baseline for comparison.

After obtaining the clustering results, for each cluster, the number of samples of each class was counted in accordance with the reference map, and then the class label with the largest number of samples was assigned to the cluster (in practice, the process of labeling clusters should be performed manually). Each experiment was repeated five times. The accuracy of each result is presented in Table 4. The classification map with the highest accuracy for each method is presented in Figure 11.

Table 4. Classification accuracy obtained by all methods.

Method	Similarity Criteria	F-Score				OA	Kappa
		Oats	Corn	Canola	Soybean		
Unconstrained	ED	71.8%	76.5%	86.1%	88.2%	83.5%	76.6%
	DTW	66.9%	73.2%	82.6%	88.0%	81.8%	73.3%
	Pearson	68.4%	52.5%	83.3%	75.9%	73.5%	62.4%
Random sampling	ED	69.5%	78.2%	84.3%	89.7%	83.6%	76.6%
	DTW	66.4%	75.9%	82.9%	88.3%	82.3%	74.1%
	Pearson	67.1%	52.9%	84.4%	81.9%	77.2%	66.6%
RSRIA	ED	74.4%	80.4%	87.6%	91.1%	86.2%	80.3%
	DTW	71.3%	79.3%	85.7%	89.4%	84.7%	77.6%
	Pearson	66.4%	36.9%	84.0%	82.0%	75.8%	63.9%
APCL	ED	75.9%	**85.5%**	89.2%	**92.0%**	**88.1%**	**82.9%**
	DTW	81.9%	83.6%	**90.0%**	89.9%	**88.1%**	82.8%
	Pearson	**82.5%**	51.0%	87.1%	82.2%	79.8%	70.2%

The four methods unconstrained clustering, random sampling, RSRIA, and APCL all achieved their best accuracy (kappa coefficient) when using ED as the similarity criterion, which were 76.6%, 76.6%, 80.3%, and 82.9%, respectively, as shown in Figure 11. APCL also achieved a kappa of 82.8% when using DTW. All the four methods had poor accuracies when using Pearson. However, the accuracy of APCL (kappa = 70.2%) was still higher than those of the other methods.

Among the four crop types, the F-score for oats was relatively low when using ED as the similarity criterion because its backscattering coefficients were close to that of other crops, as shown in Figure 7. When using Pearson as similarity criterion, the F-score for corn was substantially low for all the methods. This phenomenon was due to the relatively low shape similarity of corn's time-series curves in different incidence angles, increasing the difficulty of clustering into one group using Pearson. In addition, for the F-score of each crop type, the best accuracies of oats, corn, canola, and soybean were all achieved by APCL.

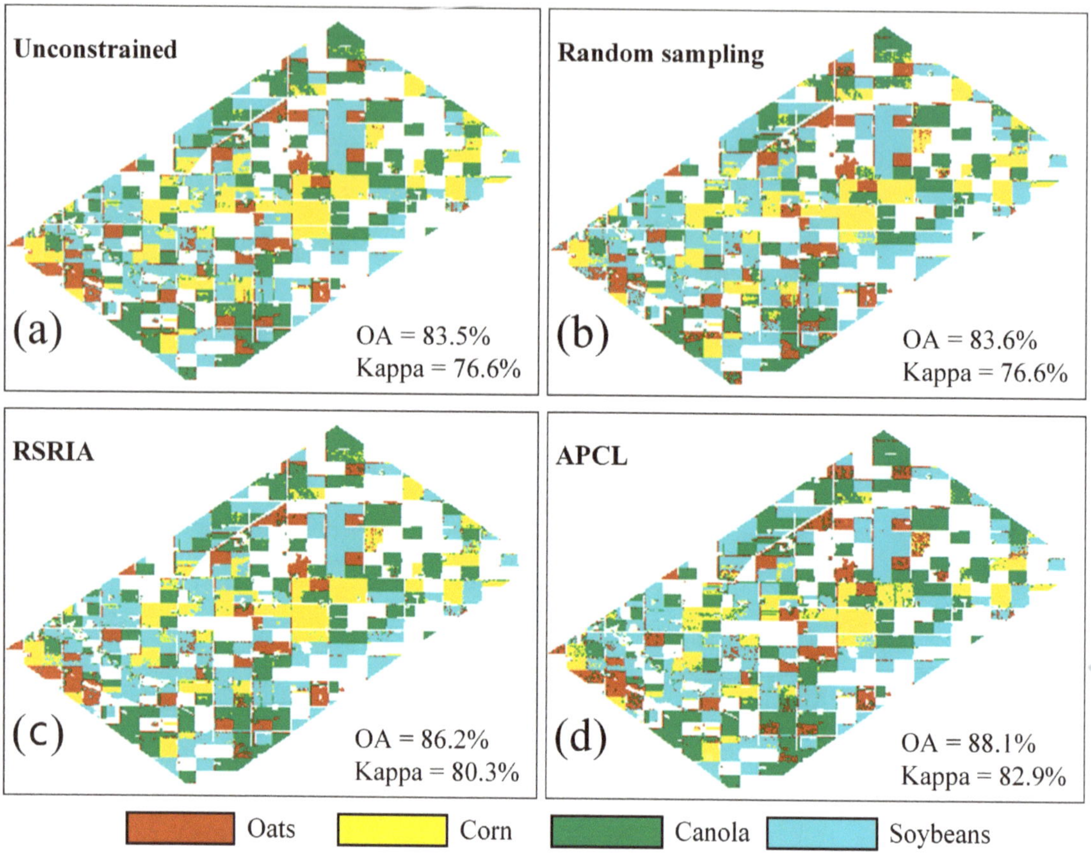

Figure 11. Crop classification maps with the highest accuracy for each method: (**a**) unconstrained clustering; (**b**) random sampling; (**c**) RSRIA; (**d**) APCL.

When the results of unconstrained clustering were used as the baseline, the constraints of random sampling minimally helped improve the accuracy, and those of RSRIA can improve the accuracy to a certain extent. This phenomenon demonstrates the effectiveness of the proposed strategy of generating CLCs within the same incidence angle region and generating MLCs between different incidence angle regions. The comparison between the results of RSRIA and APCL shows the significance of the proposed strategy of extracting candidate sample set based on silhouette coefficients. These results prove the performance and application value of APCL in the constrained clustering of Airborne SAR imagery.

4.4. Evaluation of the Informativeness of the Constraints

If the constraints can be satisfied in the unconstrained clustering results, the constrained clustering based on these constraints may not improve the clustering accuracy [36]. In other words, for a group of constraints, a large number of unsatisfied constraints in the unconstrained clustering result indicate that these constraints contain additional information. Therefore, based on the results of unconstrained clustering with PC-KMeans as the clustering method and ED as the time-series similarity criterion, the quantity of information for the pairwise constraints (6000 pairs of MLCs and 6000 pairs of CLCs) obtained by random sampling, RSRIA, and APCL was evaluated. The number of unsatisfied constraints in the unconstrained clustering results is presented in Table 5.

Table 5. Number of unsatisfied constraints in the unconstrained clustering results.

Method	Must-Link Constraint	Cannot-Link Constraint	Total
Random sampling	3089	308	3397
RSRIA	3302	304	3606
APCL	4283	1691	5974

First, the total number of unsatisfied constraints for APCL is larger than that for the other methods. Second, only a few of the CLCs of random sampling and RSRIA were unsatisfied, and many of the CLCs of APCL were unsatisfied. Therefore, the information contained in the constraints of APCL is abundant. In addition, the number of unsatisfied MLCs is substantially larger than that of the unsatisfied CLCs, which indicates that the MLCs carry additional information and are thus effective at improving the accuracy of constrained clustering. Thus, before the constraint learning, the informativeness of the different kinds of constraints should first be evaluated, and the specific constraints should then be constructed to acquire substantially useful information under limited size of constraints [38,42].

In the following, the reason for having additional unsatisfied MLCs in APCL is discussed. The clustering is markedly affected by the different incidence angles. Thus, the absolute value of the incidence angle difference between every two samples in an MLC is calculated, and then the histograms are drawn, as shown in Figure 12. Each bar in Figure 12 corresponds to the number of satisfied or unsatisfied MLCs under the specific difference of incidence angle.

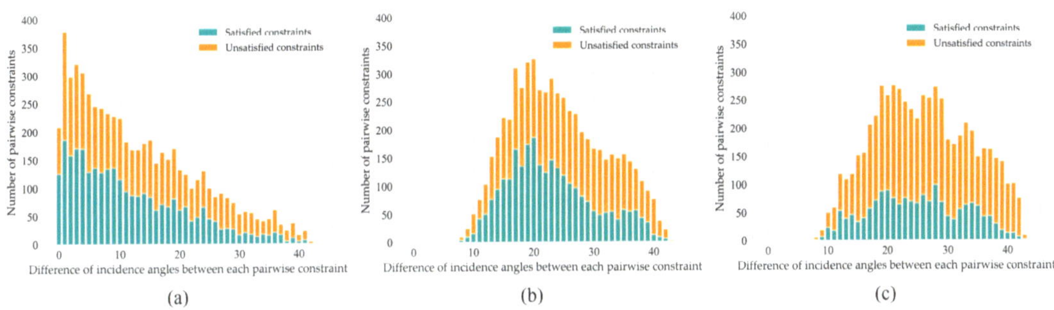

Figure 12. Numbers of satisfied and unsatisfied must-link constraints under the absolute values of the incidence angle differences: (**a**) random sampling; (**b**) RSRIA; (**c**) APCL.

In Figure 12a, most of the MLCs of random sampling show a small incidence angle difference, and half are satisfied. In Figure 12b, most of the MLCs have in an incidence angle range of 13–35°, but many remain unsatisfied. In Figure 12c, most of the MLCs have an incidence angle range of 15–39°, and the number of satisfied MLCs is significantly less than in the above results. Thus, MLCs are only minimally satisfied when the difference between incidence angles is large. This phenomenon is due to the differences in backscattering coefficients of the same crop's samples with large differences in incidence angles. Therefore, clustering these samples into one group by unconstrained clustering is difficult. Second, the proposed APCL can generate MLCs that are difficult to satisfy. The above results prove that the pairwise constraints obtained by the proposed method are more informative than that of RSRIA and random sampling.

As a semi-supervised clustering technique, constrained clustering also needs some supervised information to generate constraints. However, unlike the conventional supervised methods, supervision by instance-level constraints is more general and more realistic than specific class labels. By using knowledge, a user can specify whether pairs of samples belong to the same cluster or not even when the class labels remain unknown [36]. The poor visual effect of PolSAR images increases the difficulty in identifying the class labels of samples [29,70]. In this case, the method of generating pairwise constraints shows its advantages. However, designing the corresponding human–computer interaction system in practical application to construct constraints remains a problem that must be solved in the future.

5. Conclusions

An active pairwise constraint learning method is proposed in this paper to reduce the incidence angle effect in crop mapping. The proposed method can construct informative pairwise constraints based on the incidence angles of the samples and a non-iterative batch-mode active selection scheme. Comparison experiments on the UAVSAR images have shown the effectiveness of the proposed APCL method compared to the commonly used constraint construction methods for constrained clustering at improving the crop mapping accuracy and the superiority. However, the proposed method ignores transferring pairwise constraint information between samples to reinforce the constraints, and the future work will focus on effectively transferring constraint information. Thus, a way to pre-evaluate the optimal proportion of different types of constraints is also needed.

Author Contributions: Conceptualization: X.Q. and J.Y.; methodology: X.Q., K.S. and L.Z.; validation: X.Q. and L.Z.; investigation: X.Q., B.W. and L.Z.; resources: J.Y. and P.L.; writing of original draft: X.Q.; writing—review and editing: L.Z., J.Y., X.Q., Y.X. and K.S.; supervision: P.L. and J.Y.; project administration: J.Y. and P.L.; funding acquisition: J.Y., P.L. and B.W. All authors have read and agreed to the published version of the manuscript.

Funding: This research was funded by the National Natural Science Foundation of China (Grant No. 61971318), the Strategic Priority Research Program of the Chinese Academy of Sciences (Grant No. XDA19030201), the Shenzhen Fundamental Research Program (Grant No. JCYJ20200109150833977), the Joint Funds of the National Natural Science Foundation of China (Grant No. U2033216), and the National Natural Science Foundation of China Major Program (Grant No. 42192580).

Data Availability Statement: The data presented in this study are openly available in (UAVSAR_POLSAR) at (10.5067/7PEQV8SVR4DM).

Acknowledgments: The authors would like to present their acknowledgments to the JPL NASA and SMAPVEX 2012 teams for providing the UAVSAR images, and the NASA National Snow and Ice Data Center Distributed Active Archive Center (NSIDC DAAC) for making and providing the 2012 crop map of the Winnipeg area.

Conflicts of Interest: The authors declare no conflict of interest.

References

1. Wu, B.; Gommes, R.; Zhang, M.; Zeng, H.; Yan, N.; Zou, W.; Zheng, Y.; Zhang, N.; Chang, S.; Xing, Q.; et al. Global Crop Monitoring: A Satellite-Based Hierarchical Approach. *Remote Sens.* **2015**, *7*, 3907–3933. [CrossRef]
2. Laborde, D.; Martin, W.; Swinnen, J.; Vos, R. COVID-19 risks to global food security. *Science* **2020**, *369*, 500–502. [CrossRef] [PubMed]
3. Wu, B.; Meng, J.; Li, Q.; Yan, N.; Du, X.; Zhang, M. Remote sensing-based global crop monitoring: Experiences with China's CropWatch system. *Int. J. Digital Earth* **2014**, *7*, 113–137. [CrossRef]
4. Liang, S.; Qin, J. Data Assimilation Methods for Land Surface Variable Estimation. In *Advances in Land Remote Sensing: System, Modeling, Inversion and Application*; Liang, S., Ed.; Springer: Dordrecht, The Netherlands, 2008; pp. 313–339.
5. Wu, B.; Zhang, M.; Zeng, H.; Zhang, X.; Yan, N.; Meng, J. Agricultural monitoring and early warning in the era of big data. *J. Remote Sens.* **2016**, *20*, 1027–1037.
6. Huang, J.; Gómez-Dans, J.L.; Huang, H.; Ma, H.; Wu, Q.; Lewis, P.E.; Liang, S.; Chen, Z.; Xue, J.-H.; Wu, Y.; et al. Assimilation of remote sensing into crop growth models: Current status and perspectives. *Agric. For. Meteorol.* **2019**, *276–277*, 107609. [CrossRef]
7. Mandal, D.; Kumar, V.; Ratha, D.; Dey, S.; Bhattacharya, A.; Lopez-Sanchez, J.M.; McNairn, H.; Rao, Y.S. Dual polarimetric radar vegetation index for crop growth monitoring using sentinel-1 SAR data. *Remote Sens. Environ.* **2020**, *247*, 111954. [CrossRef]
8. Shi, H.; Yang, J.; Li, P.; Zhao, L.; Liu, Z.; Zhao, J.; Liu, W. Soil moisture estimation using two-component decomposition and a hybrid X-Bragg/Fresnel scattering model. *J. Hydrol.* **2019**, *574*, 646–659. [CrossRef]

9. Freeman, A.; Durden, S.L. A three-component scattering model for polarimetric SAR data. *IEEE Trans. Geosci. Remote Sens.* **1998**, *36*, 963–973. [CrossRef]
10. Yamaguchi, Y.; Moriyama, T.; Ishido, M.; Yamada, H. Four-component scattering model for polarimetric SAR image decomposition. *IEEE Trans. Geosci. Remote Sens.* **2005**, *43*, 1699–1706. [CrossRef]
11. Arii, M.; van Zyl, J.J.; Kim, Y. Adaptive Model-Based Decomposition of Polarimetric SAR Covariance Matrices. *IEEE Trans. Geosci. Remote Sens.* **2011**, *49*, 1104–1113. [CrossRef]
12. van Zyl, J.J.; Arii, M.; Kim, Y. Model-Based Decomposition of Polarimetric SAR Covariance Matrices Constrained for Nonnegative Eigenvalues. *IEEE Trans. Geosci. Remote Sens.* **2011**, *49*, 3452–3459. [CrossRef]
13. Cui, Y.; Yamaguchi, Y.; Yang, J.; Park, S.-E.; Kobayashi, H.; Singh, G. Three-Component Power Decomposition for Polarimetric SAR Data Based on Adaptive Volume Scatter Modeling. *Remote Sens.* **2012**, *4*, 1559–1572. [CrossRef]
14. Guo, J.; Wei, P.; Liu, J.; Jin, B.; Su, B.; Zhou, Z. Crop Classification Based on Differential Characteristics of H/α Scattering Parameters for Multitemporal Quad- and Dual-Polarization SAR Images. *IEEE Trans. Geosci. Remote Sens.* **2018**, *56*, 6111–6123. [CrossRef]
15. Niazmardi, S.; Homayouni, S.; Safari, A.; McNairn, H.; Shang, J.; Beckett, K. Histogram-based spatio-temporal feature classification of vegetation indices time-series for crop mapping. *Int. J. Appl. Earth Obs. Geoinf.* **2018**, *72*, 34–41. [CrossRef]
16. Li, H.; Zhang, C.; Zhang, S.; Atkinson, P.M. Crop classification from full-year fully-polarimetric L-band UAVSAR time-series using the Random Forest algorithm. *Int. J. Appl. Earth Obs. Geoinf.* **2020**, *87*, 102032. [CrossRef]
17. Mestre-Quereda, A.; Lopez-Sanchez, J.M.; Vicente-Guijalba, F.; Jacob, A.W.; Engdahl, M.E. Time-Series of Sentinel-1 Interferometric Coherence and Backscatter for Crop-Type Mapping. *IEEE J. Sel. Top. Appl. Earth Obs. Remote Sens.* **2020**, *13*, 4070–4084. [CrossRef]
18. Xu, J.; Zhu, Y.; Zhong, R.; Lin, Z.; Xu, J.; Jiang, H.; Huang, J.; Li, H.; Lin, T. DeepCropMapping: A multi-temporal deep learning approach with improved spatial generalizability for dynamic corn and soybean mapping. *Remote Sens. Environ.* **2020**, *247*, 111946. [CrossRef]
19. Zhang, H.; Kang, J.; Xu, X.; Zhang, L. Accessing the temporal and spectral features in crop type mapping using multi-temporal Sentinel-2 imagery: A case study of Yi'an County, Heilongjiang province, China. *Comput. Electron. Agric.* **2020**, *176*, 105618. [CrossRef]
20. Huang, X.; Reba, M.; Coffin, A.; Runkle, B.R.K.; Huang, Y.; Chapman, B.; Ziniti, B.; Skakun, S.; Kraatz, S.; Siqueira, P.; et al. Cropland mapping with L-band UAVSAR and development of NISAR products. *Remote Sens. Environ.* **2021**, *253*, 112180. [CrossRef]
21. Skakun, S.; Franch, B.; Vermote, E.; Roger, J.-C.; Becker-Reshef, I.; Justice, C.; Kussul, N. Early season large-area winter crop mapping using MODIS NDVI data, growing degree days information and a Gaussian mixture model. *Remote Sens. Environ.* **2017**, *195*, 244–258. [CrossRef]
22. Zhang, G.; Xiao, X.; Dong, J.; Kou, W.; Jin, C.; Qin, Y.; Zhou, Y.; Wang, J.; Menarguez, M.A.; Biradar, C. Mapping paddy rice planting areas through time series analysis of MODIS land surface temperature and vegetation index data. *ISPRS J. Photogramm. Remote Sens.* **2015**, *106*, 157–171. [CrossRef] [PubMed]
23. Friedman, L.; Netanyahu, N.S.; Shoshany, M. Mean shift-based clustering of remotely sensed data with agricultural and land-cover applications. *Int. J. Remote Sens.* **2013**, *34*, 6037–6053. [CrossRef]
24. Tamiminia, H.; Homayouni, S.; McNairn, H.; Safari, A. A particle swarm optimized kernel-based clustering method for crop mapping from multi-temporal polarimetric L-band SAR observations. *Int. J. Appl. Earth Obs. Geoinf.* **2017**, *58*, 201–212. [CrossRef]
25. Xiong, J.; Thenkabail, P.S.; Gumma, M.K.; Teluguntla, P.; Poehnelt, J.; Congalton, R.G.; Yadav, K.; Thau, D. Automated cropland mapping of continental Africa using Google Earth Engine cloud computing. *ISPRS J. Photogramm. Remote Sens.* **2017**, *126*, 225–244. [CrossRef]
26. Zhang, H.; Li, Q.; Liu, J.; Shang, J.; Du, X.; Zhao, L.; Wang, N.; Dong, T. Crop classification and acreage estimation in North Korea using phenology features. *GIScience Remote Sens.* **2017**, *54*, 381–406. [CrossRef]
27. Wang, S.; Azzari, G.; Lobell, D.B. Crop type mapping without field-level labels: Random forest transfer and unsupervised clustering techniques. *Remote Sens. Environ.* **2019**, *222*, 303–317. [CrossRef]
28. Konduri, V.S.; Kumar, J.; Hargrove, W.W.; Hoffman, F.M.; Ganguly, A.R. Mapping crops within the growing season across the United States. *Remote Sens. Environ.* **2020**, *251*, 112048. [CrossRef]
29. Qin, X.; Yang, J.; Li, P.; Sun, W.; Liu, W. A Novel Relational-Based Transductive Transfer Learning Method for PolSAR Images via Time-Series Clustering. *Remote Sens.* **2019**, *11*, 1358. [CrossRef]
30. McNairn, H.; Jackson, T.J.; Wiseman, G.; Bélair, S.; Berg, A.; Bullock, P.; Colliander, A.; Cosh, M.H.; Kim, S.; Magagi, R.; et al. The Soil Moisture Active Passive Validation Experiment 2012 (SMAPVEX12): Prelaunch Calibration and Validation of the SMAP Soil Moisture Algorithms. *IEEE Trans. Geosci. Remote Sens.* **2015**, *53*, 2784–2801. [CrossRef]
31. Khosravi, I.; Alavipanah, S.K. A random forest-based framework for crop mapping using temporal, spectral, textural and polarimetric observations. *Int. J. Remote Sens.* **2019**, *40*, 7221–7251. [CrossRef]
32. Wagner, W.; Noll, J.; Borgeaud, M.; Rott, H. Monitoring soil moisture over the Canadian Prairies with the ERS scatterometer. *IEEE Trans. Geosci. Remote Sens.* **1999**, *37*, 206–216. [CrossRef]
33. Menges, C.H.; Van Zyl, J.J.; Hill, G.J.E.; Ahmad, W. A procedure for the correction of the effect of variation in incidence angle on AIRSAR data. *Int. J. Remote Sens.* **2001**, *22*, 829–841. [CrossRef]

34. Ardila, J.P.; Tolpekin, V.; Bijker, W. Angular Backscatter Variation in L-Band ALOS ScanSAR Images of Tropical Forest Areas. *IEEE Geosci. Remote Sens. Lett.* **2010**, *7*, 821–825. [CrossRef]
35. Mladenova, I.E.; Jackson, T.J.; Bindlish, R.; Hensley, S. Incidence Angle Normalization of Radar Backscatter Data. *IEEE Trans. Geosci. Remote Sens.* **2013**, *51*, 1791–1804. [CrossRef]
36. Lampert, T.; Dao, T.-B.-H.; Lafabregue, B.; Serrette, N.; Forestier, G.; Crémilleux, B.; Vrain, C.; Gançarski, P. Constrained distance based clustering for time-series: A comparative and experimental study. *Data Min. Knowl. Discov.* **2018**, *32*, 1663–1707. [CrossRef]
37. Wagstaff, K.; Cardie, C. Clustering with instance-level constraints. *AAAI/IAAI* **2000**, *1097*, 577–584.
38. Davidson, I.; Ravi, S.S. The complexity of non-hierarchical clustering with instance and cluster level constraints. *Data Min. Knowl. Discov.* **2007**, *14*, 25–61. [CrossRef]
39. Wagstaff, K.; Cardie, C.; Rogers, S.; Schroedl, S. Constrained K-Means Clustering with Background Knowledge. In Proceedings of the Eighteenth International Conference on Machine Learning, San Francisco, CA, USA, 28 June–1 July 2001; pp. 577–584.
40. Asafi, S.; Cohen-Or, D. Constraints as Features. In Proceedings of the 2013 IEEE Conference on Computer Vision and Pattern Recognition, Portland, OR, USA, 23–28 June 2013; pp. 1634–1641.
41. Chen, Z.; Li, X.; Yang, F. Constrained K-means with external information. In Proceedings of the 2013 8th International Conference on Computer Science & Education, Colombo, Sri Lanka, 26–28 April 2013; pp. 490–493.
42. Lampert, T.; Lafabregue, B.; Dao, T.; Serrette, N.; Vrain, C.; Gançarski, P. Constrained Distance-Based Clustering for Satellite Image Time-Series. *IEEE J. Sel. Top. Appl. Earth Obs. Remote Sens.* **2019**, *12*, 4606–4621. [CrossRef]
43. González-Almagro, G.; Bermejo, P.S.; Suarez, J.L.; Cano, J.-R.; García, S. Monotonic Constrained Clustering: A First Approach. In *Proceedings of the Advances and Trends in Artificial Intelligence, Theory and Practices in Artificial Intelligence*; Springer: Cham, Switzerland, 2022; pp. 725–736.
44. Hazratgholizadeh, R.; Balafar, M.A.; Derakhshi, M.R.F. Active constrained deep embedded clustering with dual source. *Appl. Intell.* **2022**, 1–31. [CrossRef]
45. Peng, B.; Yao, Y.; Li, Q.; Li, X.; Lin, G.; Chen, L.; Lei, J. Clustering information-constrained 3D U-Net subspace clustering for hyperspectral image. *Remote Sens. Lett.* **2022**, *13*, 1131–1141. [CrossRef]
46. Schier, M.; Reinders, C.; Rosenhahn, B. Constrained Mean Shift Clustering. In Proceedings of the 2022 SIAM International Conference on Data Mining (SDM), Alexandria, VA, USA, 28–30 April 2022; Society for Industrial and Applied Mathematics: Philadelphia, PA, USA, 2022; pp. 235–243.
47. Zhang, X.; Zheng, Y.; Ye, X.; Peng, Q.; Wang, W.; Li, S. Clustering with implicit constraints: A novel approach to housing market segmentation. *Trans. GIS* **2022**, *26*, 585–608. [CrossRef]
48. McNairn, H.; Powers, J.; Wiseman, G. *SMAPVEX12 Land Cover Classification Map, Version 1. [Data Set]*; NASA National Snow and Ice Data Center Distributed Active Archive Center: Boulder, CO, USA, 2014. [CrossRef]
49. Chen, E.; Li, Z.; Tian, X.; Ling, F. Terrain Radiometric Correction Model and Its Validation for Space-Borne SAR Data. *Geomat. Inf. Sci. Wuhan Univ.* **2010**, *35*, 322. [CrossRef]
50. Lang, W.; Mo, L.; Yang, X.; Zhang, J.; Meng, J. Quantitative research and correction of incident angle effect for wide-swath SAR images. *J. Remote Sens.* **2013**, *17*, 1281–1294. [CrossRef]
51. Shi, H.; Lopez-Sanchez, J.M.; Yang, J.; Li, P.; Zhao, L.; Zhao, J. Contribution of Polarimetry and Multi-Incidence to Soil Moisture Estimation Over Agricultural Fields Based on Time Series of L-Band SAR Data. *IEEE J. Sel. Top. Appl. Earth Obs. Remote Sens.* **2021**, *14*, 300–313. [CrossRef]
52. NOAA National Centers for Environmental Information. *Global Surface Hourly Global Summary of the Day*. 2001. Available online: https://aquaknow.jrc.ec.europa.eu/en/content/global-surface-summary-day-gsod (accessed on 20 October 2022).
53. Zhao, L.; Yang, J.; Li, P.; Zhang, L. Seasonal inundation monitoring and vegetation pattern mapping of the Erguna floodplain by means of a RADARSAT-2 fully polarimetric time series. *Remote Sens. Environ.* **2014**, *152*, 426–440. [CrossRef]
54. Whelen, T.; Siqueira, P. Use of time-series L-band UAVSAR data for the classification of agricultural fields in the San Joaquin Valley. *Remote Sens. Environ.* **2017**, *193*, 216–224. [CrossRef]
55. Schein, A.I.; Ungar, L.H. Active learning for logistic regression: An evaluation. *Mach. Learn.* **2007**, *68*, 235–265. [CrossRef]
56. Demir, B.; Persello, C.; Bruzzone, L. Batch-Mode Active-Learning Methods for the Interactive Classification of Remote Sensing Images. *IEEE Trans. Geosci. Remote Sens.* **2011**, *49*, 1014–1031. [CrossRef]
57. Chakraborty, S.; Balasubramanian, V.; Panchanathan, S. Adaptive Batch Mode Active Learning. *IEEE Trans. Neural Netw. Learn. Syst.* **2015**, *26*, 1747–1760. [CrossRef]
58. Li, H.; Wang, Y.; Li, Y.; Xiao, G.; Hu, P.; Zhao, R. Batch mode active learning via adaptive criteria weights. *Appl. Intell.* **2020**, *51*, 3475–3489. [CrossRef]
59. Patra, S.; Bruzzone, L. A cluster-assumption based batch mode active learning technique. *Pattern Recognit. Lett.* **2012**, *33*, 1042–1048. [CrossRef]
60. Wang, Z.; Du, B.; Zhang, L.; Zhang, L. A batch-mode active learning framework by querying discriminative and representative samples for hyperspectral image classification. *Neurocomputing* **2016**, *179*, 88–100. [CrossRef]
61. Chang, C.; Huang, H. Automatic Tuning of the RBF Kernel Parameter for Batch-Mode Active Learning Algorithms: A Scalable Framework. *IEEE Trans. Cybern.* **2019**, *49*, 4460–4472. [CrossRef] [PubMed]

62. Stanitsas, P.; Cherian, A.; Morellas, V.; Papanikolopoulos, N.P. Active constrained clustering via non-iterative uncertainty sampling. In Proceedings of the 2016 IEEE/RSJ International Conference on Intelligent Robots and Systems (IROS), Daejeon, Republic of Korea, 9–14 October 2016; pp. 4027–4033. [CrossRef]
63. Rousseeuw, P.J. Silhouettes: A graphical aid to the interpretation and validation of cluster analysis. *J. Comput. Appl. Math.* **1987**, *20*, 53–65. [CrossRef]
64. MacQueen, J. Some Methods for Classification and Analysis of Multivariate Observations. In Proceedings of the 5th Berkeley Symposium on Mathematical Statistics and Probability, Los Angeles, CA, USA, 27 December 1965–7 January 1966; Volume 1, pp. 281–297.
65. Aghabozorgi, S.; Shirkhorshidi, A.S.; Wah, T.Y. Time-series clustering—A decade review. *Inf. Syst.* **2015**, *53*, 16–38. [CrossRef]
66. Sakoe, H.; Chiba, S. Dynamic programming algorithm optimization for spoken word recognition. *IEEE Trans. Acoust. Speech Signal Process.* **1978**, *26*, 43–49. [CrossRef]
67. Basu, S.; Banerjee, A.; Mooney, R.J. Active Semi-Supervision for Pairwise Constrained Clustering. In Proceedings of the 2004 SIAM International Conference on Data Mining, Lake Buena Vista, FL, USA, 22–24 April 2004; Society for Industrial and Applied Mathematics: Philadelphia, PA, USA, 2004; pp. 333–344.
68. Babaki, B. MIPKmeans. 2017. Available online: https://github.com/Behrouz-Babaki/MIPKmeans (accessed on 20 October 2022).
69. Knops, Z.F.; Maintz, J.B.A.; Viergever, M.A.; Pluim, J.P.W. Normalized mutual information based registration using k-means clustering and shading correction. *Med. Image Anal.* **2006**, *10*, 432–439. [CrossRef] [PubMed]
70. Qin, X.; Yang, J.; Zhao, L.; Li, P.; Sun, K. A Novel Deep Forest-Based Active Transfer Learning Method for PolSAR Images. *Remote Sens.* **2020**, *12*, 2755. [CrossRef]

remote sensing

Article

Convolutional Neural Networks for Agricultural Land Use Classification from Sentinel-2 Image Time Series

Alejandro-Martín Simón Sánchez [1,*], José González-Piqueras [1], Luis de la Ossa [2] and Alfonso Calera [1]

[1] Remote Sensing and GIS Group, Regional Research Institute, Campus of Albacete, University of Castilla-La Mancha, 02071 Albacete, Spain
[2] Computing Systems Department, Campus of Albacete, University of Castilla-La Mancha, 02071 Albacete, Spain
* Correspondence: alejandro.simon@uclm.es

Abstract: Land use classification (LUC) is the process of providing information on land cover and the types of human activity involved in land use. In this study, we perform agricultural LUC using sequences of multispectral reflectance Sentinel-2 images taken in 2018. LUC can be carried out using machine or deep learning techniques. Some existing models process data at the pixel level, performing LUC successfully with a reduced number of images. Part of the pixel information corresponds to multispectral temporal patterns that, despite not being especially complex, might remain undetected by models such as random forests or multilayer perceptrons. Thus, we propose to arrange pixel information as 2D yearly fingerprints so as to render such patterns explicit and make use of a CNN to model and capture them. The results show that our proposal reaches a 91% weighted accuracy in classifying pixels among 19 classes, outperforming random forest by 8%, or a specifically tuned multilayer perceptron by 4%. Furthermore, models were also used to perform a ternary classification in order to detect irrigated fields, reaching a 97% global accuracy. We can conclude that this is a promising operational tool for monitoring crops and water use over large areas.

Keywords: deep learning; remote sensing; land use classification; sentinel; time series

check for updates

Citation: Simón Sánchez, A.-M.; González-Piqueras, J.; de la Ossa, L.; Calera, A. Convolutional Neural Networks for Agricultural Land Use Classification from Sentinel-2 Image Time Series. *Remote Sens.* **2022**, *14*, 5373. https://doi.org/10.3390/rs14215373

Academic Editors: Sawaid Abbas, Janet E. Nichol, Faisal M. Qamer and Jianchu Xu

Received: 29 September 2022
Accepted: 25 October 2022
Published: 27 October 2022

1. Introduction

Agriculture is a key sector from an economic, social, and environmental point of view. Because of its high demand for water and nutrient inputs for increasing the yield, agriculture places high pressure on the surface and groundwater resources. This is an important issue for governments, policymakers, farmers, and other organizations, as estimations predict that, in order to cover the food demand by 2050, a 60% increment production will be needed [1]. Policies from the international to local scale are facing the challenge of providing natural resource managers with the tools required for the sustainability of food supply sectors and resilience to climate change [2–4]. In particular, the European and national authorities estimate the number of subsidies on the farm scale by considering farmers' reports about their practices for crop management, such as water use, fertilizers, and energy, which are properly supervised by experts who help to encourage sustainable use. In this scenario, precise crop and land use classification (LUC) are necessary for assisting in the management of sustainable natural resources [5] and also facing the consequences of climate change [6]. LUC is the basic information required for the sustainable use of land, water, energy management, and the environment [7]. Nevertheless, even gathering information about the harvested crops and their management has a high cost when it requires fieldwork.

Remote sensing is one of the main sources of information in this context. It can be applied to large areas and provides classification maps at a lower cost. The plot scale in remote sensing is limited by the pixel size of the images acquired by the sensors. However, due to the increase in both the spatiotemporal resolution of the data and processing power, improvements are becoming important in recent years [8]. Hence, Sentinel and Landsat constellations provide data to monitor plots below 0.3 ha, which is adequate for most agricultural areas [9].

Traditionally, the generation of an LUC map has consisted of the examination of multi-spectral and temporal sensor data at single positions and their surrounding areas. Therefore, a long time series of images at a high temporal frequency provides data that can be used to identify vegetation types with a high confidence level [10]. Those algorithms are based on biophysical variables that vary over time according to the specific phenology of each crop, the agronomic management, and the environmental conditions. Regarding the importance of such temporal signatures, sensors with a high temporal resolution, such as MODIS, have been used for this purpose despite their low resolution (from 250 m to 1 km depending on the data) [11,12], providing good results, since they gather the sequential information of a specific location over time [13]. The scenario is even more favorable nowadays, as there is an availability of multi-spectral remote sensing data of a relatively fine resolution (from 10 m to 60 m), provided by the Sentinel 2 constellation with a reasonable revisit frequency of 5 days [14].

The first generation of procedures for LUC maps relied on experts who built models based on indexes calculated from the bands of the satellite images, such as the temporal evolution of NDVI, which is calculated as the near-infrared (NIR) minus red reflectance, divided by the near-infrared plus red reflectance [15]. NDVI can be used to identify the land use of a single point with a reasonable computational cost [16]. The attempts and the progress in efforts to automate the process of land use classification are moving towards the use of machine learning, an area of artificial intelligence. Machine learning models use features to identify patterns in datasets. When such patterns involve a target variable, these can be used to perform tasks such as classification. Gilabert et al. showed that temporal information can be extracted from the spectral bands [17]. Experts manually fine-tuned the rules of those interpretable models based on simpler variables, according to their expertise regarding how features determine the type of crop and the knowledge about a concrete region [18]. Foerster et al. [13] and Hao et al. [19] proposed solutions based on NDVI that use decision trees and random forest models, respectively, with a pixel-based approach. J. Inglada et al. tested random forests and support vector machines in crop classification using feature extractions, such as derived indices [9]. Immitzer et al. proved that the most important bands of the Sentinel-2 images are the blue, red-edge, and shortwave infrared for crop classification in a study applying a random forest model [20]. Additionally, authors such as Hao et al. indicated that it is possible to merge datasets with a similar spatial resolution in order to enrich the time series [21]. Zhou et al. reported that optimizing the required number of images, resulting in a reduction in the time series length, had almost no impact on the accuracy of the classification [22].

On the other hand, deep learning is a recently adopted technique based on neural networks that infer those characteristics directly from data, comprising fewer pre-processing steps and improving the results, as compared to machine learning models [23]. There are many neural network types, but the convolutional (CNNs) and recurrent (RNNs) neural networks are the most useful in LUC. CNNs are mainly employed to process visual imagery, whereas RNNs are related to tasks whose input is sequential data [24,25]. As explained previously, deep learning algorithms represent a step forward, as they open the way to automatically apply models to time series of images collected from satellites throughout a crop cycle, classifying herbaceous and orchards crops and distinguishing between irrigated and non-irrigated lands. Consequently, despite not obtaining a fully automatic method of classification that is valid for all the land use cases, the use of deep learning models results in a reduction in the workload. Lyu et al. proposed an LSTM (long short-term memory, a kind of RNN) to extract spectral and temporal characteristics for change detection [26]. This work was continued by Mou et al. with the use of Conv2D layers to extract spatial features to be inputted to LSTM layers. This network, which enhanced the detection of the temporal dependency, achieved better results than detection algorithms based on spatial features alone [27]. The efficiency and the ability to ignore the shift in data so as to recognize patterns mean that the learning process of CNNs is the most suitable for image recognition [28]. Rußwurm and Körner extracted temporal features from image sequences to identify crop types with LSTMs [29]. They enhanced this model by using Conv2D layers to extract spatial features and an RNN with a bidirectional sequential encoder to introduce a sequence in a bidirectional manner [30]. Zhong et al. showed that models based on one-dimension convolutional layers (Conv1D) achieved better results than models which rely on long short-term memory layers, which achieved worse results despite being useful for analyzing sequential data. The Conv1D-based model achieved the best result among the tested classifiers [23]. Campos-Taberner et al. proposed a deep learning solution based on a recurrent neural network model with 2 Bi-LSTM layers and a fully connected layer to classify 10 crops using Sentinel-2 images and their bands, in addition to NDVI and enhanced NDVI [31]. Portalés-Julià et al. assessed Sentinel-2's capabilities in identifying abandoned crops, achieving the best results with the Bi-LSTM neural networks and random forest by classifying two major classes: active or abandoned crops and eight subclasses [32]. Ruiz et al. presented a classification model using CNNs with very-high-resolution aerial orthoimages from the Spanish plan of aircraft orthophotography (PNOA) and calculated the NDVI from a Sentinel-2 level 2A image time series to determine the type of soil, according to six different classes and whether they were abandoned or not [33]. Amani et al. showed that models firstly trained offline can be used on cloud platforms and applied to classify available online data, taking advantage of their satellite imagery and geospatial datasets on the planetary scale [34].

In this article, we propose a model for land use classification that requires few images and offers good results in the area of our focus. While some previous models work at the image level, carrying out segmentation and requiring large datasets, we consider that: (i) a pixel can contain enough information to successfully perform classification; (ii) part of this information lies in multispectral temporal patterns; and (iii) although such patterns are not especially complex, existing works based on random forest might not capture them, whereas others models, such as LTSM, are unnecessarily complex and require larger sequences. Based on these assumptions, we propose an approach that arranges the information corresponding to a pixel as a 2D yearly fingerprint and uses a convolution-based model (CNN) for the prediction. This approach can render multispectral temporal patterns more explicit and improve the classification. In fact, the state-of-the-art algorithms for time series classification are based on convolutions [35,36]. We also added a problem-specific process of oversampling, trying to deal with variations in phenology. We used this approach to perform (i) the classification of the main crop classes, focusing on herbaceous and woody crops, and (ii) discrimination between irrigated and non-irrigated areas.

Therefore, a relevant contribution of this work is the consequent improvement in pixel-based land use classification with a small number of images by representing data as a 2D fingerprint and using a CNN model. We tested our proposal in a well-known agricultural area in the Mancha Oriental aquifer in Spain. Improvement regarding these issues is of great interest, as land use information is a basic input for water accounting [37], water footprint estimations for environmental management [38], and yield prediction modelling [39].

2. Materials and Methods

Figure 1 summarizes, in a workflow, the content of this section, showing how the satellite and labeled data are selected, gathered, and later processed to build, assess, and employ the model.

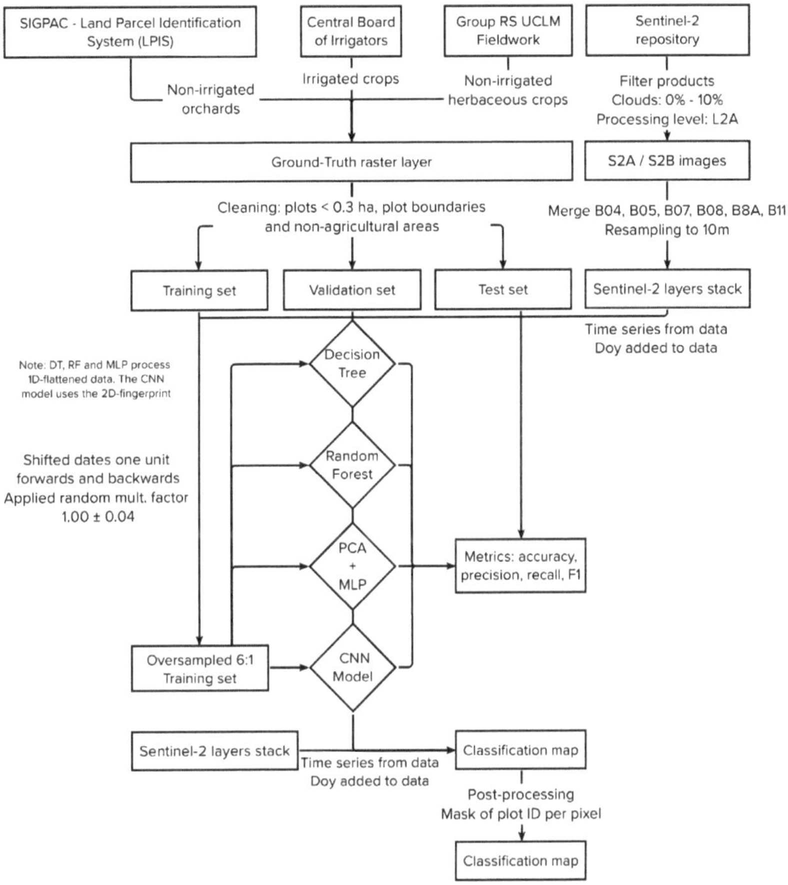

Figure 1. Classification workflow.

2.1. Input Data

Our study takes as its basis a set of 24 images from the year 2018, corresponding to different dates from March to October, as shown in Table 1. The images were downloaded from the Copernicus Open Access Hub server, and we considered only those in which the percentage of clouds was below 10%, setting a processing level of L2A. This criterion reduces the noise introduced by the presence of clouds and shade on the surface. This set of images was used to generate spectral-temporal signatures.

Table 1. Dates of the T30SWJ granule set of 24 Sentinel 2 images at level L2A, selected with a maximum of 10% cloud coverage in the study area.

2018							
March	**April**	**May**	**June**	**July**	**August**	**September**	**October**
27	14, 26	4, 16	13, 18, 23	3, 8, 13, 18, 23, 28	2, 12, 17, 22, 27	1, 13, 21	3, 6

Sentinel 2 data requires pre-processing, insofar as the information on the bands is stored in different files that have different resolutions. Thus, bands with the lowest resolution, which can range between 20 and 60 m, need to be resampled to obtain a resolution of 10 m so that they can all be merged into a single tiff file for each date (Table 2). In addition, those pixels belonging to plots with areas under 0.3 ha were excluded from the study because of the limitation of the resolution of the satellite.

Table 2. Pre-processed resampled tiff image correspondence with Sentinel-2 bands.

#	Sentinel-2 Band	Original Res.	Target Res.	Wavelength (μm)
1	B02	10 m	10 m	0.490
2	B03	10 m	10 m	0.560
3	B04	10 m	10 m	0.665
4	B08	10 m	10 m	0.842
5	B05	20 m	10 m	0.705
6	B06	20 m	10 m	0.740
7	B07	20 m	10 m	0.783
8	B8A	20 m	10 m	0.865
9	B11	20 m	10 m	1.610
10	B12	20 m	10 m	2.190
11	B01	60 m	10 m	0.443
12	B09	60 m	10 m	0.945

As deep learning enables the identification of more patterns from the raw features of the input data, the experts' need for the introduction of new features built using original data with their knowledge, such as NDVI, is not particularly relevant, although it may lead to an improvement in some cases. As a matter of fact, this characteristic results in a clear advantage when compared to traditional methods [40]. In this proposal, the information is managed at the pixel level, so that each one of them is treated as a single and independent entity that is formed by the bands of information for the 24 selected dates. Moreover, considering previous works that stated the importance of spectral information [41], the number of bands used in this practical case was reduced from 12 bands to 6 bands, without any impact on the performance of the model, in order to optimize the computational resources in terms of the memory usage. The chosen bands, ordered by relevance, are B4, B8, B11, B8A, B7, and B5. Figure 2 shows both the NDVI sequence and the 2D fingerprint at a pixel level that corresponds to the land use of forage irrigated crops. The NDVI sequence is an index that can be interpreted by agronomic experts and allows them to build models according to their knowledge of the phenology of the crops, as the curve is correlated with the different growing stages [42,43], although the classification process using only NDVI does not consider all the available relevant information. In contrast, the representation of data at a pixel level as a 2D fingerprint carries the most important information, and it can be processed as an image by employing algorithms, such as CNNs, to efficiently find patterns and relationships between the bands and the different dates.

2.2. Area of Study

The area of study (Figure 3) is located in Albacete, Spain. For this area, the Sentinel-2 granule corresponds to T30SWJ. This location was selected due to the variety of land use, the availability of images with a low percentage of clouds, and the high expertise in land use classification regarding this area, with many national and international projects having been carried out in this test location and accurate field information being available from recent years [44,45].

The ground truth data are composed of information corresponding to the year 2018. They were provided by the following sources: (i) the Central Board of Irrigators, which contributed the irrigated plot data, obtained from fieldwork; (ii) the non-irrigated data provided by the Remote Sensing and GIS Group (Albacete, University of Castilla-La Mancha), based on their fieldwork; and (iii) SIGPAC, the national Land Parcel Identification System in Spain, which provided non-irrigated woody crop field information based on the farmers' reports. SIGPAC information is reliable, since it corresponds to long-term crops that are intended to be cultivated over many years. In order to facilitate the assessments by scientists, agronomists and technicians in the area, according to the experience derived from previous classification projects focusing on agronomic and sustainable criteria and considering the aim to deal with water management, the categories shown in Table 3 were selected, distinguishing between irrigated and non-irrigated crops.

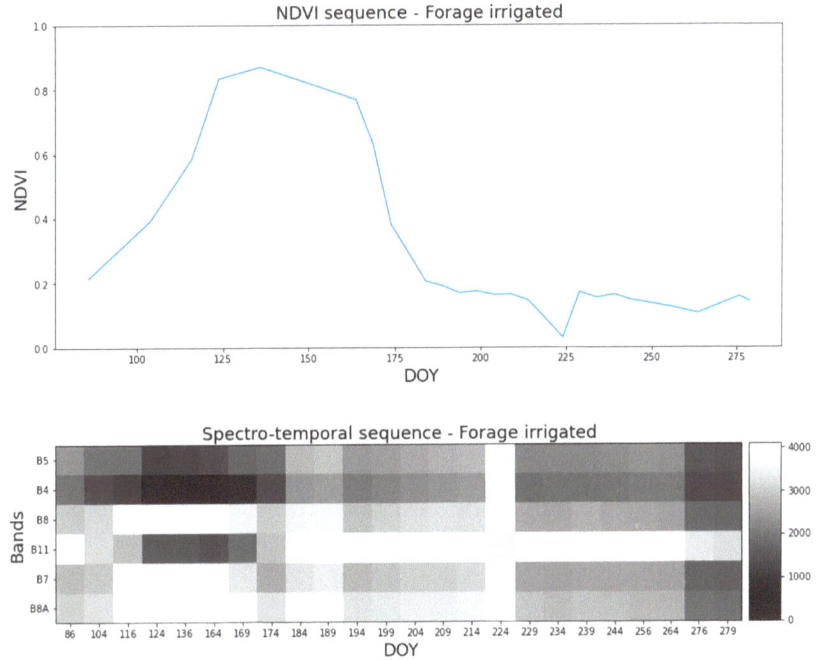

Figure 2. NDVI and 2D fingerprint representation of a forage irrigated pixel. Day 224 indicates the presence of a cloud.

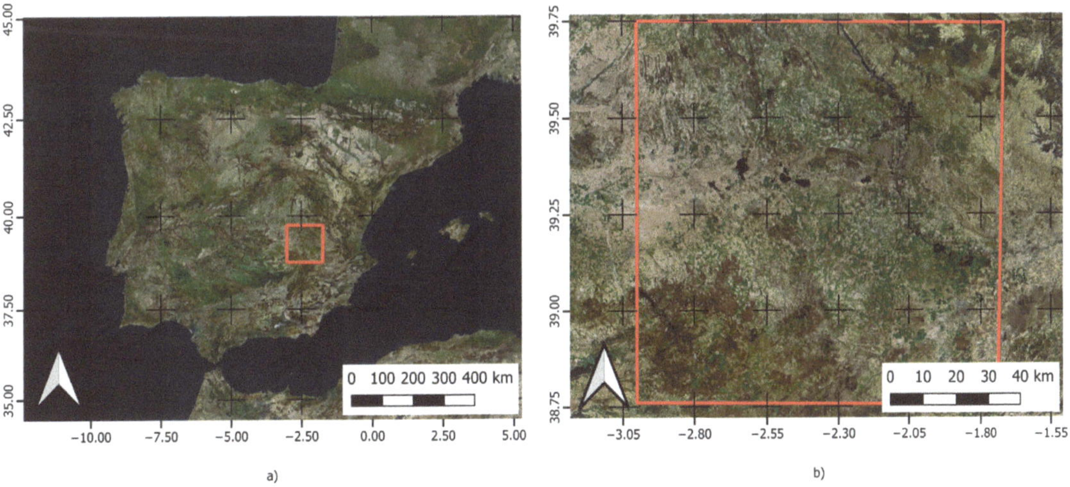

Figure 3. (**a**) Location of the experimental area in the Iberian Peninsula. (**b**) View of the experimental area (coordinates in EPSG: 4326).

Table 3. Land use classification categories.

#	Class	Hectares	%
1	Forage irrigated	4331	2.75
2	Spring irrigated	28,473	18.05
3	Spring irrigated Chinese garlic	6483	4.11
4	Spring irrigated purple garlic and horticulture	2852	1.81
5	Late spring irrigated	1842	1.17
6	Summer irrigated, low cover and horticulture	5572	3.53
7	Summer irrigated, high coverage	7160	4.54
8	Irrigated alfalfa	5365	3.40
9	Double crop	6904	4.38
10	Non-irrigated	7275	4.61
11	Bare soil	1098	0.70
13	Irrigated vineyard	9704	6.15
15	Non-irrigated vineyard	37,263	23.62
16	Irrigated olive trees	753	0.48
18	Non-irrigated olive trees	9266	5.87
19	Irrigated shell fruit trees	6368	4.04
21	Non-irrigated shell fruit trees	14,915	9.45
22	Irrigated fruit trees	175	0.11
24	Non-irrigated fruit trees	1972	1.25
	Total	157,771	

2.3. Selected Models: Decision Tree, Random Forest, Multilayer Perceptron, and Neural Networks

Some existing works carry out LUC by means of a set of rules built and tuned by experts [18]. These can be represented as a decision tree. Therefore, the first method selected to test the proposed classification scheme was the automatic learning of decision trees, where nodes represent conditions for the variables and leaves correspond to classes. These models are popular because of their low learning complexity, being carried out with a greedy algorithm, and because they essentially perform a selection of relevant features, discarding the rest [46].

Depending on the problem, decision trees can under- or overfit the training data. In general, ensemble models are frequently used in machine learning because they can deal with both problems at the same time. One of the most popular ensembles is the random forest, which combines bagging and randomly built decision trees [47].

Neural networks (NN) are models inspired by the structure of biological neural networks. An artificial neuron is a simple processing unit characterized by a set of weights. It receives a finite set of values—i.e., as many as there are weights—as the input, creates a weighted linear combination of them, and applies an activation function to the resulting value to generate an output. In their most basic form, neural networks (NN) are composed of several layers of artificial neurons arranged as an input layer, an output layer, and several hidden layers. Each neuron in the input layer processes the input data, producing an output value. Then, for all the remaining neurons, the input values correspond to the outputs produced in the previous layer. Lastly, the values generated by the neuron of the output layer are predefined labels that correspond to the labels or target variable. This is called forward propagation. Activation functions, such as sigmoid or hyperbolic tangent, together with the use of hidden layers, allow neural networks to represent complex, non-linear decision functions [48]. Multilayer perceptrons (MLPs) are the most basic type of fully connected feedforward neural network [49]. In this work, MLPs were first tested on their own but did not obtain acceptable results. Finally, the performance of this model was remarkably improved by applying principal component analysis (PCA) [50]. This method captures linear dependences among the input features and projects data onto a set of components, retaining most of its variance while reducing dimensionality and noise, which, in this case, can be produced by the presence of clouds [51].

The last developed model is based on the use of CNNs to extract the temporal-multispectral patterns from time series. As stated, the data were prepared as 2D images which contain temporal and multispectral data on each of their axes. Therefore, this enables the application of two-dimensional CNNs, allowing them to identify those patterns in a 2D array. This model (Figure 4) is composed of six initial Conv2D layers, which receive input data, with dimensions 24 dates $\times$ 7 bands, and reduce the number of parameters for the next layers. Usually, pooling layers are used after each convolution to reduce the spatial dimension of the feature maps, which was not considered in this concrete case so as to preserve the information. Convolutions of 1×7, 3×3, and 1×5 are considered in this model; thus, it can identify the changes over time for each band and the relationships between them. Finally, the flattened output of the Conv2D layers is used as an input for the final four dense hidden layers, in which L2 regularization is also applied to avoid the overfitting of the data.

For this work, we used the machine learning methods implemented in Scikit-Learn. This is a state-of-the-art machine learning method, used in python, as it provides many tools for tasks such as the use of models for classification, regression, clustering, dimensionality reduction, etc. [52]. We also used TensorFlow to implement the convolutional neural networks. Its particular purpose is focused on deep neural networks, and it provides tools and options that can be used to configure and build deep learning models [53].

2.4. Data Preparation: Hold-Out, Oversampling, and Post-Processing

The whole dataset is composed of sequential data corresponding to the evolution of the spectrum of each pixel over a year. We used a hold-out procedure [54], dividing the data into one subset for the model training; another, the validation set, for the model tunning; and the last, the test set, for the model evaluation. Table 4 shows the sizes of the resulting sets. Although larger training datasets help us to both detect more complex patterns and prevent overfitting, learning curves do not show improvement if the proportion of data used for training is increased above 1% of the data. On the other hand, test sets are usually smaller than training sets because of the lack of data. However, this leads to estimations affected by randomness and noise. In this scenario, models can be evaluated with the whole set of (remaining) pixels. This renders the estimation of the performance indicators far more robust.

We also improved the training with synthetic data. For the purpose of capturing the variations in the phenology over different years due to environmental and management conditions, we also propose a methodology based on oversampling, generating new instances by assigning the readings corresponding to pixels of the previous or later capture date. In addition to this, the data were replicated and multiplied by a random factor of 1.00 ± 0.04 to introduce an acceptable range of variation, which can represent the scenarios at different times or locations. Moreover, when the data are loaded, the day of the year is normalized using the sine function, so that it has a temporal reference, with values between zero and one. Thus, the first and last days of the year have similar values near zero.

Additionally, since our focus was on pixel-based classification, a procedure was proposed to integrate it into a plot level. This post-processing step consisted of overlaying the plot vector layer with the raster image of the classification, assigning the majority class to the pixels corresponding to each plot. The procedure consisted of checking every plot. When the majority class among the classified pixels for that plot is over 40%, it proceeds to verify the rest of the classes of the classified pixels. In a case where the previous condition holds, and the pixel representation of a minority class is under 25%, those underrepresented pixels, which are considered to be misclassified, change their classification value to the majority class of the pixels in that plot.

2.5. Evaluation

The classes are unbalanced in the dataset, as can be observed from the areas and the percentages in Table 3. This dataset was treated by selecting a minimum number of instances of each class for the training set and considering the weights of each class so as to adjust their importance during the training. In this way, the models do not overfit the most representative classes, and their ability to predict the classes in which there are fewer samples is improved [55].

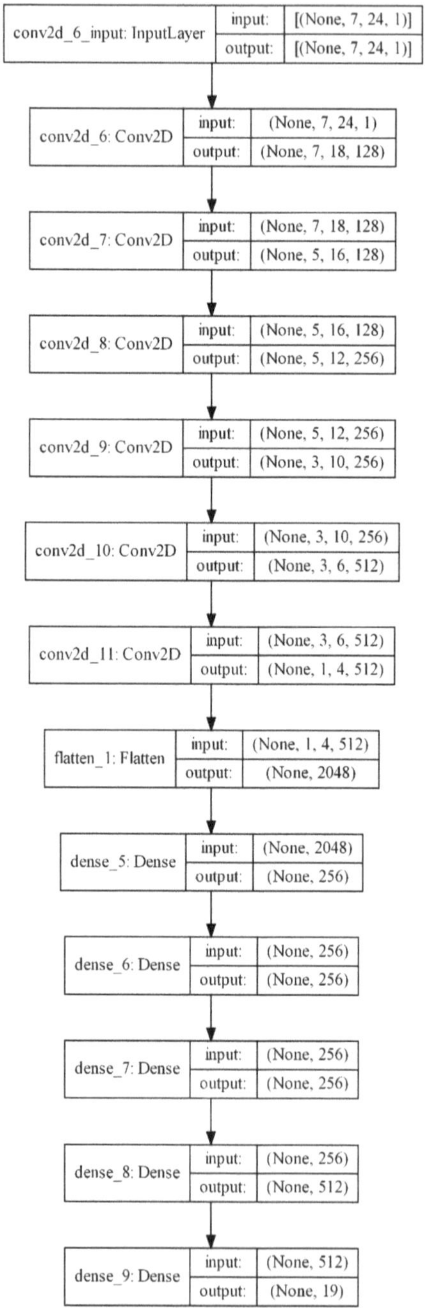

Figure 4. Representation of the model architecture.

Table 4. Dataset divided for land use classification.

Crop and Irrigation Soil Classification		
	Pixels	%
Training set	119,364	1.00
Validation set	834,753	6.97
Test set	11,030,736	92.03

In order to assess the results, the following metrics were applied to the validation and test sets [56,57]: (i) *accuracy*, which shows the rate of correct predictions (*CP*) made by the model and is the ratio between the number of *CP* divided by the number of total cases (*TC*); (ii) *recall*, which is the number of total correctly predicted positive results (*TP*) divided by the total number of cases which should be detected as positive (sum of the false negatives, *FN* and *TP*); (iii) *precision*, which expresses the rate between the *TP* and the sum of the *TP* and false negatives (*FN*); and the (iv) *F1-score*, which is the harmonic mean between the *precision* and *recall*, with values within the range of [0, 1]. The *recall*, *precision*, and *F1-Score* are defined for the binary classification and are calculated for each class, considering the class in question as positive and the rest as negative. The summary results reported for each model correspond to the weighted average of the metrics, whose weights correspond to the number of pixels of each class.

$$Accuracy = \frac{CP}{TC}, \tag{1}$$

$$Recall = \frac{TP}{(FN + TP)}, \tag{2}$$

$$Precision = \frac{TP}{(FP + TP)}, \tag{3}$$

$$F_1 Score = 2 \cdot \frac{Precision \cdot Recall}{Precision + Recall}. \tag{4}$$

3. Results

The selected models were applied to the test set to assess them and validate their performance according to the statistics proposed. Thus far, this section has shown how traditional machine and deep learning models perform in these tasks.

3.1. Decision Tree

The report in Table 5 shows the expected behavior regarding the low complexity of this algorithm, which is not capable of training a robust model. Considering each class, the *F1* is slightly higher in the herbaceous irrigated crops than in the herbaceous non-irrigated crops, bare soil, and orchards. There are specific classes, such as bare soil or irrigated fruit trees, that are correctly detected, as the *recall* shows a higher value. However, they are not well classified, as there is confusion with regard to the other classes, as the *precision* is low. This results in a low *F1* because those previous metrics are not balanced. According to these metrics and the achieved 73% overall accuracy, this model lacks robustness.

3.2. Random Forest

Random forest has a great modelling power but, at the same time, it maintains a low variance and does not usually produce overfitting. The report in Table 6 shows the expected improvement compared to the decision tree. Thus, the overall accuracy reaches 83%, and the average *F1* also reaches 83%. However, the assessment needs to be performed considering each class. The *F1* is above 85% in the herbaceous irrigated crops, whereas the model is not as accurate in distinguishing the bare soil or herbaceous non-irrigated crops. Even so, it reaches an 80% *F1* for those two classes. Regarding the orchards, these results are one step ahead of the previous ones, despite the fact that there is still room for refinement. The *F1* of the irrigated herbaceous crops is over 90%, high enough to be used to identify those classes with acceptable confidence in an application scenario.

Table 5. Land use classification: decision tree.

Class	Precision	Recall	F1	Support
Forage irrigated	0.61	0.68	0.64	343,643
Spring irrigated	0.91	0.90	0.91	2,373,450
Spring irrigated Chinese garlic	0.82	0.83	0.82	546,212
Spring irrigated purple garlic and horticulture	0.88	0.90	0.89	235,297
Late spring irrigated	0.64	0.73	0.68	141,147
Summer irrigated, low cover and horticulture	0.84	0.82	0.83	455,175
Summer irrigated, high coverage	0.89	0.87	0.88	585,448
Irrigated alfalfa	0.94	0.91	0.93	445,925
Double crop	0.87	0.80	0.83	572,968
Herbaceous non-irrigated	0.62	0.58	0.60	569,945
Bare soil	0.49	0.71	0.58	77,114
Irrigated vineyard	0.54	0.51	0.53	734,295
Non-irrigated vineyard	0.74	0.72	0.73	2,111,911
Irrigated olive trees	0.41	0.57	0.48	43,715
Non-irrigated olive trees	0.38	0.34	0.36	321,381
Irrigated shell fruit trees	0.51	0.51	0.51	514,900
Non-irrigated shell fruit trees	0.50	0.58	0.54	888,792
Irrigated fruit trees	0.20	0.78	0.32	6974
Non-irrigated fruit trees	0.14	0.11	0.12	62,444
Accuracy			0.73	11,030,736
Weighted avg.	0.74	0.73	0.74	

Table 6. Land use classification: random forest.

Class	Precision	Recall	F1	Support
Forage irrigated	0.90	0.80	0.85	343,643
Spring irrigated	0.95	0.95	0.95	2,373,450
Spring irrigated Chinese garlic	0.89	0.95	0.92	546,212
Spring irrigated purple garlic and horticulture	0.97	0.96	0.97	235,297
Late spring irrigated	0.97	0.80	0.87	141,147
Summer irrigated, low cover and horticulture	0.91	0.93	0.92	455,175
Summer irrigated, high coverage	0.97	0.93	0.95	585,448
Irrigated alfalfa	0.97	0.97	0.97	445,925
Double crop	0.96	0.91	0.93	572,968
Herbaceous non-irrigated	0.79	0.74	0.76	569,945
Bare soil	0.86	0.79	0.82	77,114
Irrigated vineyard	0.75	0.52	0.61	734,295
Non-irrigated vineyard	0.74	0.87	0.80	2,111,911
Irrigated olive trees	0.87	0.64	0.74	43,715
Non-irrigated olive trees	0.63	0.34	0.44	321,381
Irrigated shell fruit trees	0.78	0.64	0.70	514,900
Non-irrigated shell fruit trees	0.57	0.73	0.64	888,792
Irrigated fruit trees	0.68	0.89	0.77	6974
Non-irrigated fruit trees	0.35	0.14	0.20	62,444
Accuracy			0.83	11,030,736
Weighted avg.	0.83	0.83	0.83	

3.3. Multilayer Perceptron with PCA

The multilayer perceptron (MLP) algorithm was applied on its own, without preprocessing or cleaning the data, and we found that the model performed poorly. The application of PCA to the spectro-temporal sequences resulted in a notable gain in the performance of the MLP in the training process, with the advantage of its ability to reduce the number of features per sample and the noise of the dataset. As shown in Table 7, the most basic neural network outperforms the results from the decision tree and random forest, with an 87% weighted *F1*. Observing the *F1* in each class, it can be inferred that both the *precision* and the *recall* are balanced, with promising results in the case of herbaceous irrigated crops.

Table 7. Land use classification: MLP + PCA.

Class	Precision	Recall	F1	Support
Forage irrigated	0.86	0.85	0.86	343,643
Spring irrigated	0.96	0.94	0.95	2,373,450
Spring irrigated Chinese garlic	0.95	0.95	0.95	546,212
Spring irrigated purple garlic and horticulture	0.97	0.98	0.97	235,297
Late spring irrigated	0.88	0.92	0.90	141,147
Summer irrigated, low cover and horticulture	0.93	0.93	0.93	455,175
Summer irrigated, high coverage	0.96	0.94	0.95	585,448
Irrigated alfalfa	0.97	0.97	0.97	445,925
Double crop	0.93	0.94	0.94	572,968
Herbaceous non-irrigated	0.81	0.80	0.81	569,945
Bare soil	0.86	0.95	0.90	77,114
Irrigated vineyard	0.72	0.70	0.71	734,295
Non-irrigated vineyard	0.86	0.85	0.85	2,111,911
Irrigated olive trees	0.70	0.83	0.76	43,715
Non-irrigated olive trees	0.70	0.55	0.62	321,381
Irrigated shell fruit trees	0.78	0.80	0.79	514,900
Non-irrigated shell fruit trees	0.72	0.79	0.76	888,792
Irrigated fruit trees	0.39	0.96	0.56	6974
Non-irrigated fruit trees	0.22	0.33	0.26	62,444
Accuracy			0.87	11,030,736
Weighted avg.	0.87	0.87	0.87	

3.4. Convolutional Neural Network

The report on the crop classification (Table 8) with the use of deep learning shows a substantial improvement compared to the previous results. In this case, the CNNs behaved as expected by recognizing temporal patterns in the time series of the reflectance bands, represented as a 2D array. In other words, the temporal fingerprint per pixel is treated as an image. This raises the overall accuracy to 91%, performing better than the other models. The highest performance is obtained in irrigated herbaceous crops, with the *F1* scores ranging from 93% to 98%. As for the non-irrigated herbaceous crops and bare soil, the *F1* is around 90%, significantly higher than the rest of the presented models. Regarding orchards, it shows the best performance in the detection of vineyard and shell fruit trees, with an *F1* between 80% and 89%. In the case of olive trees, the detection is better, with an F1 of 81% if irrigation is present and 65% if not. The *F1* drops dramatically in the classes of fruit trees, not surpassing 67%. This is due to the lower number of pixels of these classes in the dataset. This issue will be explained further in the discussion. A sample of the classification is included in Figure 5. Figure 6 shows the results of the classification in this area compared to the available ground truth data.

Considering the lower F1 in the categories corresponding to orchards, it is worth verifying this in the confusion matrix (Figure 7). As can be observed in detail, those cases of misclassification mainly refer to crops of the same two major aggregated categories, herbaceous crops and orchards, with no mismatch between them. Moreover, it is also relevant that irrigated crops tend to be misclassified, with other irrigated crops and non-irrigated crops repeating this same pattern. Therefore, we considered using the same architecture to build a model with the sole purpose of identifying irrigated soil as an aggregated category.

Table 8. Land use classification CNN. Pixel-based.

Class	Precision	Recall	F1	Support
Forage irrigated	0.93	0.93	0.93	343,643
Spring irrigated	0.97	0.97	0.97	2,373,450
Spring irrigated Chinese garlic	0.97	0.96	0.97	546,212
Spring irrigated purple garlic and horticulture	0.98	0.98	0.98	235,297
Late spring irrigated	0.95	0.96	0.95	141,147
Summer irrigated, low cover and horticulture	0.96	0.95	0.96	455,175
Summer irrigated, high coverage	0.97	0.96	0.97	585,448
Irrigated alfalfa	0.98	0.98	0.98	445,925
Double crop	0.96	0.96	0.96	572,968
Herbaceous non-irrigated	0.89	0.89	0.89	569,945
Bare soil	0.89	0.98	0.93	77,114
Irrigated vineyard	0.82	0.79	0.80	734,295
Non-irrigated vineyard	0.89	0.89	0.89	2,111,911
Irrigated olive trees	0.73	0.91	0.81	43,715
Non-irrigated olive trees	0.66	0.64	0.65	321,381
Irrigated shell fruit trees	0.87	0.87	0.87	514,900
Non-irrigated shell fruit trees	0.82	0.82	0.82	888,792
Irrigated fruit trees	0.52	0.97	0.67	6974
Non-irrigated fruit trees	0.31	0.41	0.36	62,444
Accuracy			0.91	11,030,736
Weighted avg.	0.91	0.91	0.91	

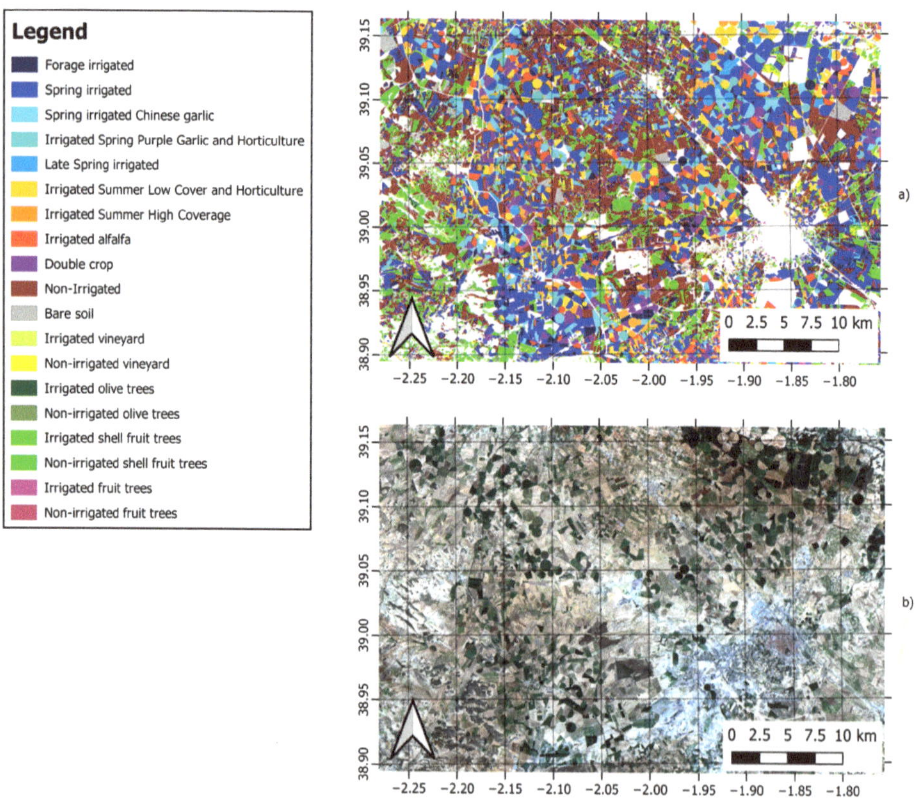

Figure 5. (**a**) Land use classification among 19 classes. (**b**) View with RGB bands from Sentinel-2 image in Albacete, Spain (coordinates in EPSG: 4326).

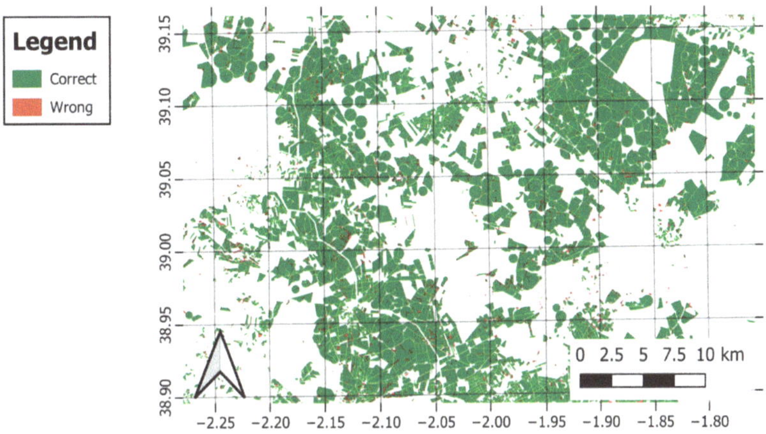

Figure 6. Classification results compared to ground truth data (coordinates in EPSG: 4326).

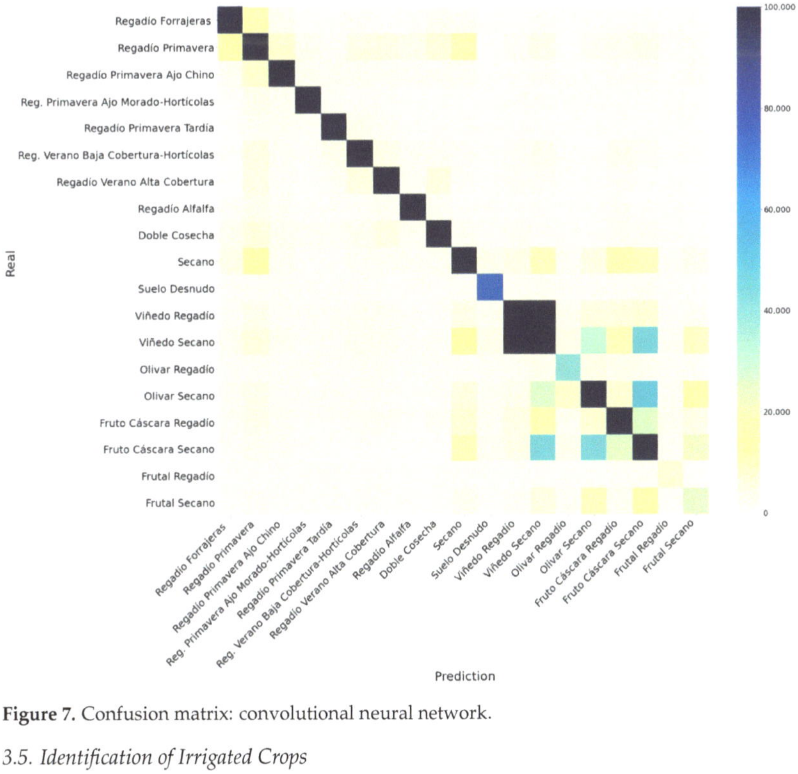

Figure 7. Confusion matrix: convolutional neural network.

3.5. *Identification of Irrigated Crops*

As can be observed, the performance of the model based on CNNs in detecting irrigated crops separately seems to be high. The original categories were grouped into three classes, irrigated, non-irrigated, and bare ground, to build and train another model with the same architecture. The results (Table 9) for the identification of the irrigated soil are outstanding, since the most relevant class is the one corresponding to irrigated crops, and 96% of pixels are detected, as the *recall* shows. At the same time, the model prediction is right 97% of the time, when it classifies a pixel as irrigated. A sample of the classification is included in Figure 8. Both the overall accuracy and the average *F1-score* are 0.95.

Table 9. Classification of irrigated, non-irrigated, and bare ground. Pixel-based.

Class	Precision	Recall	F1	Support
Irrigated	0.97	0.96	0.96	6,999,149
Non-irrigated	0.93	0.95	0.94	3,954,473
Bare ground	0.92	0.98	0.95	77,114
Accuracy			0.95	11,030,736
Weighted avg.	0.95	0.95	0.95	

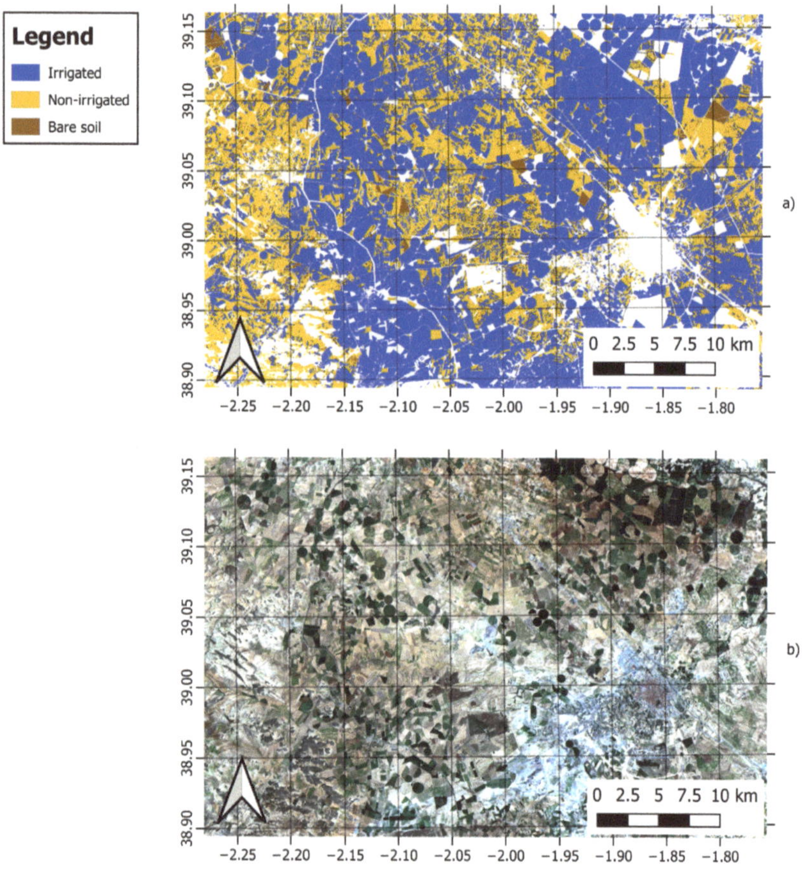

Figure 8. (**a**) Land use classification per irrigation management. (**b**) View with RGB bands from Sentinel-2 image in Albacete-Spain (Coordinates in EPSG: 4326).

3.6. Analysis at the Plot Level

Since the results of the deep neural network with the use of convolutional layers offer the best statistics among all the tested models, the previously exposed post-processing algorithm was applied to the classification generated using the CNN, thereby integrating the pixel-based classification into a plot level. This resulted in a slight overall improvement (Table 10), which is mainly noticeable for the orchards, with a higher F1-score, whereas it is slightly lower in the case of the herbaceous irrigated crops. A sample of the classification is included in Figure 9.

Table 10. Land use classification CNN. Post-processing.

Class	Precision	Recall	F1	Support
Forage irrigated	0.91	0.90	0.91	343,643
Spring irrigated	0.94	0.97	0.95	2,373,450
Spring irrigated Chinese garlic	0.96	0.93	0.95	546,212
Spring irrigated purple garlic and horticulture	0.96	0.95	0.96	235,297
Late spring irrigated	0.93	0.84	0.88	141,147
Summer irrigated, low cover and horticulture	0.94	0.91	0.93	455,175
Summer irrigated, high coverage	0.97	0.93	0.95	585,448
Irrigated alfalfa	0.96	0.97	0.96	445,925
Double crop	0.92	0.94	0.93	572,968
Herbaceous non-irrigated	0.93	0.93	0.93	569,945
Bare soil	0.94	1.00	0.97	77,114
Irrigated vineyard	0.88	0.84	0.86	734,295
Non-irrigated vineyard	0.89	0.93	0.91	2,111,911
Irrigated olive trees	0.84	0.94	0.89	43,715
Non-irrigated olive trees	0.76	0.64	0.69	321,381
Irrigated shell fruit trees	0.92	0.93	0.92	514,900
Non-irrigated shell fruit trees	0.87	0.86	0.87	888,792
Irrigated fruit trees	0.72	0.99	0.83	6974
Non-irrigated fruit trees	0.45	0.43	0.44	62,444
Accuracy			0.92	11,030,736
Weighted avg.	0.91	0.92	0.91	

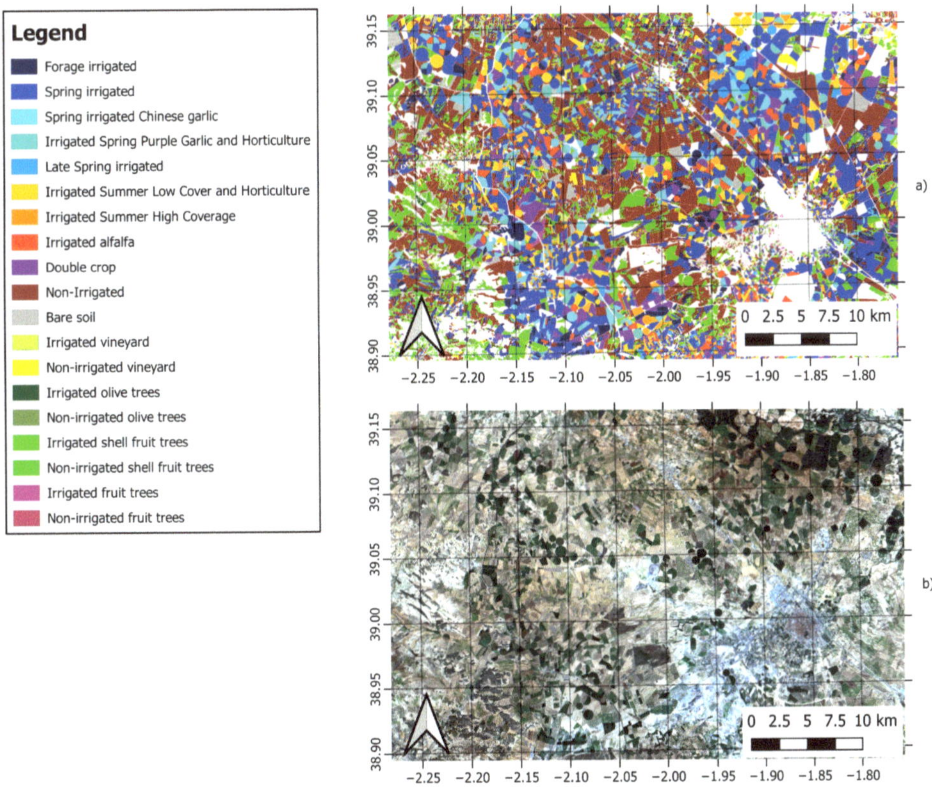

Figure 9. (**a**) Land use classification among 19 classes. Post-processing. (**b**) View with RGB bands from Sentinel-2 image in Albacete, Spain (coordinates in EPSG: 4326).

After applying post-processing, in the case of the classification of irrigated crops, the results were slightly improved compared to those obtained using the base algorithm, as shown in Table 11.

Table 11. Classification of irrigated, non-irrigated, and bare ground. Post-processing.

Class	Precision	Recall	F1	Support
Irrigated	0.98	0.97	0.98	6,999,149
Non-irrigated	0.95	0.96	0.96	3,954,473
Bare ground	0.96	1.00	0.98	77,114
Accuracy			0.97	11,030,736
Weighted avg.	0.97	0.97	0.97	

The *recall* shows a 97% accuracy in detecting the irrigated pixels, whereas 98% of the pixels classified within the irrigated category actually correspond to that class. A sample of the classification is included in Figure 10.

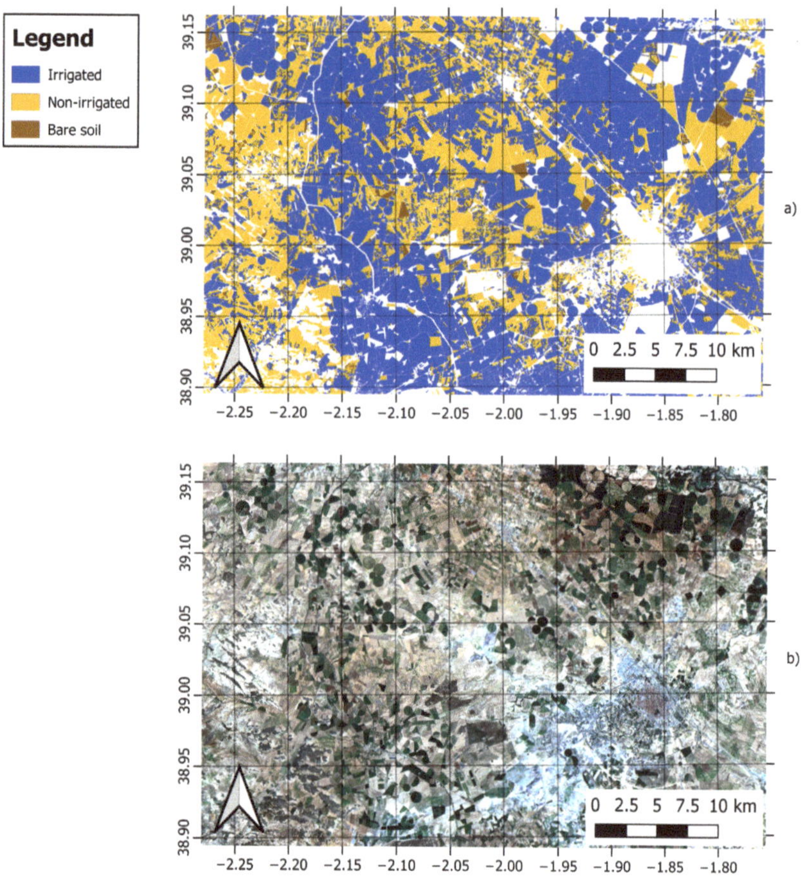

Figure 10. (**a**) Land use classification per irrigation management. Post-processing. (**b**) View with RGB bands from Sentinel-2 image in Albacete, Spain (Coordinates in EPSG: 4326).

4. Discussion

The core information for the crop classification comes from the Sentinel 2A and 2B constellations of the Copernicus program, which, nowadays, provide a high temporal and spatial resolution. A total of 24 images gathered on different dates from March to October were used. Despite observing a considerable time gap between May and June in this application, such a situation did not affect the performance of the model, as shown by the results. Additional data from different sources could be added, such as Landsat or national orthophotography program images and local sensors, so as to provide biophysical variables. For operational purposes, according to the computational resources, and to ensure the future scalability to other areas, Sentinel-2 sensors provide enough information for a successful classification for the purpose of pursuing an automatic process.

Our main objective was to represent the sequential data as a 2D spectro-temporal fingerprint of each pixel, which is particularly interesting with respect to its processing with machine learning. Therefore, it was tested using several algorithms, including CNNs, reducing the number of parameters for the fully connected layers and thus obtaining an enhancement of the speed and a reduction in the complexity of the model used to train and evaluate the data. The process of classification, designed to generate a land use cover, considers only the multispectral and temporal dimensions of the information at the pixel level.

The convolutional neural network output was compared to those generated using other, simpler algorithms in global terms (Table 12) and compared by considering each separate class (Table 13). CNNs were proved to outperform any other model applied to the experimental data considered in this study with a similar prediction time. Since the training phase can only take place once the model has been updated, the improvement in global accuracy highly compensates for the training time, which is important for generating reliable land use classification maps. Both the overall accuracy and weighted average F1 are affected by all the results achieved for every class. In general, the performance of all the models considered is high. However, considering the resolution of the satellite images, we expected this to result in a slightly lower performance in the case of orchards. The mismatches between these classes correspond mainly to the same aggregated categories distinguishing between irrigated and non-irrigated crops. Considering the importance of identifying irrigated crops, a specific model for this purpose was considered, using the same architecture, to identify the irrigated land use, showing a 95% accuracy and a 95% F1, whereas it increased to 97% for both metrics when post-processing was applied.

Table 12. Statistics for the comparison of the five selected models used to identify 19 classes.

Model	Metrics					
	19 Classes Accuracy	19 Classes Weighted Avg. F1	Irrigation Detection Accuracy	Irrigation Detection Weighted Avg. F1	Fitting Time (s)	Prediction Time Per Million Samples (s)
Decision Tree	73%	74%			448	0.60
Random Forest	83%	83%			257	12.74
PCA + MLP	87%	87%			1505	11.69
CNN	91%	91%	95%	95%	11,576	13.76
CNN w/Postprocessing	92%	91%	97%	97%		

Table 13. Comparison of F1s for each model per class.

Class	F1			
	DT	RF	PCA + MLP	CNN
Forage irrigated	0.64	0.85	0.86	0.93
Spring irrigated	0.91	0.95	0.95	0.97
Spring irrigated Chinese garlic	0.82	0.92	0.95	0.97
Spring irrigated purple garlic and horticulture	0.89	0.97	0.97	0.98
Late spring irrigated	0.68	0.87	0.90	0.95

Table 13. *Cont.*

Class	F1			
	DT	RF	PCA + MLP	CNN
Summer irrigated, low cover and horticulture	0.83	0.92	0.93	0.96
Summer irrigated, high coverage	0.88	0.95	0.95	0.97
Irrigated alfalfa	0.93	0.97	0.97	0.98
Double crop	0.83	0.93	0.94	0.96
Herbaceous non-irrigated	0.60	0.76	0.81	0.89
Bare soil	0.58	0.82	0.90	0.93
Irrigated vineyard	0.53	0.61	0.71	0.80
Non-irrigated vineyard	0.73	0.80	0.85	0.89
Irrigated olive trees	0.48	0.74	0.76	0.81
Non-irrigated olive trees	0.36	0.44	0.62	0.65
Irrigated shell fruit trees	0.51	0.70	0.79	0.87
Non-irrigated shell fruit trees	0.54	0.64	0.76	0.82
Irrigated fruit trees	0.32	0.77	0.56	0.67
Non-irrigated fruit trees	0.12	0.20	0.26	0.36

According to the proposals shown in Table 14, different types of models can be used in land use classification. These include traditional models based on decision trees and random forests, which constitute one of the most successful machine learning methods and deep learning models based on the use of recurrent neural networks, such as long short-term memory and convolutional neural networks.

Table 14. Previous proposals' overall performance.

Proposal	RS	Features	Model	Accuracy	Classes
Ours	Sentinel-2	BOA ref.	Conv2D	91%	19
Portalés-Julià et al. [32]	Sentinel-2	Ref. and BSI	Bi-LSTM	98.2%	9
Campos-Taberner et al. [31]	Sentinel-2	Ref. and NDVI	Bi-LSTM	98.6%	16
Fan et al. [58]	Sentinel-2	BOA ref.	RF	96–98%	9
Zhong et al. [23]	Landsat	EVI	Conv1D	85.5%	14
Rußwurm and Körner [30]	Sentinel-2	TOA ref.	Bi-ConvLSTM	89.6%	17
Hao et al. [19]	MODIS	Phenological metrics	RF	89%	6
Foerster et al. [13]	Landsat	NDVI	DT	73%	11

Similar works applied to a set of images with pixel sizes larger than the plot showed a low accuracy of 50% in crop classification in some areas because of the presence of trees in the fields and the lack of resolution [9]. In light of this, the input information was filtered with a minimum threshold of 0.3 ha per plot, as explained in the methodology.

Hao et al. proposed a model which employs the phenological features obtained from a MODIS time series, whose resolution is lower than that of Landsat and Sentinel-2 images. They aimed to classify a reduced group of six classes, achieving an overall accuracy of 89% with a model based on random forest [19].

As Fan et al. demonstrated [58], the data volume of the Sentinel 2 images is large compared to the data provided by other satellites because of the medium-high resolution of the sensors. Therefore, we considered a similar solution, optimizing the volume of the training, while obtaining data to enable the model to learn without making the training set too large to be processed. Their model consists of a random forest that classifies the land use among nine classes. In this respect, it is necessary to point out that, when comparing algorithms, a smaller number of classes leads to simpler and more accurate models. In our proposal, we used CNNs inspired by Zhong et al.'s work [23], consisting of Conv1D layers that can be used to extract temporal patterns from an enhanced vegetation index (EVI) time series for 14 classes, whereas our model used Conv2D layers to extract temporal and multispectral patterns from the 2D fingerprint so as to classify 19 classes.

Rußwurm and Körner proposed a model whose input is derived from a Sentinel-2 top-of-atmosphere (TOA) time series and a maximum cloud level of 80%. They employed a Bi-ConvLSTM model, which obtained an 89.6% global accuracy in classifying 17 herbaceous crops, considering the spatial and temporal distribution [30]. In contrast, our proposal establishes a threshold of 10% for the presence of clouds in the image in order to consider it acceptable and employs a bottom-of-atmosphere (BOA) time series to perform the classification of both herbaceous crops and orchards, considering only the temporal distribution per pixel.

Campos-Taberner et al. presented a study which aimed to explain deep learning, in which they employed a bidirectional LSTM model [31]. Portalés-Julià et al. made improvements on their previous work, in which they continued using Bi-LSTM models [32], obtaining a 94.3% accuracy using random forests and an over 98% accuracy in their study area using Bi-LSTM networks. This may be the result of both the fact that the model did not discriminate irrigated areas and classified fewer classes. The use of LSTMs leads to higher computational costs in terms of time and resources compared to CNNs, which we used in our proposal with the aim of achieving an efficient model with a reduced number of parameters and a similar performance [36].

5. Conclusions

This work offered a comparison of different deep learning algorithms, which were used to identify 19 agricultural land use classes in the area of Albacete, Spain, a well-known testing area for the use of traditional decision tree algorithms. For this purpose, each pixel was characterized as a 2D fingerprint using a sequence of one-granule Sentinel 2 multispectral data over the year of 2018, developing traditional machine learning and deep neural networks models in order to classify the land use as herbaceous and orchards crops and consider whether they are irrigated or not. The best results were achieved by the deep neural networks containing two-dimensional convolutional layers (Conv2D), which outperformed the other accepted classifiers, such as random forests and multilayer perceptron, whose results were good in previous works. This model, based on CNNs, performs a pixel-based classification that analyzes the multispectral and temporal components, proving that it can obtain a high overall accuracy of 91%, regardless of the spatial distribution, by identifying multispectral-temporal patterns in the 2D pixel fingerprint. The accuracy of the model is higher in case of the herbaceous crops than the orchards, a which is even more notable if irrigation is applied. Additionally, the model we built to detect irrigated areas detects 97% of the irrigated areas with a 98% *precision*. Considering these achievements, a model based on two-dimensional convolutional layers shows promising results and potential to be applied to the area in question. This model is trained for a specific area; thus, the crops or the characteristics of other locations are not within its knowledge. For that purpose, additional possibilities for future work can be conceived in two different ways. The first option is to re-train the model with new samples from the other area, so that its weight changes to fit the new data. A second option is to take the layers of this model as a base in order to build a new one. These choices would allow for the application of transfer learning and take advantage of this previous work. This also implies an improvement, since experts in agronomy are currently performing heavy manual work in order to make classifications, which is a task that can take months to complete [18].

Author Contributions: The conceptualization, methodology, and validation were conducted by all the authors. Software, L.d.l.O. and A.-M.S.S.; resources and data curation, A.C., J.G.-P. and A.-M.S.S.; writing—original draft preparation, A.-M.S.S.; writing—review and editing, J.G.-P. and L.d.l.O.; visualization, A.-M.S.S. and L.d.l.O.; supervision, L.d.l.O. and J.G.-P.; project administration and funding acquisition, J.G.-P. and L.d.l.O. All authors have read and agreed to the published version of the manuscript.

Funding: This research was funded by project REXUS (H2020 EU, ref. 101003632), IRENE (Spanish Ministry of Research, PID2020-113498RB-C21), the Predoctoral Grant from UCLM (2020-PREDUCLM-16149), ERDF, A Way of Making Europe (SBPLY/21/180501/000148), and EO_TIME (Spanish Ministry of Research, PCI2018-093140).

Institutional Review Board Statement: Not applicable.

Informed Consent Statement: Not applicable.

Acknowledgments: Special thanks go to David Sánchez Pérez, Irene López Arellano, and the Central Board of Irrigators for providing field data for the validation.

Conflicts of Interest: The authors declare no conflict of interest.

References

1. Dubois, O.; Faurès, J.; Felix, E.; Flammini, A.; Hoogeveen, J.; Pluschke, L.; Puri, M.; Ünver, O. The Water-Energy-Food Nexus. A New Approach in Support of Food Security and Sustainable Agriculture | Policy Support and Governance | Food and Agriculture Organization of the United Nations. Available online: https://www.fao.org/policy-support/tools-and-publications/resources-details/es/c/421718/ (accessed on 16 November 2021).
2. Deng, L.; Guo, S.; Yin, J.; Zeng, Y.; Chen, K. Multi-objective optimization of water resources allocation in Han River basin (China) integrating efficiency, equity and sustainability. *Sci. Rep.* **2022**, *12*, 798. [CrossRef] [PubMed]
3. Drakvik, E.; Kogevinas, M.; Bergman, Å.; Devouge, A.; Barouki, R.; on behalf of the HERA (Health and Environment Research Agenda) Consortium; Kogevinas, M.; Bergman, Å.; Drakvik, E. Priorities for research on environment, climate and health, a European perspective. *Environ. Health* **2022**, *21*, 37. [CrossRef] [PubMed]
4. Elliot, T.; Torres-Matallana, J.A.; Goldstein, B.; Babí Almenar, J.; Gómez-Baggethun, E.; Proença, V.; Rugani, B. An expanded framing of ecosystem services is needed for a sustainable urban future. *Renew. Sustain. Energy Rev.* **2022**, *162*, 112418. [CrossRef]
5. Matton, N.; Canto, G.S.; Waldner, F.; Valero, S.; Morin, D.; Inglada, J.; Arias, M.; Bontemps, S.; Koetz, B.; Defourny, P. An Automated Method for Annual Cropland Mapping along the Season for Various Globally-Distributed Agrosystems Using High Spatial and Temporal Resolution Time Series. *Remote Sens.* **2015**, *7*, 13208–13232. [CrossRef]
6. Gómez, C.; White, J.C.; Wulder, M.A. Optical remotely sensed time series data for land cover classification: A review. *ISPRS J. Photogramm. Remote Sens.* **2016**, *116*, 55–72. [CrossRef]
7. Leemhuis, C.; Thonfeld, F.; Näschen, K.; Steinbach, S.; Muro, J.; Strauch, A.; López, A.; Daconto, G.; Games, I.; Diekkrüger, B. Sustainability in the Food-Water-Ecosystem Nexus: The Role of Land Use and Land Cover Change for Water Resources and Ecosystems in the Kilombero Wetland, Tanzania. *Sustainability* **2017**, *9*, 1513. [CrossRef]
8. Sánchez, J.M.; Galve, J.M.; González-Piqueras, J.; López-Urrea, R.; Niclòs, R.; Calera, A. Downscaling MODIS land surface temperature to Sentinel-2 spatial resolution in the Barrax test site. *SPIE* **2019**, *11174*, 317–323.
9. Inglada, J.; Arias, M.; Tardy, B.; Hagolle, O.; Valero, S.; Morin, D.; Dedieu, G.; Sepulcre, G.; Bontemps, S.; Defourny, P.; et al. Assessment of an Operational System for Crop Type Map Production Using High Temporal and Spatial Resolution Satellite Optical Imagery. *Remote Sens.* **2015**, *7*, 12356–12379. [CrossRef]
10. Vuolo, F.; Neuwirth, M.; Immitzer, M.; Atzberger, C.; Ng, W.-T. How much does multi-temporal Sentinel-2 data improve crop type classification? *Int. J. Appl. Earth Obs. Geoinf.* **2018**, *72*, 122–130. [CrossRef]
11. Carrao, H.; Gonçalves, P.; Caetano, M. Contribution of multispectral and multitemporal information from MODIS images to land cover classification. *Remote Sens. Environ.* **2008**, *112*, 986–997. [CrossRef]
12. Friedl, M.A.; McIver, D.K.; Hodges, J.C.F.; Zhang, X.Y.; Muchoney, D.; Strahler, A.H.; Woodcock, C.E.; Gopal, S.; Schneider, A.; Cooper, A.; et al. Global land cover mapping from MODIS: Algorithms and early results. *Remote Sens. Environ.* **2002**, *83*, 287–302. [CrossRef]
13. Foerster, S.; Kaden, K.; Förster, M.; Itzerott, S. Crop type mapping using spectral-temporal profiles and phenological information. *Comput. Electron. Agric.* **2012**, *89*, 30–40. [CrossRef]
14. Revill, A.; Florence, A.; MacArthur, A.; Hoad, S.P.; Rees, R.M.; Williams, M. The Value of Sentinel-2 Spectral Bands for the Assessment of Winter Wheat Growth and Development. *Remote Sens.* **2019**, *11*, 2050. [CrossRef]
15. Huang, S.; Tang, L.; Hupy, J.P.; Wang, Y.; Shao, G. A commentary review on the use of normalized difference vegetation index (NDVI) in the era of popular remote sensing. *J. For. Res.* **2021**, *32*, 1–6. [CrossRef]
16. Silleos, N.; Misopolinos, N.; Perakis, K. Relationships between remote sensing spectral indices and crops discrimination. *Geocarto Int.* **1992**, *7*, 41–51. [CrossRef]
17. Gilabert, M.A.; González-Piqueras, J.; García-Haro, J. Acerca de los Indices de Vegetación. *Rev. Teledetec.* **1997**, *8*, 10.
18. Conrad, C.; Fritsch, S.; Zeidler, J.; Rücker, G.; Dech, S. Per-Field Irrigated Crop Classification in Arid Central Asia Using SPOT and ASTER Data. *Remote Sens.* **2010**, *2*, 1035–1056. [CrossRef]
19. Hao, P.; Zhan, Y.; Wang, L.; Niu, Z.; Shakir, M. Feature Selection of Time Series MODIS Data for Early Crop Classification Using Random Forest: A Case Study in Kansas, USA. *Remote Sens.* **2015**, *7*, 5347–5369. [CrossRef]
20. Immitzer, M.; Vuolo, F.; Atzberger, C. First experience with Sentinel-2 data for crop and tree species classifications in central Europe. *Remote Sens.* **2016**, *8*, 166. [CrossRef]

21. Hao, P.; Wang, L.; Niu, Z.; Aablikim, A.; Huang, N.; Xu, S.; Chen, F. The Potential of Time Series Merged from Landsat-5 TM and HJ-1 CCD for Crop Classification: A Case Study for Bole and Manas Counties in Xinjiang, China. *Remote Sens.* **2014**, *6*, 7610–7631. [CrossRef]
22. Zhou, F.; Aining, Z.; Townley-Smith, L. A data mining approach for evaluation of optimal time-series of MODIS data for land cover mapping at a regional level. *ISPRS J. Photogramm. Remote Sens.* **2013**, *84*, 114–129. [CrossRef]
23. Zhong, L.; Hu, L.; Zhou, H. Deep learning based multi-temporal crop classification. *Remote Sens. Environ.* **2019**, *221*, 430–443. [CrossRef]
24. Valueva, M.V.; Nagornov, N.N.; Lyakhov, P.A.; Valuev, G.V.; Chervyakov, N.I. Application of the residue number system to reduce hardware costs of the convolutional neural network implementation. *Math. Comput. Simul.* **2020**, *177*, 232–243. [CrossRef]
25. Li, X.; Wu, X. Constructing Long Short-Term Memory based Deep Recurrent Neural Networks for Large Vocabulary Speech Recognition. *arXiv* **2015**, arXiv:Computer Science/1410.4281. [CrossRef]
26. Lyu, H.; Lu, H.; Mou, L. Learning a Transferable Change Rule from a Recurrent Neural Network for Land Cover Change Detection. *Remote Sens.* **2016**, *8*, 506. [CrossRef]
27. Mou, L.; Bruzzone, L.; Zhu, X.X. Learning Spectral-Spatial-Temporal Features via a Recurrent Convolutional Neural Network for Change Detection in Multispectral Imagery. *IEEE Trans. Geosci. Remote Sens.* **2019**, *57*, 924–935. [CrossRef]
28. Krizhevsky, A.; Sutskever, I.; Hinton, G.E. ImageNet classification with deep convolutional neural networks. *Commun. ACM* **2012**, *2*, 1097–1105. [CrossRef]
29. Rußwurm, M.; Korner, M. Temporal Vegetation Modelling Using Long Short-Term Memory Networks for Crop Identification from Medium-Resolution Multi-spectral Satellite Images. In Proceedings of the 2017 IEEE Conference on Computer Vision and Pattern Recognition Workshops (CVPRW), Honolulu, HI, USA, 21 July 2017; pp. 1496–1504.
30. Rußwurm, M.; Körner, M. Multi-Temporal Land Cover Classification with Sequential Recurrent Encoders. *ISPRS Int. J. Geo Inf.* **2018**, *7*, 129. [CrossRef]
31. Campos-Taberner, M.; García-Haro, F.J.; Martínez, B.; Gilabert, M.A. Deep learning for agricultural land use classification from Sentinel-2. *Rev. Teledetec.* **2020**, *56*, 35–48. [CrossRef]
32. Portalés-Julià, E.; Campos-Taberner, M.; García-Haro, F.J.; Gilabert, M.A. Assessing the sentinel-2 capabilities to identify abandoned crops using deep learning. *Agronomy* **2021**, *11*, 654. [CrossRef]
33. Ruiz, L.A.; Almonacid-Caballer, J.; Crespo-Peremarch, P.; Recio, J.A.; Pardo-Pascual, J.E.; Sánchez-García, E. Automated classification of crop types and condition in a mediterranean area using a fine-tuned convolutional neural network. *Int. Soc. Photogramm. Remote Sens.* **2020**, *43*, 1061–1068. [CrossRef]
34. Amani, M.; Kakooei, M.; Moghimi, A.; Ghorbanian, A.; Ranjgar, B.; Mahdavi, S.; Davidson, A.; Fisette, T.; Rollin, P.; Brisco, B.; et al. Application of Google Earth Engine Cloud Computing Platform, Sentinel Imagery, and Neural Networks for Crop Mapping in Canada. *Remote Sens.* **2020**, *12*, 3561. [CrossRef]
35. Dempster, A.; Petitjean, F.; Webb, G.I. ROCKET: Exceptionally fast and accurate time series classification using random convolutional kernels. *Data Min. Knowl. Discov.* **2019**, *34*, 1454–1495. [CrossRef]
36. Weytjens, H.; De Weerdt, J. Process Outcome Prediction: CNN vs. LSTM (with Attention). *Int. Conf. Bus. Process Manag.* **2020**, *397*, 321–333. [CrossRef]
37. Murmu, S.; Biswas, S. Application of Fuzzy Logic and Neural Network in Crop Classification: A Review. *Aquat. Procedia* **2015**, *4*, 1203–1210. [CrossRef]
38. Garrido-Rubio, J.; González-Piqueras, J.; Campos, I.; Osann, A.; González-Gómez, L.; Calera, A. Remote sensing–based soil water balance for irrigation water accounting at plot and water user association management scale. *Agric. Water Manag.* **2020**, *238*, 106236. [CrossRef]
39. Campos, I.; González-Gómez, L.; Villodre, J.; González-Piqueras, J.; Suyker, A.E.; Calera, A. Remote sensing-based crop biomass with water or light-driven crop growth models in wheat commercial fields. *Field Crops Res.* **2018**, *216*, 175–188. [CrossRef]
40. Anuradha, T.; Tigadi, A.; Ravikumar, M.; Nalajala, P.; Hemavathi, S.; Dash, M. Feature Extraction and Representation Learning via Deep Neural Network. In Proceedings of the Computer Networks, Big Data and IoT; Pandian, A.P., Fernando, X., Haoxiang, W., Eds.; Springer Nature: Singapore, 2022; pp. 551–564.
41. Campos-Taberner, M.; García-Haro, F.J.; Martínez, B.; Izquierdo-Verdiguier, E.; Atzberger, C.; Camps-Valls, G.; Gilabert, M.A. Understanding deep learning in land use classification based on Sentinel-2 time series. *Sci. Rep.* **2020**, *10*, 17188. [CrossRef]
42. González-Piqueras, J.; Rubio, E.; Calera, A.; Moratalla, A. Intensive field campaigns in the framework of demeter project. *AIP Conf. Proc.* **2006**, *852*, 67–74.
43. Calera, A.; González-Piqueras, J.; Melia, J. Monitoring barley and corn growth from remote sensing data at field scale. *Int. J. Remote Sens.* **2004**, *25*, 97–109. [CrossRef]
44. Sánchez, J.M.; Galve, J.M.; González-Piqueras, J.; López-Urrea, R.; Niclòs, R.; Calera, A. Monitoring 10-m LST from the combination MODIS/Sentinel-2, validation in a high contrast semi-arid agroecosystem. *Remote Sens.* **2020**, *12*, 1453. [CrossRef]
45. Solano-Correa, Y.T.; Bovolo, F.; Bruzzone, L.; Fernandez-Prieto, D. A Method for the Analysis of Small Crop Fields in Sentinel-2 Dense Time Series. *IEEE Trans. Geosci. Remote Sens.* **2020**, *58*, 2150–2164. [CrossRef]
46. Breiman, L.; Friedman, J.; Stone, C.J.; Olshen, R.A. *Classification and Regression Trees*; Taylor & Francis: New York, NY, USA, 1984; ISBN 978-0-412-04841-8.
47. Breiman, L. Random Forests. *Mach. Learn.* **2001**, *45*, 5–32. [CrossRef]

48. Matheus, C.J.; Hohensee, W.E. Learning in artificial neural systems. *Comput. Intell.* **1987**, *3*, 283–294. [CrossRef]
49. Bebis, G.; Georgiopoulos, M. Feed-forward neural networks. *IEEE Potentials* **1994**, *13*, 27–31. [CrossRef]
50. Multilayer Perceptron. Available online: https://deepai.org/machine-learning-glossary-and-terms/multilayer-perceptron (accessed on 15 December 2021).
51. Bro, R.; Smilde, A.K. Principal component analysis. *Anal. Methods* **2014**, *6*, 2812–2831. [CrossRef]
52. Scikit-Learn: Machine Learning in Python—Scikit-Learn 1.0.1 Documentation. Available online: https://scikit-learn.org/stable/ (accessed on 20 December 2021).
53. TensorFlow Core | Machine Learning for Beginners and Experts. Available online: https://www.tensorflow.org/overview (accessed on 1 January 2022).
54. Ripley, B.D. *Pattern Recognition and Neural Networks*; Cambridge University Press: Cambridge, UK, 1996; ISBN 978-0-521-71770-0.
55. King, G.; Zeng, L. Logistic Regression in Rare Events Data. *Polit. Anal.* **2001**, *9*, 137–163. [CrossRef]
56. Glossary of Terms Journal of Machine Learning. Available online: http://robotics.stanford.edu/~{}ronnyk/glossary.html (accessed on 20 December 2021).
57. Zhao, C.; Yang, J.; Shi, H.; Chen, T. Transforming approach for assessing the performance and applicability of rice arsenic contamination forecasting models based on regression and probability methods. *J. Hazard. Mater.* **2022**, *424*, 127375. [CrossRef]
58. Fan, J.; Zhang, X.; Zhao, C.; Qin, Z.; De Vroey, M.; Defourny, P. Evaluation of crop type classification with different high resolution satellite data sources. *Remote Sens.* **2021**, *13*, 911. [CrossRef]

remote sensing

MDPI

Article

Paddy Rice Phenological Mapping throughout 30-Years Satellite Images in the Honghe Hani Rice Terraces

Jianbo Yang [1,2,3], Jianchu Xu [1,2,4], Ying Zhou [5,*], Deli Zhai [6], Huafang Chen [1,2], Qian Li [1,2,3] and Gaojuan Zhao [1,2]

1 Center for Mountain Futures, Kunming Institute of Botany, Chinese Academy of Sciences, Kunming 650201, China
2 Department of Economic Plants and Biotechnology, Yunnan Key Laboratory for Wild Plant Resources, Kunming Institute of Botany, Chinese Academy of Sciences, Kunming 650201, China
3 University of Chinese Academy of Sciences, Beijing 100049, China
4 CIFOR-ICRAF China Program, World Agroforestry (ICRAF), Kunming 650201, China
5 Institute of Agricultural Resources and Regional Planning, Chinese Academy of Agricultural Sciences, Beijing 100081, China
6 CAS Key Laboratory of Tropical Forest Ecology, Xishuangbanna Tropical Botanical Garden, Chinese Academy of Sciences, Mengla 666303, China
* Correspondence: zhouying@caas.cn

Abstract: The Honghe Hani Rice Terraces represent the coexistence between natural and cultural systems. Despite being listed as a World Heritage Site in 2013, certain natural and anthropogenic factors have changed land use/land cover, which has led to a reduction in the size of the paddy rice area. It is difficult to accurately assess these changes due to the lack of historical maps of paddy rice croplands with fine spatial resolution. Therefore, we integrated a random forest classifier and phenological information to improve mapping accuracy and stability. We then mapped the historical distribution of land use/land cover in the Honghe Hani Rice Terraces from 1989–1991 to 2019–2021 using the Google Earth Engine. Finally, we analyzed the driving forces of land use types in the Honghe Hani Rice Terraces. We found that: (1) forests, shrubs or grasslands, and other croplands could be discriminated from paddy rice during the flooding and transplanting period, and water bodies and buildings could also be discriminated from paddy rice during the growing and harvesting period. (2) Inputting phenological feature data improved mapping accuracy and stability compared with single phenological periods. (3) In the past thirty years, 10.651%, 8.810%, and 5.711% of paddy rice were respectively converted to forests, shrubs or grasslands, and other croplands in the Honghe Hani Rice Terraces. (4) Lower agricultural profits and drought led to problems in identifying the driving mechanisms behind paddy rice distribution changes. This study demonstrates that phenological information can improve the mapping accuracy of rice terraces. It also provides evidence for the change in the size of the rice terrace area and associated driving forces in Southwest China.

Keywords: Honghe Hani Rice Terraces; Landsat; land use/land cover; phenology; Google Earth Engine

check for updates

Citation: Yang, J.; Xu, J.; Zhou, Y.; Zhai, D.; Chen, H.; Li, Q.; Zhao, G. Paddy Rice Phenological Mapping throughout 30-Years Satellite Images in the Honghe Hani Rice Terraces. *Remote Sens.* **2023**, *15*, 2398. https://doi.org/10.3390/rs15092398

Academic Editor: Mario Cunha

Received: 8 March 2023
Revised: 22 April 2023
Accepted: 2 May 2023
Published: 4 May 2023

1. Introduction

Paddy rice is an important caloric source that feeds more than half of the world's population [1]. Rice fields are also vital temporary and anthropogenic wetlands that support habitats for wildlife [2]. Mapping temporal-spatial changes in paddy rice plays a crucial role in food security and ecological conservation [3]. However, applying satellite images to map paddy rice distribution presents several challenges. The similarity of spectral features between paddy rice and other crops or vegetation types throughout the year is one of the challenges for mapping paddy rice [4]. On the one hand, the spectral changes of deciduous plants, other crops, and paddy rice show a consistent positive correlation trend with precipitation and temperature changes [5]. On the other hand, different cultivation conditions, e.g., the input of nitrogen fertilizers or irrigation, also affect the spectral change of paddy rice [6]. These problems bring challenges when discriminating paddy rice from other crops or vegetation types. Therefore, determining the differences in spectral features between other crops and vegetation types at different growing phases or cultivation conditions is the key to paddy rice mapping.

The availability of satellite images with suitable time windows is another challenge for paddy rice mapping. In general, paddy rice phenology can be split into three periods: (1) the flooding and transplanting rice period (FTP); (2) the rice growing period (GP); and (3) the fallow period after harvest (FP) [7,8]. Previous studies used phenological features to map and discriminate paddy rice from other land use types by detecting its surface water and green vegetation change [8–11]. However, due to cloud cover, cloud shadow, and fog contamination, it is difficult to acquire high-quality images or suitable time window images for detecting the phenological phase of paddy rice [3,11,12]. This also limits the application of classifiers to map paddy rice based on spectral bands and vegetation indices of phenological features [3]. Therefore, a few previous studies have integrated classifiers and phenological information to map paddy rice distribution and assess mapping performance.

Historical mapping is an effective tool to monitor or assess the temporal-spatial change in land use/land cover (LULC) and ecosystem services [13,14]. Many historical LULC maps of paddy rice croplands have been produced based on satellite images. However, coarse-resolution data might lead to uncertainty in the LULC change process [14,15], making it challenging to map paddy rice and LULC at the landscape scale. Landsat data have provided finer spatial resolution and temporal coverage from 1984 to the present day [16], so Landsat has been widely used to map paddy rice at regional and national scales. We found that most current historical mappings of paddy rice based on Landsat images lacked information about other land use types, leading to problems in identifying the driving mechanisms behind paddy rice distribution changes.

Ethnic minorities, primarily the Hani people, have shaped the steep, mid-mountain slopes of the Ailao Mountains into rice terraces in a subtropical monsoon climate over the past 1300 years [17,18]. The Hani culture believes that natural environments have their divine owners, and that the Hani people serve as custodians with the right to use these natural environments, but also a duty to protect them [18,19]. Based on the traditional cultural system of "vernacular knowledge", the Hani people maintain the balance between the local natural environment and traditional agricultural systems [20]. Thus, the Hani people have developed a sustainable landscape composed of forests, water systems, villages, and terraces [21,22]. This harmonious coexisting system of the Honghe Hani Rice Terraces (HHRT) maintains local ecosystem services and rice productivity while providing aesthetic diversity and cultural value for humans [18,20,22]. Consequently, the HHRT was designated as a Globally Important Agricultural Heritage System by UNFAO in 2010 and as a World Heritage Site by UNESCO in 2013 [20,21,23]. Mapping temporal-spatial changes of paddy rice in the HHRT is vital to local landscape planning and supporting heritage conservation.

Traditional rice terraces worldwide face various challenges from natural and anthropogenic factors [24,25]. Economic benefits, labor shortages, and food habit changes have led to rice terraces decreasing by 40% in Japan [26,27]. Soil erosion is also threatening the quality of most terraces [28]. Meteorological disasters, organic fertilizer misuse, pests, and soil erosion are the primary reasons that have led to rice terrace degeneration and loss at Ifugao in the Philippines [29]. The HHRT is also facing a series of problems: conversion of paddy rice areas to dry crops [30], farmer outmigration [31], landslide risk [17], climate change [18], etc. These problems are causing traditional culture and paddy rice area losses in the HHRT [31,32]. Most of these studies were conducted by questionnaire surveys or field monitoring, resulting in a lack of temporal-spatial quantified evidence to reveal driving forces in the rice terraces.

Here, we will investigate the traditional ecological knowledge used to acquire phenological information on paddy rice in HHRT and phenological-based paddy rice to produce historical mapping in HHRT using Landsat time-series images and the Google Earth Engine (GEE). In this study, we will address four objectives: (1) assess the separability between paddy rice and other land use types at different phenological periods; (2) assess the accuracy and stability of paddy rice mapping between three inputs of phenological information; (3) update the historical maps of LULC for HHRT from 1989–1991 to 2019–2021 using GEE and phenological information; and (4) analyze the driving forces of LULC in HHRT from 1989–1991 to 2019–2021.

2. Materials and Methods

2.1. Study Area

The HHRT (22°30′–23°30′N, 101°30′–103°30′E), distributed across four counties (Yuanyang, Honghe, Jinping, Lvchun), covers 978.071 km^2 in the southern Yunnan Province, China (Figure 1). The terraces are located on the south edge of the Ailao Mountains, which range in elevation from 339 m to 2865 m. Due to the lower latitudes, the study area has a typical 'stereoscopic' climate. The average temperature in the study area is 25 °C. The Honghe hot-dry valley in the northern part of the Ailao Mountains has the highest recorded temperature (42 °C), and the lowest recorded temperature on the highest mountain is 11.6 °C [21]. The major climate in the study area is the subtropical monsoon. Precipitation is mainly influenced by the southwest monsoon from the Indian Ocean, and the southern area of the Honghe River has the highest annual rainfall (1397.6 mm) in Yunnan Province [21]. River valleys with high evaporation and abundant rain often cause the study area to have dense fog and clouds, seriously affecting paddy rice mapping.

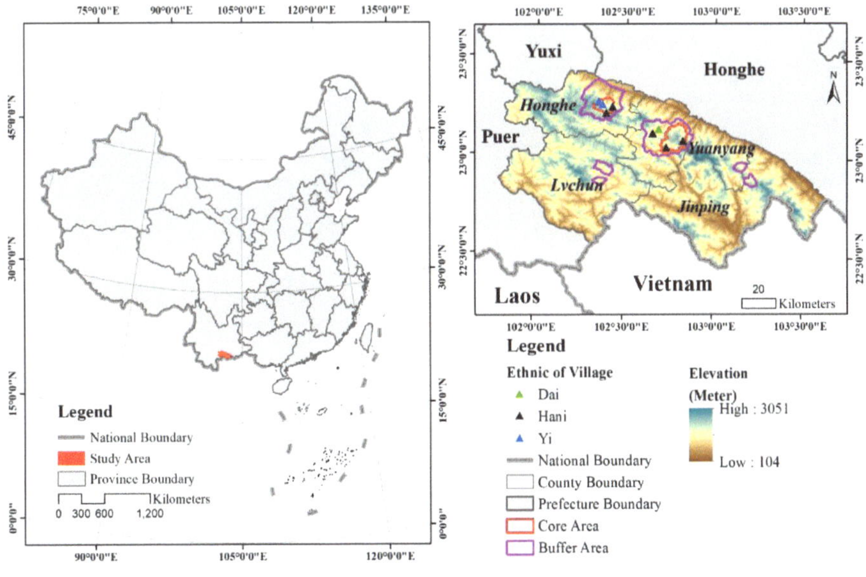

Figure 1. The location of the HHRT in China and its elevation ranges.

2.2. Phenological Features of Paddy Rice in the HHRT

2.2.1. Traditional Ecological Knowledge Investigation

We used semi-structured interviews based on the recall method to collect traditional ecological knowledge (TEK) of paddy rice phenology. We investigated TEK from eight villages in Honghe and Yuanyang Counties (Figure 1), which mainly included five Hani ethnic minority villages, two Yi ethnic minority villages, and one Dai ethnic minority village in the HHRT. Each local village head invited one or two elders to answer our TEK questions. These elders were all above 60 years old and could speak Mandarin or the local dialects. Some previously served as the heads of paddy rice production or were priests (咪谷) in their villages. Each interview was composed of four people in a relatively closed room, including an interviewer who was responsible for asking questions, an interviewer who recorded and supplemented questions, a participant who answered our questions, and a translator who was responsible for conveying the correct meaning between Mandarin or the local dialect and their ethnic minority language. The questions included the planting history of rice species in the HHRT, the traditional paddy rice farming methods in the HHRT, the phenological features and dates of each paddy rice production sector, and the conventional technology and equipment used in paddy rice production. The interviewer did not follow the list of TEK questions during the semi-structured interviews but allowed the interviewees to freely answer our questions according to their experiences and perceptions [33]. This helped to increase reciprocal interactions between interviewers and participants [34].

2.2.2. Traditional Paddy Rice Ecological Knowledge

The Hani people have plenty of TEK for paddy rice production. Local governments encourage HHRT farmers to plant traditional rice species to protect traditional cultures. Although the Hani Terrace zones have high enough temperatures and rainfall, they only plant single rice [23]. Therefore, Table 1 and Figure 2 show that the rice phenology in HHRT could be separated into three periods: (1) FTP; (2) GP; and (3) harvesting period (HP). The phenological dates used by the interviewees to answer our questions are based on the lunar calendar. In the past 20 years, the Gregorian calendar has been, on average, 36 days earlier than the lunar calendar.

Table 1. The Hani lunar calendar and related farming activities.

<table>
<tr><th>Month of Lunar Calendar</th><th>Meaning of Hani Nationality</th><th>Farming Activities</th><th>Vegetation Indices Changes of Paddy Rice</th></tr>
<tr><td>January</td><td>The month of creatures awakening</td><td>Nursery rice seedlings</td><td rowspan="3">NDVI lower than 0.4; EVI, NDSVI, and LSWI lower than 0.2</td></tr>
<tr><td>February</td><td rowspan="2">The season for transplanting seedlings</td><td rowspan="2">Transplant seedlings and hoeing in terraces</td></tr>
<tr><td>March</td></tr>
<tr><td>April</td><td rowspan="2">The season of leisure</td><td rowspan="2">hoeing, hunting, repair farm tools</td><td rowspan="3">NDVI from 0.4 increased to 0.8; EVI, NDSVI, and LSWI from 0.2 increased to 0.7, 0.55, and 0.4, respectively.</td></tr>
<tr><td>May</td></tr>
<tr><td>June</td><td>The harvest preparation month</td><td>hoeing, repair farm tools, prepare for the harvest</td></tr>
<tr><td>July</td><td>The month of rice growing</td><td rowspan="2">Autumn harvest, fallow</td><td rowspan="3">NDVI decreased to lower than 0.4; EVI, NDSVI, and LSWI decreased to lower than 0.2.</td></tr>
<tr><td>August</td><td>The month of rice maturation</td></tr>
<tr><td>September</td><td>The alternate month of the new year and the last year</td><td>Harvest late rice, fallow</td></tr>
<tr><td>October</td><td>The first month of new year</td><td rowspan="3">Fallow, plough the paddy lands, repair ridges and farm tools</td><td rowspan="3">NDVI lower than 0.4; EVI, NDSVI, and LSWI lower than 0.2</td></tr>
<tr><td>November</td><td>The month of creature hibernation</td></tr>
<tr><td>December</td><td>The month of seed germination</td></tr>
</table>

Notes: October of the lunar calendar is the first month of the new year for the Hani nationality.

As shown in Table 1, the lunar calendar month of January is the awakened month of creatures in the Hani nationality [35]. In January of the lunar calendar, the day of the year (DOY) is from 36 to 66, and the Hani people nurse rice seedlings and flood the paddies. February and March of the lunar calendar (DOY from 66 to 126) are the time for transplanting seedlings. The DOY of the FTP in the lunar calendar is from 36 to 126. In this period, the paddy rice NDVI values are lower than 0.4, and the EVI, NDSVI, and LSWI values are lower than 0.2. At the same time, the paddy rice values of NDVI, EVI, and NDSVI are lower than those of forests and shrubs or grasslands, but the paddy rice value of NDSVI is lower than that of other croplands. The paddy rice values of LSWI were lower than those of forests but higher than those of shrubs or grasslands, and other croplands (Figure 2).

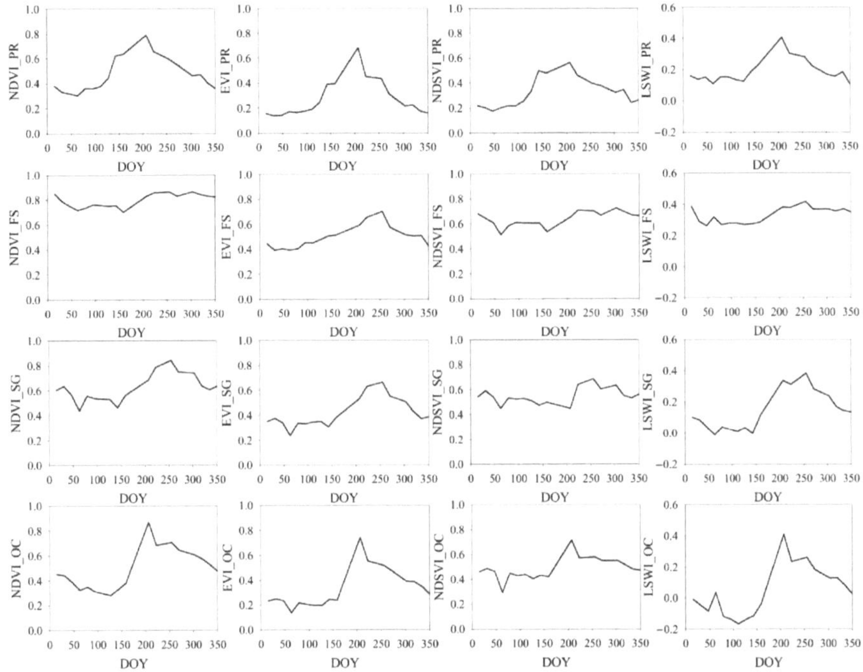

Figure 2. The time series of NDVI, EVI, NDSVI, and LSWI indices for paddy rice (PR), forests (FS), shrubs or grasslands (SG), and other croplands (OC) calculated from Landsat 8 images in 2019: 50 points for each land use type.

The work of the Hani people from April to June of the lunar calendar is hoeing, repairing farm tools, and preparing to harvest (Table 1). This period can be defined as the GP of paddy rice and the DOY is from 126 to 216. In this period, the paddy rice value of NDVI increases from 0.4 to 0.8, EVI increases from 0.2 to 0.7, NDSVI increases from 0.2 to 0.55, and LSWI increases from 0.2 to 0.4 (Figure 2). At the same time, the vegetation indices of forests, shrubs or grasslands, and other croplands also show an increasing trend during GP.

The major rice harvesting period is from late July to August of the lunar calendar. A part of the autumn harvest will last until early September of the lunar calendar (Table 1). This period can be defined as HP and the DOY is 216 to 300. Therefore, the paddy rice values of NDVI, EVI, NDSVI, and LSWI show a decreasing trend but are higher than those of the FTP.

After the harvest period, the paddy rice in the HHRT enters the FP. The Hani people celebrate the rice harvest from the first Loong Day to the first Monkey Day in October of the lunar calendar. In addition, the paddy lands are fallowed from October to December of the lunar calendar, and the Hani people plough the paddy lands and repair ridges and farm tools (Table 1). However, the HHRT paddy flooding differs greatly from other places. The local interviewees said that most rice paddies are flooded after harvest to the transplant period. The DOY of the fallow and flooding period is from 300 to 36 of the following year. In this period, the paddy rice values of NDVI, EVI, and NDSVI are lower than those of forests, shrubs or grasslands, and other croplands. The paddy rice value of LSWI is lower than that of forests but higher than that of shrubs or grasslands, and other croplands (Figure 2).

2.3. Landsat Data Pre-Processing

Image pre-processing, phenological detection, classification, and validation were analysed by Google Earth Engine (GEE) Python API and geemap [36]. Given that Landsat satellites have several advantages, such as a long observation history, fine spatial resolution, and free access, Landsat satellite datasets are widely used to monitor LULC change and map historical datasets [37]. We selected Surface Reflectance Tier 1 (SRT1) products of the Thematic Mapper Sensor (TM) and the Operational Land Imager (OLI) to map the land use types from 1989 to 2011 and 2013 to 2021, respectively. Because the band numbers, wavelength ranges, and spatial resolution of the Enhanced Thematic Mapper Plus (ETM+) are similar to those of TM, a previous study used TM images and ETM+ images to map the historical distribution of paddy rice in Japan [4]. In this study, we combined TM images from 2010–2011 and ETM+ images from 2012 into an image collection from 2010 to 2012 to make up for missed Landsat images in 2012. A total of 1456 Landsat images were used in this study from 1989 to 2021 (Table S1), and the path/row of Landsat satellites in the HHRT are 128/44, 128/45, 129/44, 129/45, and 130/44.

Both Landsat datasets underwent atmospheric correction and orthorectification. Then, the CFmask algorithm was used to reduce the effect of clouds and cloud shadows on the images. Previous studies have proven that the normalized difference vegetation index (NDVI), enhanced vegetation index (EVI), and land surface water index (LSWI) can help improve the mapping accuracy of paddy rice [8,9,38]. The normalized difference senescent vegetation index (NDSVI) can help detect vegetation senescence caused by leaf water loss [39]. Finally, we calculated four spectral indices and added these indices into spectral bands based on the GEE platform, including NDVI [40], EVI [41], NDSVI [39], and LSWI [8]:

$$\mathrm{NDVI} = \frac{\rho_{nir} - \rho_{red}}{\rho_{nir} + \rho_{red}} \tag{1}$$

$$\mathrm{EVI} = 2.5 \times \frac{\rho_{nir} - \rho_{red}}{\rho_{nir} + 6 \times \rho_{red} - 7.5 \times \rho_{blue} + 1} \tag{2}$$

$$\mathrm{NDSVI} = \frac{\rho_{swir1} - \rho_{red}}{\rho_{swir1} + \rho_{red}} \tag{3}$$

$$\mathrm{LSWI} = \frac{\rho_{nir} - \rho_{swir1}}{\rho_{nir} + \rho_{swir1}} \tag{4}$$

where ρ_{blue}, ρ_{red}, ρ_{nir}, and ρ_{swir1} are the reflectance of the blue, red, near-infrared (NIR), and shortwave-infrared 1 (SWIR1) bands in Landsat images, respectively.

2.4. Ground Reference Data

We collected reference data using a stratified sampling method based on Yang's products [14]. A total of 1000 points of interest (POIs) were collected between 1999 and 2019. The POIs in 1999 and 2019 were used to map the land use types of Landsat TM images and Landsat OLI images, respectively. In this study, we defined six land use types, including paddy rice, forests, shrubs or grasslands, other croplands, water bodies, and buildings. To ensure that each land use type had enough reference data for the training and testing datasets, each one was equally allocated 50 POIs. In addition, we assigned the remaining 700 POIs based on the land use area of Yang's dataset in 1999 and 2019. Yang's dataset was based on GEE, and Landsat images mapped annual LULC data in China from 1990 to 2019 [14]. This dataset has six land use types in the study area: forests, shrubs, grasslands, croplands, buildings, and water bodies. Therefore, we merged the shrubs and grasslands of Yang's dataset as a new land use type in our classification scheme, and the POIs of paddy rice were incorporated into the croplands. We randomly allocated POIs of each land use type in 1999 and 2019 by Yang's dataset and the number of POIs. Because the average overall accuracy of Yang's dataset is 79.30% [14], we double-checked all of the POIs on Google Earth with high spatial resolution. In the end, there were 118 and 97 POIs for paddy rice, 555 and 528 POIs for forests, 127 and 126 POIs for shrubs or grasslands, 97 and 141 for other croplands, 52 and 53 POIs for water bodies, and 51 and 55 POIs for buildings in 1999 and 2019, respectively.

Finally, we randomly selected 50 POIs of each land use type to extract the time series values of NDVI, EVI, NDSVI, and LSWI in 2019 to map the phenology features (Figure 2) and to calculate the Jeffries–Matusita distance (JMD) separability values between indices and spectral bands of the six land use types [42]. The basic idea of JMD is that if spectral or index values become more different between various land use types, the land use types are discriminated more easily [43]. The value of JMD ranges from 0 to 2, and when the value reaches 2, the land use types can be better discriminated [43]. The JMD is calculated using the equation [43,44]:

$$J_{ij} = 2 \times \left(1 - e^{-B_{ij}}\right) \tag{5}$$

in which

$$B_{ij} = \frac{1}{8} \times \left(u_i - u_j\right)^2 \times \frac{2}{v_i^2 + v_j^2} + \frac{1}{2} \times \ln\left(\frac{v_i^2 + v_j^2}{2v_i v_j}\right) \tag{6}$$

where J_{ij} is the value of JMD, B_{ij} is the value of Bhattacharyya distance, and u_i and v_i are the means and the variance of adjacent segments of class i, respectively. We calculated the JMD in R v.4.0.2.

2.5. Classification & Data Input

Random forest has higher stability, lower noise and overfitting, and higher accuracy than other machine learning algorithms [45,46]. Therefore, the random forest classifier has been widely used in remote sensing classification. The principle of the random forest classifier is based on multiple decision trees and bootstrap aggregation to train reference data [47,48]. The number of trees in the random forest is the number of repeat times in the ensemble or the number of "trees" (decision trees) in the "forest". To reduce overfitting and maintain mapping accuracy in this study, the number of trees was set to 500 in the random forest classifier [46,49]. A random sub-selection of 70% of the POIs was used as the training dataset to map the land use, and the remaining 30% of the POIs were used as the testing dataset to assess the mapping accuracy.

Previous studies have proven that phenology-based approaches can help improve paddy rice mapping accuracy [3,4,8,9]. However, monsoonal climatic conditions in lower latitude areas that lead to persistent cloud limits the detection of paddy rice phenological features during GP [11]. To ensure we had enough Landsat images with lower cloud covers (<5%) to analyze the driving forces of paddy rice area changes, we calculated median values of Landsat spectral bands and vegetation indices during the FTP (DOY from 1 to 126) and the growing and harvesting period (GHP, DOY from 126 to 300) data input within three-year periods [4], which helped to detect the phenological features of paddy rice. We obtained 11 Landsat image collections of three-year periods: 1989–1991, 1992–1994, 1995–1997, 1998–2000, 2001–2003, 2004–2006, 2007–2009, 2010–2012, 2013–2015, 2016–2018, and 2019–2021. Among them, except for 2001–2003 (cloud cover = 27.455%) and 2013–2015 (cloud cover = 5.771%), the cloud covers of the other Landsat image collections are lower than 4%. Finally, we compared three different input datasets to assess the performance of phenological features for LULC mapping improvement. The phenological input dataset (PID) consisted of the vegetation indices (NDVI, EVI, NDSVI, and LSWI) and Landsat spectral bands (B4–7) in the time series of the FTP and GHP. Another two input datasets consisted of the vegetation indices (NDVI, EVI, NDSVI, and LSWI) and Landsat spectral bands (B4–7) in the FTP (FID) and GHP (GID), respectively. The workflow is shown in Figure 3.

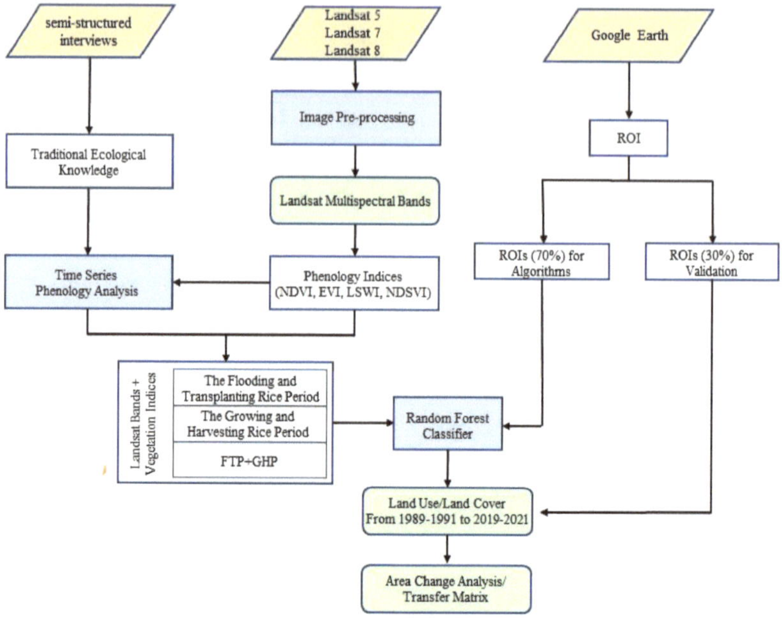

Figure 3. The land use/land cover mapping workflow in the HHRT.

2.6. Validation

To assess the accuracy between different input data, we used K-fold cross-validation of random sub-selections of 70% of POIs to map LULC in 2019–21, and the remaining 30% of POIs to assess the accuracy of thematic maps [45]. Then, we randomly selected 300 samples for accuracy validation from six periods (1989–1991, 1995–1997, 1998–2000, 2004–2006, 2010–2012, and 2013–2015). We checked these samples by visual interpretation with Landsat images and Google Earth [14]. Finally, we utilized the F1 score (FS), producer's accuracy (PA), user's accuracy (UA), overall accuracy (OA), and kappa coefficient (KC) to determine the accuracy of the land use maps [14].

2.7. Area Changes and Driving Force Analysis

We divided the Landsat images from 1989 to 2022 into collections of three-year periods to cover the gaps caused by clouds and cloud shadows [4,11]. In this study, we found that PID achieved the highest mapping accuracy and stability. Therefore, PID was used to map the land use distribution from 1989–1991 to 2019–2021. Because cloud cover percentage in 2001–2003 was 27.455%, we did not use the land use map in 2001–2003 to analyze driving forces. Instead, we used ArcGIS 10.5 software (version: 10.5.0.6491) and linear regression to compare the differences in land use change trends and land use transfer between HHRT and outside the HHRT from 1989–1991 to 2019–2021. Previous studies revealed that economic developments and climate change are the major driving forces impacting paddy rice area changes in China [50,51]. Therefore, economic data (GDP and proportion of primary industry) and climate data (annual temperature and annual precipitation) were used to assess the driving forces of land use types in four countries from 1989–1991 to 2019–2021 by Pearson correlation analysis in R. Among them, we collected Yunnan Statistical Yearbook economic data from 1989 to 2022 and climate data from local meteorological departments from 1989 to 2021. Finally, we randomly interviewed 37 local people in 8 villages to support driving force analysis. The average age of interviewees was 47 years old. The questionnaires included household income sources, agricultural production, employment planning, forest coverage changes, and water yield changes.

3. Results

3.1. The Separability Analysis in Two Phenological Periods

The vegetation indices and Landsat spectral bands in the FTP performed better when discriminating paddy rice from forests, shrubs or grasslands, and other croplands (Figure 4 and Table 2). The JMD values of NDVI, EVI, and NDSVI between paddy rice with forests and paddy rice with shrubs or grasslands during the FTP are more than 1. This meant that paddy rice with forests and paddy rice with shrubs or grasslands could be discriminated in the FTP by adding the vegetation indices NDVI, EVI, and NDSVI (Figure 4 and Table 2). In addition, B6 (1.059), NDSVI (0.994), B7 (0.979), and LSWI (0.842) show better separability performances than other indices or bands to discriminate paddy rice and other croplands (Figure 4 and Table 2). Furthermore, during the FTP, only B6 (1.006) can better discriminate paddy rice and water bodies, and LSWI (1.092) and B7 (1.247) can better discriminate paddy rice and buildings (Table 2).

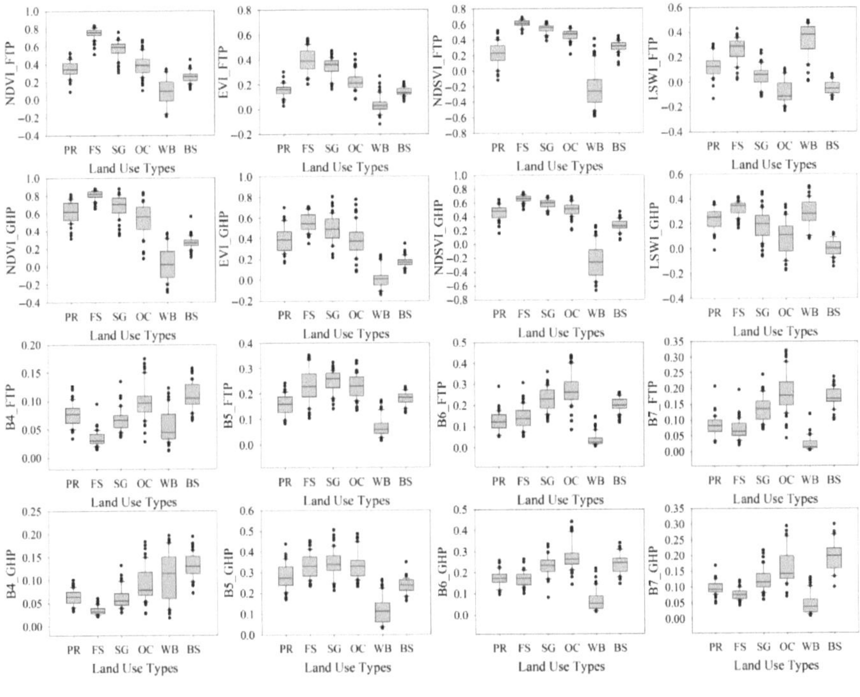

Figure 4. The annual mean of different land use types in four indices and spectral bands. FTP is the flooding and transplanting rice period, GHP is the growing and harvesting period, B4 is red band, B5 is NIR band, B6 is SWIR 1 band, B7 is SWIR 2 band, PR is paddy rice, FS is forests, SG is shrubs or grasslands, OC is other croplands, WB is water bodies, and BS is buildings.

The vegetation indices and Landsat bands in GHP performed better when discriminating paddy rice from water bodies and buildings (Figure 4 and Table 2). The vegetation indices NDVI (1.700), EVI (1.509), NDSVI (1.755) and Landsat bands B5 (1.136) and B6 (1.070) could be used to discriminate paddy rice and water bodies during GHP. At the same time, the vegetation indices of NDVI (1.607), EVI (1.148), LSWI (1.465) and Landsat bands of B4 (1.394) and B7 (1.305) performed better in discriminating paddy rice and buildings during GHP (Figure 4 and Table 2). However, only NDVI, NDSVI, and B4 during GHP could be used to discriminate paddy rice and forests. No vegetation indices or Landsat bands during GHP could be used to discriminate paddy rice with shrubs or grasslands, and paddy rice with other croplands (Table 2).

Table 2. Results of the Jeffries–Matusita distance (JMD) between paddy rice (PR) and forests (FS), shrubs or grasslands (SG), other croplands (OC), water bodies (WB), and buildings (BS).

Indices or Bands	FTP					GHP				
	PR vs. FS	PR vs. SG	PR vs. OC	PR vs. WB	PR vs. BS	PR vs. FS	PR vs. SG	PR vs. OC	PR vs. WB	PR vs. BS
NDVI	1.934	1.034	0.094	0.710	0.338	1.029	0.068	0.148	1.700	1.607
EVI	1.444	1.379	0.368	0.751	0.117	0.547	0.158	0.026	1.509	1.148
NDSVI	1.707	1.445	0.994	0.991	0.338	1.188	0.516	0.030	1.755	0.850
LSWI	0.458	0.193	0.842	0.833	1.092	0.398	0.111	0.504	0.088	1.465
B4	1.118	0.062	0.339	0.237	0.462	1.038	0.057	0.650	0.793	1.394
B5	0.531	0.952	0.618	0.968	0.186	0.175	0.209	0.155	1.136	0.325
B6	0.035	0.856	1.059	1.006	0.749	0.007	0.428	0.803	1.070	0.594
B7	0.037	0.527	0.979	0.814	1.247	0.304	0.269	0.798	0.764	1.305

3.2. Phenological Information Improved Mapping Accuracy

By adding phenological information, the mapping accuracy and stability performed better than the other two single phenological periods. The OA and KC of PID on average are 0.871 (SD = 0.011) and 0.813 (SD = 0.015), which are higher than 0.859 (SD = 0.021) and 0.794 (SD = 0.027) of FID on average and 0.796 (SD = 0.016) and 0.695 (SD = 0.023) of GID on average, respectively. In addition, Figure 5 shows that the PA, UA, and FS of PID are higher than those of FID in paddy rice, forests, shrubs or grasslands, and water bodies. However, the UA and FS of PID are lower than those of FID in other croplands and buildings. The PA, UA, and FS of PID are higher than those of GID in each land use type. Figure 5 also shows that PID resulted in the highest stability of PA, UA, and FS compared with the other two single phenological periods. Moreover, Figure 5 also shows that the GID had better stability of PA, UA, and FS than the FID, and the PA, UA, and FS of water bodies in the GID are higher than those in the FID. Therefore, adding phenological information helps improve mapping accuracy and stability.

3.3. Paddy Rice Decreased from the 1990s to 2020s

We selected the PID using the random forest classifier to map the land use changes from 1989–1991 to 2019–2021. The average OA and KC in six periods were 88.721 ± 2.798% and 80.387 ± 4.716%, respectively (Table S2).

Paddy rice, shrubs or grasslands, and other croplands show a decreasing trend from 1989–1991 to 2019–2021 (Figures 6 and 7). The loss proportions of paddy rice, shrubs or grasslands, and other croplands in HHRT are lower than those out of HHRT, but the increased proportion of forests in HHRT is higher than that outside the HHRT. Figure 7 and Table 3 show that 24.724% and 42.529% of paddy rice, 0.468% and 8.484% of shrubs or grasslands, and 51.943% and 56.879% of other croplands decreased in HHRT and outside the HHRT from 1989–1991 to 2019–2021, respectively. The forests increased to 31.572% and 17.027% in HHRT and outside the HHRT from 1989–1991 to 2019–2021, respectively. In addition, the area of paddy rice in the HHRT decreased from 13.659% in 1989–1991 to 5.566% in 2010–2012, then increased to 13.880% in 2013–2015, but decreased to 10.282% in 2019-2021. Although the HHRT was listed as a World Heritage Site in 2013, the paddy rice area decreased from 2013–2015 to 2019–2021 (Figure 7). The area of paddy rice outside the HHRT decreased from 6.347% in 1989–1991 to 2.511% in 2010–2012, then increased to 5.675% in 2013–2015, but decreased to 3.648% in 2019–2021.

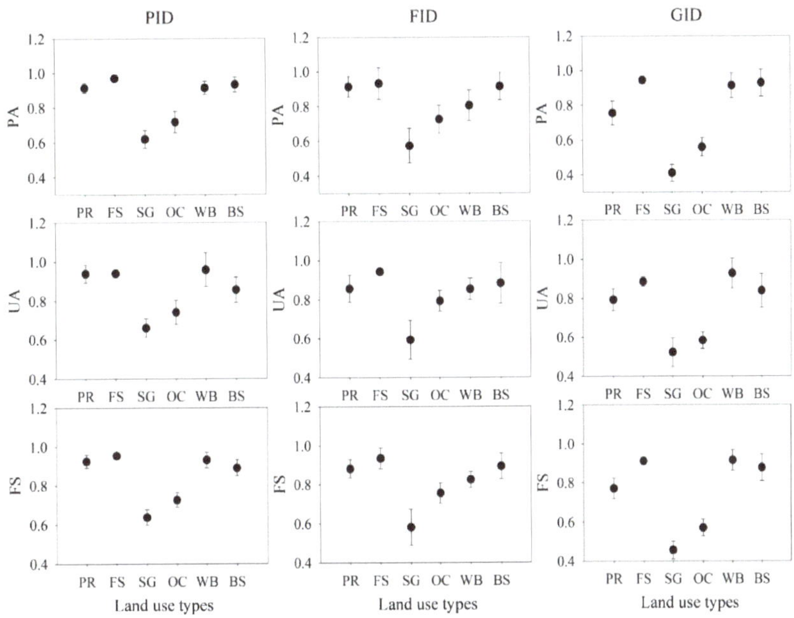

Figure 5. Accuracy assessment with producer's accuracy (PA), user's accuracy (UA), and F1 score (FS) in 2019–2021 from three kinds of data input: phenological features input dataset (PID); FTP input dataset (FID); and GHP input dataset (GID). PR is paddy rice, FS is forests, SG is shrubs or grasslands, OC is other croplands, WB is water bodies, and BS is buildings.

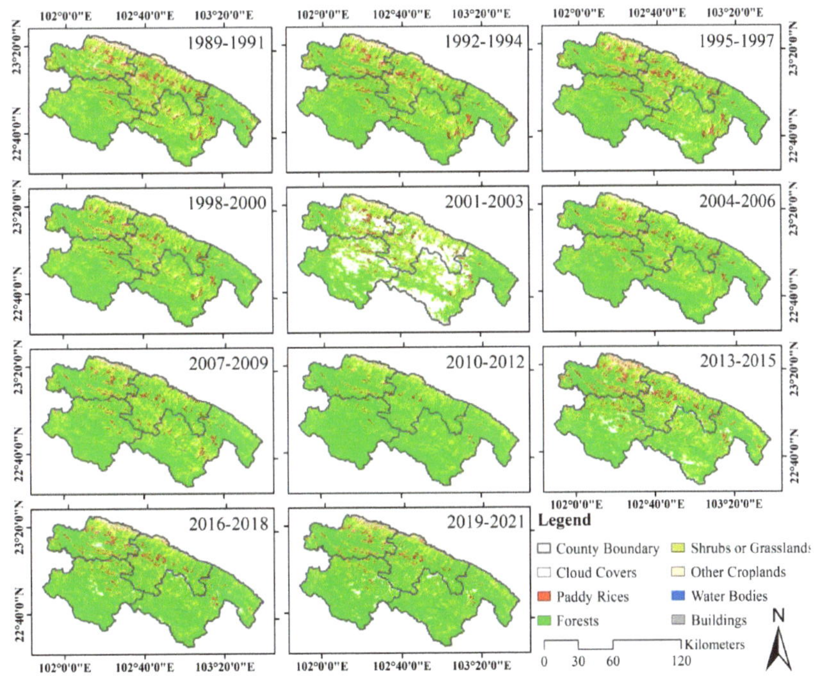

Figure 6. The land use cover changes in the study area from 1989–1991 to 2019–2021.

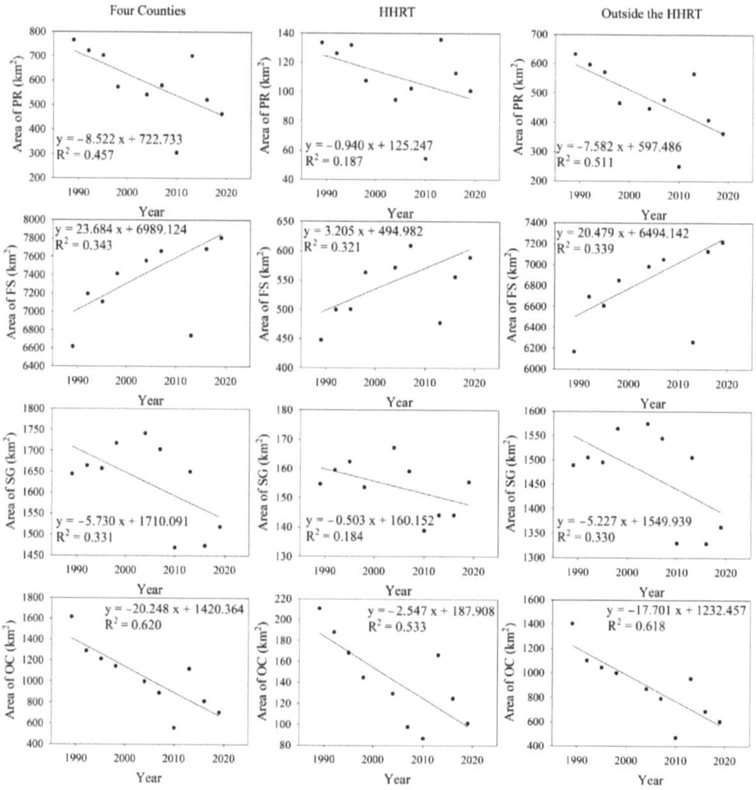

Figure 7. Area changes and trends of paddy rice (PR), forests (FS), shrubs or grasslands (SG), and other croplands (OC) from 1989–1991 to 2019–2021.

Paddy rice is the major land contributor converted to other land use types from 1989–1991 to 2019–2021. Figure 6 and Table 3 show that 10.651% and 22.963%, 8.810% and 11.886%, and 5.711% and 6.640% of paddy rice areas were converted to forests, shrubs or grasslands, and other croplands in the HHRT and outside the HHRT from 1989–1991 to 2019–2021, respectively. At the same time, 34.602% of other croplands and 36.517% of shrubs or grasslands were converted to forests outside the HHRT, and 30.597% of other croplands and 38.976% of shrubs or grasslands were converted to forests in the HHRT. In addition, 23.822% and 24.615% of other croplands were converted to shrubs or grasslands in the HHRT and outside the HHRT from 1989–1991 to 2019–2021, respectively. Therefore, paddy rice and other croplands were the primary sources of the forests and shrubs or grasslands.

3.4. The Driving Factors of LULC

The Pearson correlation analysis and questionnaires revealed that climate and economic factors affected paddy rice area changes (Figure 8). Our questionnaires show that 7 household interviewees rely on agriculture for their income with an average income of 11,125 yuan per year. Working as a migrant worker is the main income source for 27 household interviewees with an average income of 50,763 yuan per year. Regarding willingness to cultivate paddy rice, 16 interviewees hoped the next generation of their family would continue to grow rice and inherit rice terrace culture. Due to the lower agricultural profits, 11 interviewees were unwilling for the next generation to engage in agricultural production. In addition, Pearson correlation analysis shows a negative correlation between GDP and paddy rice (r = −0.493) but a positive correlation between the proportion of primary industry and paddy rice (r = 0.704). The results of the questionnaires indicated that searching for higher incomes might be an important reason for local farmers abandoning agricultural lands. At the same time, we investigated the reasons for rice production changes. Fifteen interviewees thought new rice varieties improved rice production while four interviewees thought that precipitation reduction and water shortages decreased rice production. Figure 8 shows a negative correlation between annual temperature and paddy rice (r = −0.728) but a positive correlation between annual precipitation and paddy rice (r = 0.587).

Table 3. The transfer matrix of land use and land cover in four counties, the HHRT, and outside the HHRT in 1989–1991 and 2019–2021 (km^2).

			1989–1991					
		Land Use Types	**Paddy Rice**	**Forests**	**Shrubs or Grasslands**	**Other Croplands**	**Water Bodies**	**Buildings**
2019–2021	Four counties	Paddy rice	260.243	124.580	16.231	41.091	5.538	14.199
		Forests	282.653	5629.271	1033.406	685.266	1.672	16.710
		Shrubs or Grasslands	102.447	438.720	421.858	529.110	0.764	18.378
		Other Croplands	90.305	137.487	135.016	326.121	1.158	13.950
		Water Bodies	8.045	8.346	0.926	3.036	13.283	4.629
		Buildings	16.243	16.889	11.022	23.117	1.485	17.310
	HHRT	Paddy rice	68.315	17.112	2.508	9.658	0.562	2.225
		Forests	31.323	367.917	91.377	77.892	0.166	1.392
		Shrubs or Grasslands	14.261	31.550	40.941	66.429	0.032	1.540
		Other Croplands	17.277	13.469	16.272	52.263	0.054	1.639
		Water Bodies	0.693	0.234	0.062	0.173	0.207	0.184
		Buildings	1.550	2.443	2.339	4.134	0.027	2.843
	Out ofHHRT	Paddy rice	191.935	107.459	13.722	31.425	4.976	11.962
		Forests	251.326	5261.489	942.090	607.408	1.506	15.317
		Shrubs or Grasslands	88.190	407.126	380.884	462.603	0.732	16.830
		Other Croplands	73.027	124.008	118.721	273.767	1.105	12.308
		Water Bodies	7.352	8.111	0.864	2.863	13.078	4.444
		Buildings	14.690	14.443	8.682	18.974	1.456	14.457

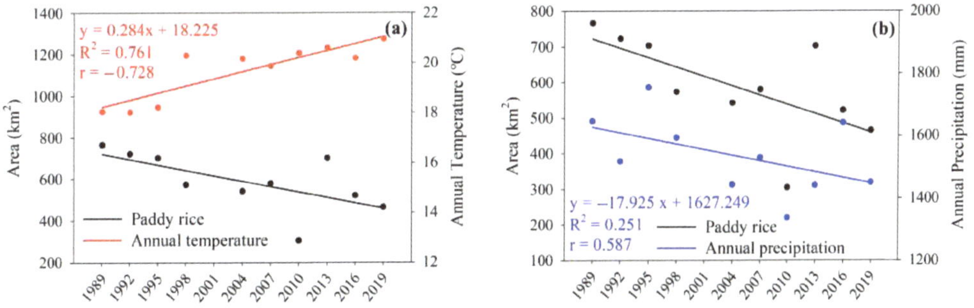

Figure 8. *Cont.*

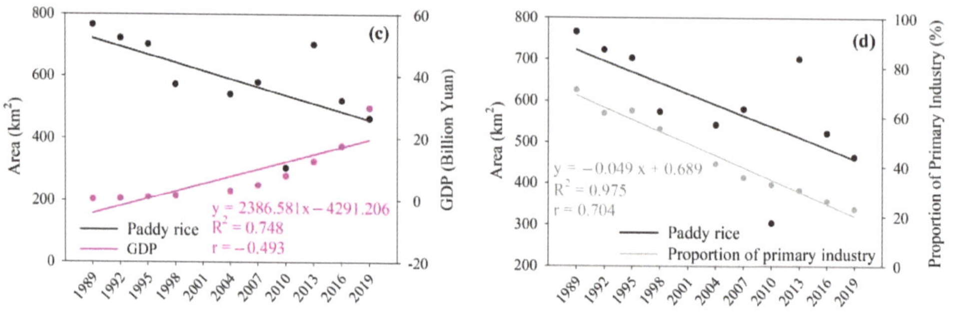

Figure 8. The correlation coefficient (r) between paddy rice and its driving forces. (**a**–**d**) are the correlation coefficients between rice paddy with annual temperature, annual precipitation, GDP, and proportion of primary industry in four counties from 1989–1991 to 2019–2021, respectively.

4. Discussion

4.1. Phenological Features Improve Paddy Rice Mapping

Our study illustrated that adding phenological information can help improve paddy rice mapping accuracy [3,4,8–11]. Forests, water bodies, and buildings did not have obvious phenological features (Figures 2 and 4), and vegetation indices or spectral bands can help to discriminate these land use types from paddy rice at different phenological periods. The vegetation indices of forests, e.g., NDVI and NDSVI, are much higher than those of paddy rice in any period (Figure 4). In particular, the JMD of forests and paddy rice in NDVI and NDSVI are higher than 1.7 during the FTP (Table 2), which can be used to discriminate forests and paddy rice at FTP. With paddy rice growing at GHP, the vegetation indices of paddy rice are higher than those of water bodies and buildings. Therefore, vegetation indices can be used to identify water bodies and paddy rice, and buildings and paddy rice at GHP [8,9].

However, deciduous vegetation, such as shrubs or grasslands, other croplands, and paddy rice, show a similar annual variation [52]. Therefore, it is difficult to identify paddy rice and other deciduous vegetation types at GHP using vegetation indices and Landsat spectral bands (Table 2 and Figure 4). Because shrubs or grasslands have plant cover during the FTP, the vegetation indices of shrubs or grasslands are higher than those of other croplands and paddy rice (Table 2 and Figure 4) [5]. Previous studies used vegetation indices or bands that could capture water information at transplanting periods to identify paddy rice and other croplands [8–10,53–55]. In our study, Figure 4 shows that SWIR (B6–7) has higher separability to discriminate other croplands and paddy rice because SWIR can detect water and water absorption [39,56]. Therefore, the calculated NDSVI and LSWI based on SWIR, can be used to discriminate other croplands and paddy rice at the FTP by detecting water differences in leaf tissues and surface moisture, respectively [8,39].

Our study found that single phenological images show a lower accuracy but had overfitting problems when mapping paddy rice or land use types. This limited feature in input datasets might be related to the lower accuracy [57]. Schulz et al. (2021) reported that incorporating time series data of phenological information can help address the overfitting problems of crop monitoring or mapping [58]. By adding phenological information of vegetation indices and Landsat spectral bands, we achieved higher mapping accuracy and stability of paddy rice and land use types than either FID or GID. Therefore, we recommend using phenological information to map paddy rice or deciduous vegetation.

4.2. The Area Changes of Paddy Rice in the Hani Terraces

The loss of paddy rice is caused by several factors [50,51], including economic development, climate, policy, and technology. Economic development has been a major factor impacting paddy rice area losses in southern China since the early 1980s [50,59]. In contrast, industrialization and urbanization have replaced paddy rice in built-up and residential areas in South China in the past 30 years [60,61]. Lower agricultural profits might be the major driver of paddy rice area losses in the HHRT [32]. The development of the secondary and tertiary industry brings a large number of employment opportunities and increased incomes. This leads to farm laborers migrating to cities for higher profits. Zhang et al. (2017b, 2020) reported that young adults prefer to find a nonfarm job in cities because a family's annual paddy rice profit is much lower than the income from nonfarm jobs in cities [31,62]. Therefore, parts of paddy rice areas have been transferred to other croplands or left uncultivated (Table 3). The remaining labor force in rice terraces mainly consists of elderly people or uneducated women [23]. The HHRT has diverse tourism resources that could recall young entrepreneurs back to the HHRT, and support rural tourism development [20,22]. In addition, the rice-fish-duck integrated farming model not only maintains paddy rice areas and ecosystem services but can also help to improve local economic incomes [23]. Therefore, we recommend that the HHRT develop the rural tourism industry and rice-fish-duck integrated farming model to increase incomes and maintain paddy rice areas.

The climate also has an impact on the paddy rice area changes in the HHRT. Previous studies reported that climate change impacted the growing period and distribution of paddy rice [51,63]. However, continuous severe drought events are an important driving factor impacting paddy rice losses in HHRT [18]. Yang et al. (2019) reported that continuous drought in the Yunnan Province mainly occurred in 1988–1990, 2003–2007, and 2010–2015 [64]. Jiao et al. (2012) also reported that severe drought events caused terraces to dry-up in 2005 and 2010 [18]. In addition, Abbas et al. (2014) and Nichol and Abbas (2015) reported that croplands and shrublands suffered more seriously in dry conditions than forests in the Yunnan Province due to the severe drought event in 2010 [65,66]. These studies confirmed our research that a reduction in rainfall led to the paddy rice area in the HHRT in 2004–2006 (94.607 km^2) and 2010–2012 (54.517 km^2) being smaller than that in other periods (Figures 6 and 8). The improvement of hydraulic facilities can help reduce the impact of drought on paddy rice.

Other factors could also lead to paddy rice losses in HHRT. For example, the implementation of the Grain for Green Program might lead to paddy rice being converted to forests. Landslide risk might lead to temporary or permanent loss of paddy rice areas [17].

4.3. Applications

Our method can help researchers map the crops and vegetation distribution in finer-scale Landsat data. Phenological dates, especially rice flooding and transplanting dates, are widely used to map paddy rice distribution based on Landsat images [4,10–12]. In this study, we acquired paddy rice phenological dates from semi-structured interviews and questionnaire interviews. This not only helped to acquire the specific phenological dates but also helped to explain the reasons for the paddy rice area change. Moreover, because of cloud and fog contamination, it is difficult to acquire enough high-quality images in different phenological periods for mapping paddy rice in the study area [4,12]. GEE provides users with high-performance computing capabilities, massive raster data, and advanced algorithms [16]. Therefore, we merged the images of the three-year time intervals based on GEE to acquire the specific phenological dates with lower cloud and fog contamination. Finally, our results only rely on Landsat spectral bands and vegetation indices, so they can be widely used to map the historical distributions of paddy rice with phenological features.

5. Conclusions

Historical maps of paddy rice are an important tool to support heritage conservation and management in the HHRT. Here, we acquired the phenological date of paddy rice by semi-structured interviews, and compared the separability of paddy rice and other land use types. We found that paddy rice could be discriminated from other land use types in different phenological periods of paddy rice. By adding phenological features, the PID achieved a higher mapping accuracy and stability than any single phenological period.

Therefore, we recommend adding phenological information to map paddy rice or other crops. Moreover, we mapped the LULC with three-year periods in HHRT from 1989–1991 to 2019–2021 based on GEE and phenological information. The results show that paddy rice is the major source that was converted to other land use types, and the area losses of paddy rice in the HHRT are lower than those outside the HHRT. The questionnaires and Pearson correlation analysis revealed that lower agricultural profits and drought are the major drivers leading to paddy rice losses in the HHRT. The development of the rural tourism industry and rice-fish-duck integrated farming model could increase employment opportunities and incomes which can help to recall young entrepreneurs back to the HHRT. Thus, our thematic maps can help researchers and local governments in landscape planning and heritage conservation, and support the driving force analysis of traditional rice terrace areas.

Supplementary Materials: The following supporting information can be downloaded at: https://www.mdpi.com/article/10.3390/rs15092398/s1, Table S1: The Landsat image numbers in different phenological periods and Landsat sensors from 1989–1991 to 2019–2021; Table S2: The mapping accuracy in six periods; Supplementary information A: Paddy rice mapping code in 1998–2000.

Author Contributions: Conceptualization, J.Y. and J.X.; Methodology, J.Y., Y.Z. and D.Z.; Software, J.Y. and H.C.; Validation, J.Y., H.C. and Q.L.; Investigation, J.Y., J.X., Y.Z., Q.L. and G.Z.; Writing—Original Draft Preparation, J.Y.; Writing—Review & Editing, J.Y. and D.Z.; Supervision, J.X. and Y.Z.; Funding Acquisition, J.X. and Y.Z. All authors have read and agreed to the published version of the manuscript.

Funding: This research was funded by Study on Green System of agricultural production in Hani Terrace Area (grant number Y2021YJ12), Yunnan Provincial Science and Technology Department (grant number 202205AM070007).

Data Availability Statement: The data presented in this study are available on request from the corresponding author.

Acknowledgments: This research is part of Honghe Agroforestry Ecosystem Research Station of Yunnan Province, and supported by China Agricultural Green Development Association. We thanks Zhihao Qin for his valuable comments, thanks Beinn Purvis, Eyalira Jacob Okal, and Hui Cao for language editing, and thanks Haiqin Qin, Yunfen Xu, and Rui Xiao for questionnaire investigation.

Conflicts of Interest: The authors declare no conflict of interest.

References

1. Kuenzer, C.; Knauer, K. Remote Sensing of Rice Crop Areas—A Review. *Int. J. Remote Sens.* **2013**, *34*, 2101–2139. [CrossRef]
2. Elphick, C.S. Functional Equivalency between Rice Fields and Seminatural Wetland Habitats. *Conserv. Biol.* **2000**, *14*, 181–191. [CrossRef]
3. Dong, J.; Xiao, X. Evolution of regional to global paddy rice mapping methods: A review. *ISPRS J. Photogramm. Remote Sens.* **2016**, *119*, 214–227. [CrossRef]
4. Carrasco, L.; Fujita, G.; Kito, K.; Miyashita, T. Historical mapping of rice fields in Japan using phenology and temporally aggregated Landsat images in Google Earth Engine. *ISPRS J. Photogramm. Remote Sens.* **2022**, *191*, 277–289. [CrossRef]
5. Ibrahim, S.A.; Kaduk, J.; Tansey, K.; Balzter, H.; Lawal, U.M. Detecting phenological changes in plant functional types over West African savannah dominated landscape. *Int. J. Remote Sens.* **2021**, *42*, 567–594. [CrossRef]
6. Song, S.; Gong, W.; Zhu, B.; Huang, X. Wavelength selection and spectral discrimination for paddy rice, with laboratory measurements of hyperspectral leaf reflectance. *ISPRS J. Photogramm. Remote Sens.* **2011**, *66*, 672–682. [CrossRef]
7. Toan, T.L.; Ribbes, F.; Li-Fang, W.; Floury, N.; Kung-Hau, D.; Jin Au, K.; Fujita, M.; Kurosu, T. Rice crop mapping and monitoring using ERS-1 data based on experiment and modeling results. *IEEE Trans. Geosci. Remote Sens.* **1997**, *35*, 41–56. [CrossRef]
8. Xiao, X.; Boles, S.; Liu, J.; Zhuang, D.; Frolking, S.; Li, C.; Salas, W.; Moore, B. Mapping paddy rice agriculture in southern China using multi-temporal MODIS images. *Remote Sens. Environ.* **2005**, *95*, 480–492. [CrossRef]
9. Xiao, X.; Boles, S.; Frolking, S.; Li, C.; Babu, J.Y.; Salas, W.; Moore, B. Mapping paddy rice agriculture in South and Southeast Asia using multi-temporal MODIS images. *Remote Sens. Environ.* **2006**, *100*, 95–113. [CrossRef]
10. Dong, J.; Xiao, X.; Kou, W.; Qin, Y.; Zhang, G.; Li, L.; Jin, C.; Zhou, Y.; Wang, J.; Biradar, C.; et al. Tracking the dynamics of paddy rice planting area in 1986–2010 through time series Landsat images and phenology-based algorithms. *Remote Sens. Environ.* **2015**, *160*, 99–113. [CrossRef]

11. Kontgis, C.; Schneider, A.; Ozdogan, M. Mapping rice paddy extent and intensification in the Vietnamese Mekong River Delta with dense time stacks of Landsat data. *Remote Sens. Environ.* **2015**, *169*, 255–269. [CrossRef]
12. Jiang, M.; Xin, L.; Li, X.; Tan, M.; Wang, R. Decreasing Rice Cropping Intensity in Southern China from 1990 to 2015. *Remote Sens.* **2019**, *11*, 35. [CrossRef]
13. Johnson, D.M. Using the Landsat archive to map crop cover history across the United States. *Remote Sens. Environ.* **2019**, *232*, 111286. [CrossRef]
14. Yang, J.; Huang, X. The 30m annual land cover dataset and its dynamics in China from 1990 to 2019. *Earth Syst. Sci. Data* **2021**, *13*, 3907–3925. [CrossRef]
15. Sulla-Menashe, D.; Gray, J.M.; Abercrombie, S.P.; Friedl, M.A. Hierarchical mapping of annual global land cover 2001 to present: The MODIS Collection 6 Land Cover product. *Remote Sens. Environ.* **2019**, *222*, 183–194. [CrossRef]
16. Gorelick, N.; Hancher, M.; Dixon, M.; Ilyushchenko, S.; Thau, D.; Moore, R. Google Earth Engine: Planetary-scale geospatial analysis for everyone. *Remote Sens. Environ.* **2017**, *202*, 18–27. [CrossRef]
17. Gao, X.; Roder, G.; Jiao, Y.; Ding, Y.; Liu, Z.; Tarolli, P. Farmers' landslide risk perceptions and willingness for restoration and conservation of world heritage site of Honghe Hani Rice Terraces, China. *Landslides* **2020**, *17*, 1915–1924. [CrossRef]
18. Jiao, Y.; Li, X.; Liang, L.; Takeuchi, K.; Okuro, T.; Zhang, D.; Sun, L. Indigenous ecological knowledge and natural resource management in the cultural landscape of China's Hani Terraces. *Ecol. Res.* **2012**, *27*, 247–263. [CrossRef]
19. Yang, J.; Wang, Y.-C.; Wang, D.; Guo, L. Application of Traditional Knowledge of Hani People in Biodiversity Conservation. *Sustainability* **2018**, *10*, 4555. [CrossRef]
20. Chan, J.H.; Iankova, K.; Zhang, Y.; McDonald, T.; Qi, X. The role of self-gentrification in sustainable tourism: Indigenous entrepreneurship at Honghe Hani Rice Terraces World Heritage Site, China. *J. Sustain. Tour.* **2016**, *24*, 1262–1279. [CrossRef]
21. UNESCO. *Cultural Landscape of Honghe Hani Rice Terraces*; UNESCO: Paris, France, 2013.
22. Gu, H.; Jiao, Y.; Liang, L. Strengthening the socio-ecological resilience of forest-dependent communities: The case of the Hani Rice Terraces in Yunnan, China. *Forest Policy Econ.* **2012**, *22*, 53–59. [CrossRef]
23. Yuan, Y.; Xu, G.; Shen, N.; Nie, Z.; Li, H.; Zhang, L.; Gong, Y.; He, Y.; Ma, X.; Zhang, H.; et al. Valuation of Ecosystem Services for the Sustainable Development of Hani Terraces: A Rice-Fish-Duck Integrated Farming Model. *Int. J. Environ. Res. Public Health* **2022**, *19*, 8549. [CrossRef] [PubMed]
24. Zhang, Y.; He, L.; Li, X.; Zhang, C.; Qian, C.; Li, J.; Zhang, A. Why are the Longji Terraces in Southwest China maintained well? A conservation mechanism for agricultural landscapes based on agricultural multi-functions developed by multi-stakeholders. *Land Use Pol.* **2019**, *85*, 42–51. [CrossRef]
25. Yu, M.; Li, Y.; Luo, G.; Yu, L.; Chen, M. Agroecosystem composition and landscape ecological risk evolution of rice terraces in the southern mountains, China. *Ecol. Indic.* **2022**, *145*, 109625. [CrossRef]
26. Sarker, A. Economics of underproduction: A polycentric approach for a depopulated commons in Japan. *Ecol. Econ.* **2020**, *171*, 106597. [CrossRef]
27. Yasushi, M.; Masaya, S.; Eisei, M.; Kumiko, T. When do rice terraces become rice terraces? *Paddy Water Environ.* **2019**, *17*, 323–330. [CrossRef]
28. Arnáez, J.; Lana-Renault, N.; Lasanta, T.; Ruiz-Flaño, P.; Castroviejo, J. Effects of farming terraces on hydrological and geomorphological processes. A review. *Catena* **2015**, *128*, 122–134. [CrossRef]
29. Joshi, R.C.; Matchoc, O.R.O.; Bahatan, R.G.; Pena, F.A.D. Farmers' knowledge, attitudes and practices of rice crop and pest management at Ifugao Rice Terraces, Philippines. *Int. J. Pest Manag.* **2000**, *46*, 43–48. [CrossRef]
30. Yang, L.; Liu, M.; Lun, F.; Yuan, Z.; Zhang, Y.; Min, Q. An Analysis on Crops Choice and Its Driving Factors in Agricultural Heritage Systems—A Case of Honghe Hani Rice Terraces System. *Sustainability* **2017**, *9*, 1162. [CrossRef]
31. Zhang, Y.; Min, Q.; Zhang, C.; He, L.; Zhang, S.; Yang, L.; Tian, M.; Xiong, Y. Traditional culture as an important power for maintaining agricultural landscapes in cultural heritage sites: A case study of the Hani terraces. *J. Cult. Herit.* **2017**, *25*, 170–179. [CrossRef]
32. Zhang, Y.; Li, X.; Min, Q. How to balance the relationship between conservation of Important Agricultural Heritage Systems (IAHS) and socio-economic development? A theoretical framework of sustainable industrial integration development. *J. Clean. Prod.* **2018**, *204*, 553–563. [CrossRef]
33. Kallio, H.; Pietilä, A.-M.; Johnson, M.; Kangasniemi, M. Systematic methodological review: Developing a framework for a qualitative semi-structured interview guide. *J. Adv. Nurs.* **2016**, *72*, 2954–2965. [CrossRef] [PubMed]
34. Zhang, J.-Q.; Mammides, C.; Corlett, R.T. Reasons for the Survival of Tropical Forest Fragments in Xishuangbanna, Southwest China. *Forests* **2020**, *11*, 159. [CrossRef]
35. Yang, J. Adaptation Mechanism on Climate Change of Traditional Agriculture Management in Hani Rice Terraces. Ph.D. Thesis, Beijing Forestry University, Beijing, China, 2015.
36. Wu, Q. geemap: A Python package for interactive mapping with Google Earth Engine. *J. Open Source Softw.* **2020**, *5*, 2305. [CrossRef]
37. Wulder, M.A.; White, J.C.; Loveland, T.R.; Woodcock, C.E.; Belward, A.S.; Cohen, W.B.; Fosnight, E.A.; Shaw, J.; Masek, J.G.; Roy, D.P. The global Landsat archive: Status, consolidation, and direction. *Remote Sens. Environ.* **2016**, *185*, 271–283. [CrossRef]
38. Zhang, G.; Xiao, X.; Dong, J.; Xin, F.; Zhang, Y.; Qin, Y.; Doughty, R.B.; Moore, B. Fingerprint of rice paddies in spatial–temporal dynamics of atmospheric methane concentration in monsoon Asia. *Nat. Commun.* **2020**, *11*, 554. [CrossRef]

39. Qi, J.; Marsett, R.; Heilman, P.; Bieden-bender, S.; Moran, S.; Goodrich, D.; Weltz, M. RANGES improves satellite-based information and land cover assessments in southwest United States. *Eos Trans. Am. Geophys. Union* **2002**, *83*, 601–606. [CrossRef]
40. Defries, R.S.; Townshend, J.R.G. NDVI-derived land cover classifications at a global scale. *Int. J. Remote Sens.* **1994**, *15*, 3567–3586. [CrossRef]
41. Huete, A.; Didan, K.; Miura, T.; Rodriguez, E.P.; Gao, X.; Ferreira, L.G. Overview of the radiometric and biophysical performance of the MODIS vegetation indices. *Remote Sens. Environ.* **2002**, *83*, 195–213. [CrossRef]
42. Thomas, I.L.; Ching, N.P.; Benning, V.M.; D'Aguanno, J.A. Review Article A review of multi-channel indices of class separability. *Int. J. Remote Sens.* **1987**, *8*, 331–350. [CrossRef]
43. Jung, R.; Ehlers, M. Comparison of two feature selection methods for the separability analysis of intertidal sediments with spectrometric datasets in the German Wadden Sea. *Int. J. Appl. Earth Obs. Geoinf.* **2016**, *52*, 175–191. [CrossRef]
44. Wang, Y.; Qi, Q.; Liu, Y. Unsupervised Segmentation Evaluation Using Area-Weighted Variance and Jeffries-Matusita Distance for Remote Sensing Images. *Remote Sens.* **2018**, *10*, 1193. [CrossRef]
45. Yang, J.; Xu, J.; Zhai, D.-L. Integrating Phenological and Geographical Information with Artificial Intelligence Algorithm to Map Rubber Plantations in Xishuangbanna. *Remote Sens.* **2021**, *13*, 2793. [CrossRef]
46. Belgiu, M.; Drăguţ, L. Random forest in remote sensing: A review of applications and future directions. *ISPRS J. Photogramm. Remote Sens.* **2016**, *114*, 24–31. [CrossRef]
47. Breiman, L. Random Forests. *Mach. Learn.* **2001**, *45*, 5–32. [CrossRef]
48. Blickensdörfer, L.; Schwieder, M.; Pflugmacher, D.; Nendel, C.; Erasmi, S.; Hostert, P. Mapping of crop types and crop sequences with combined time series of Sentinel-1, Sentinel-2 and Landsat 8 data for Germany. *Remote Sens. Environ.* **2022**, *269*, 112831. [CrossRef]
49. Rodriguez-Galiano, V.F.; Ghimire, B.; Rogan, J.; Chica-Olmo, M.; Rigol-Sanchez, J.P. An assessment of the effectiveness of a random forest classifier for land-cover classification. *ISPRS J. Photogramm. Remote Sens.* **2012**, *67*, 93–104. [CrossRef]
50. Zhang, G.; Xiao, X.; Biradar, C.M.; Dong, J.; Qin, Y.; Menarguez, M.A.; Zhou, Y.; Zhang, Y.; Jin, C.; Wang, J.; et al. Spatiotemporal patterns of paddy rice croplands in China and India from 2000 to 2015. *Sci. Total Environ.* **2017**, *579*, 82–92. [CrossRef]
51. Dong, J.; Xiao, X.; Zhang, G.; Menarguez, M.A.; Choi, C.Y.; Qin, Y.; Luo, P.; Zhang, Y.; Moore, B. Northward expansion of paddy rice in northeastern Asia during 2000–2014. *Geophys. Res. Lett.* **2016**, *43*, 3754–3761. [CrossRef]
52. Bridhikitti, A.; Overcamp, T.J. Estimation of Southeast Asian rice paddy areas with different ecosystems from moderate-resolution satellite imagery. *Agric. Ecosyst. Environ.* **2012**, *146*, 113–120. [CrossRef]
53. Pan, X.-Z.; Uchida, S.; Liang, Y.; Hirano, A.; Sun, B. Discriminating different landuse types by using multitemporal NDXI in a rice planting area. *Int. J. Remote Sens.* **2010**, *31*, 585–596. [CrossRef]
54. Kordelas, G.A.; Manakos, I.; Aragonés, D.; Díaz-Delgado, R.; Bustamante, J. Fast and Automatic Data-Driven Thresholding for Inundation Mapping with Sentinel-2 Data. *Remote Sens.* **2018**, *10*, 910. [CrossRef]
55. Verma, A.; Kumar, A.; Lal, K. Kharif crop characterization using combination of SAR and MSI Optical Sentinel Satellite datasets. *J. Earth Syst. Sci.* **2019**, *128*, 230. [CrossRef]
56. Du, Y.; Zhang, Y.; Ling, F.; Wang, Q.; Li, W.; Li, X. Water Bodies' Mapping from Sentinel-2 Imagery with Modified Normalized Difference Water Index at 10-m Spatial Resolution Produced by Sharpening the SWIR Band. *Remote Sens.* **2016**, *8*, 354. [CrossRef]
57. Barr, M. A Novel Technique for Segmentation of High Resolution Remote Sensing Images Based on Neural Networks. *Neural Process. Lett.* **2020**, *52*, 679–692. [CrossRef]
58. Schulz, C.; Holtgrave, A.-K.; Kleinschmit, B. Large-scale winter catch crop monitoring with Sentinel-2 time series and machine learning–An alternative to on-site controls? *Comput. Electron. Agric.* **2021**, *186*, 106173. [CrossRef]
59. Xin, F.; Xiao, X.; Dong, J.; Zhang, G.; Zhang, Y.; Wu, X.; Li, X.; Zou, Z.; Ma, J.; Du, G.; et al. Large increases of paddy rice area, gross primary production, and grain production in Northeast China during 2000–2017. *Sci. Total Environ.* **2020**, *711*, 135183. [CrossRef]
60. Liu, J.; Zhang, Z.; Xu, X.; Kuang, W.; Zhou, W.; Zhang, S.; Li, R.; Yan, C.; Yu, D.; Wu, S.; et al. Spatial patterns and driving forces of land use change in China during the early 21st century. *J. Geogr. Sci.* **2010**, *20*, 483–494. [CrossRef]
61. Ghose, B. Food security and food self-sufficiency in China: From past to 2050. *Food Energy Secur.* **2014**, *3*, 86–95. [CrossRef]
62. Zhang, Y.; He, L.; Liu, L.; Min, Q. Labor productivity of small-scale agriculture and its influence on agricultural landscape conservation in mountainous areas in China: A case study of rice farming in Hani terraced region. *Environ. Sci. Pollut. Res.* **2020**, *27*, 39795–39806. [CrossRef]
63. Wang, X.; Ciais, P.; Li, L.; Ruget, F.; Vuichard, N.; Viovy, N.; Zhou, F.; Chang, J.; Wu, X.; Zhao, H.; et al. Management outweighs climate change on affecting length of rice growing period for early rice and single rice in China during 1991–2012. *Agric. For. Meteorol.* **2017**, *233*, 1–11. [CrossRef]
64. Yang, C.; Tuo, Y.; Ma, J.; Zhang, D. Spatial and Temporal Evolution Characteristics of Drought in Yunnan Province from 1969 to 2018 Based on SPI/SPEI. *Water Air Soil Pollut.* **2019**, *230*, 269. [CrossRef]

65. Abbas, S.; Nichol, J.E.; Qamer, F.M.; Xu, J. Characterization of Drought Development through Remote Sensing: A Case Study in Central Yunnan, China. *Remote Sens.* **2014**, *6*, 4998–5018. [CrossRef]
66. Nichol, J.E.; Abbas, S. Integration of remote sensing datasets for local scale assessment and prediction of drought. *Sci. Total Environ.* **2015**, *505*, 503–507. [CrossRef]

remote sensing

MDPI

Article

Regional Accuracy Assessment of 30-Meter GLC_FCS30, GlobeLand30, and CLCD Products: A Case Study in Xinjiang Area

Jingpeng Liu [1], **Yu Ren** [2] and **Xidong Chen** [1,3,*]

[1] College of Surveying and Geo-Informatics, North China University of Water Resources and Electric Power, Zhengzhou 450046, China; 201904306@stu.ncwu.edu.cn

[2] College of Geography and Environmental Sciences, Northwest Normal University, Lanzhou 730070, China; 2021222961@nwnu.edu.cn

[3] Future Urbanity & Sustainable Environment (FUSE) Lab, Division of Landscape Architecture, Department of Architecture, Faculty of Architecture, The University of Hong Kong, Hong Kong SAR 999007, China

* Correspondence: xdchenrs@hku.hk

Abstract: With the development of remote sensing technology, a number of fine-resolution (30-m) global/national land cover (LC) products have been developed. However, accuracy assessments for the developed LC products are commonly conducted at global and national scales. Due to the limited availability of representative validation observations and reference data, knowledge relating to the accuracy and applicability of existing LC products on a regional scale is limited. Since Xinjiang, China, exhibits diverse surface cover and fragmented urban landscapes, existing LC products generally have high classification uncertainty in this region. This makes Xinjiang suitable for assessing the accuracy and consistency of exiting fine-resolution land cover products. In order to improve knowledge of the accuracy of existing fine-resolution LC products at the regional scale, Xinjiang province was selected as the case area. First, we employed an equal-area stratified random sampling approach with climate, population density, and landscape heterogeneity information as constraints, along with the hexagonal discrete global grid system (HDGGS) as basic sampling grids to develop a high-density land cover validation dataset for Xinjiang (HDLV-XJ) in 2020. This is the first publicly available regionally high-density validation dataset that can support analysis at a regional scale, comprising a total of 20,932 validation samples. Then, based on the generated HDLV-XJ dataset, the accuracies and consistency among three widely used 30-m LC products, GLC_FCS30, GlobeLand30, and CLCD, were quantitatively evaluated. The results indicated that the CLC_FCS30 exhibited the highest overall accuracy (88.10%) in Xinjiang, followed by GlobeLand30 (with an overall accuracy of 83.58%) and CLCD (81.57%). Moreover, through a comprehensive analysis of the relationship between different environmental conditions and land cover product performance, we found that GlobeLand30 performed best in regions with high landscape fragmentation, while GLC_FCS30 stood out as the most outstanding product in areas with uneven proportions of land cover types. Our study provides a novel insight into the suitability of these three widely-used LC products under various environmental conditions. The findings and dataset can provide valuable insights for the application of existing LC products in different environment conditions, offering insights into their accuracies and limitations.

Keywords: land cover; remote sensing; validation dataset; accuracy assessment; consistency analysis; stratified random sampling

Citation: Liu, J.; Ren, Y.; Chen, X. Regional Accuracy Assessment of 30-Meter GLC_FCS30, GlobeLand30, and CLCD Products: A Case Study in Xinjiang Area. *Remote Sens.* **2024**, *16*, 82. https://doi.org/10.3390/rs16010082

Academic Editors: Sawaid Abbas, Janet E. Nichol, Faisal M. Qamer and Jianchu Xu

Received: 26 October 2023
Revised: 20 December 2023
Accepted: 21 December 2023
Published: 25 December 2023

1. Introduction

Land cover (LC) monitoring plays a pivotal role in ecological environment governance, agricultural and forestry management, urban and rural planning, and ecological conservation. The rapid development of computer technology and the continuous sharing of medium- to fine-resolution remote sensing data, such as from Landsat and Sentinel, have ushered the world into an era of fine-resolution remote sensing [1,2]. When conducting large-scale macro-scale land cover research, a resolution of 30 m is considered optimal [3,4]. Moreover, this resolution has an advantageous data size, enabling more efficient data processing and storage.

In recent years, many 30-m fine-resolution LC products have been produced. For example, Chen et al. (2015) developed the well-known GlobeLand30 using a pixel-object-knowledge-based (POK) classification strategy based on multi-temporal Landsat and HJ-1A/B imagery [3], Zhang et al. (2021) generated a global-scale 30-m Global Land Cover with Fine Classification System (GLC_FCS30) product based on the spectral library of land cover and time-series Landsat data [1], and Yang et al. (2021) utilized Landsat data and the random forest classifier to generate the annual China Land Cover Dataset (CLCD) based on the Google Earth Engine (GEE) [5]. Similarly, Gong et al. (2013) produced the Finer Resolution Observation and Monitoring of Global Land Cover (FROM_GLC) product using the Landsat thematic mapper (TM) and enhanced thematic mapper Plus (ETM+) data [6]. The United States Geological Survey (USGS) collaborated with multiple federal agencies to produce the 30-m National Land Cover Database (NLCD), which covers the entire United States [7]. These large-scale fine-resolution LC products provide adequate information to observe the fundamental characteristics and trends of land cover.

However, the accuracies reported for the abovementioned large-scale LC products are generally towards global and national scales (Table 1). For instance, the accuracy of GlobeLand30 was evaluated using 38,644 globally distributed validation samples [3]. Similarly, the accuracy of GLC_FCS30 was assessed using 44,043 samples that were distributed across the globe [1]. For the assessment of CLCD accuracy, a total of 5463 validation points within China were used [5]. Nevertheless, due to the differences in scene complexity and data quality across regions, the accuracy of land cover products inevitably varies from region to region [3,8–10]. In some complex areas, the accuracy of LC products is even much lower than the overall accuracy of the report [11]. For example, the overall accuracy of GlobeLand30 is reported to be 83.5% worldwide. However, its accuracy significantly drops to approximately 50% in Africa [12]. Similarly, in the Central Asian region, the accuracy of GlobeLand30 was found to be only about 46% [13]. Cui et al. (2023) collected thirteen sets of global or national-scale land cover datasets. Through the visual interpretation of high-resolution images, ground "truth" samples were collected to evaluate the data accuracy across Northeast China [14]. Accordingly, the quantitative evaluation of the accuracy and consistency of existing 30-m land cover products at the regional scale is of great significance for the scientific use of these products.

Table 1. Summary of existing accuracy evaluation studies of 30-m LC products.

Validated LC Products	Validation Area	Sample Quantity	Literature
GLC_FCS30	Globe	44,043	Zhang et al. (2021) [1]
CLCD	China	5463	Yang et al. (2021) [5]
GlobeLand30	Globe	38,644	Chen et al. (2015) [3]
FROM-GLC30	Globe	38,664	Zhao et al. (2014) [15]
GlobeLand30, FROM-GLC30, and GLC_FCS30	Globe	79,112	Zhao et al. (2023) [16]

The validation of land cover products is typically conducted by randomly selecting validation points and visually interpreting the labels of each point using high-resolution imagery for comparison with the product [17]. Therefore, in order to accurately evaluate the accuracy of land cover products in a specific region, a scientific high-density validation dataset is a crucial prerequisite [10,15,18–20]. The construction of a high-density verification dataset in a scientific manner, with sufficient quantities of representative samples covering different categories in each region, is a crucial issue. The research conducted by Sahr (2011) demonstrated that dividing the study area into equal-sized grids and assigning validation points at the grid level ensures coverage of sample points in all regions [21]. Zhao et al. (2023) demonstrated that considering multiple factors such as climate and population when allocating sampling points and employing a stratified random sampling approach can result in a more reasonable distribution of validation points [16]. To ensure an adequate number of samples in complex areas and appropriate placement of sampling points in homogeneous regions, this study employs a multi-indicator equal-area stratified random sampling method [15]. This approach not only increases the sample size of rare land cover categories but also effectively reduces the standard deviation of accuracy assessment [16]. Additionally, the use of equal-area stratified random sampling ensures the balance of different categories during the sample selection process, thereby enhancing the representativeness and reliability of the assessment results.

To this end, we first employed an equal-area stratified random sampling approach with climate, population density, and landscape heterogeneity information as constraints, along with the hexagonal discrete global grid system (HDGGS) as basic sampling grids to develop a High-Density Land cover Validation dataset in Xinjiang (HDLV-XJ). Then, the consistency and accuracy of three widely used 30-m resolution LC products (GLC_FCS30, CLCD, GlobeLand30) in 2020 were analyzed based on the developed validation dataset. Additionally, the suitability of these three LC products under different environment conditions was for the first time analyzed and revealed. The results of this study can provide an important reference value for the accuracy assessment of existing LC products on regional scales and for the application of LC products in specific scenarios. Additionally, our HDLV-XJ dataset can provide valuable data support for regional-scale research.

2. Study Area and Data

2.1. Study Area

The study area chosen for this research was the Xinjiang Uyghur Autonomous Region, situated between 73°40′ and 96°18′ east longitude and 34°25′ and 48°10′ north latitude (Figure 1). As the largest provincial-level administrative region in China, Xinjiang encompasses an extensive land area, accounting for over one-sixth of the country's total land area, with a precise measurement of 1,664,897 km^2 [22]. Located in the Northwestern part of China, Xinjiang lies in the hinterland of the Eurasian continent and is considered one of the provinces farthest from the ocean within China. Xinjiang has a typical temperate continental climate that has only a few and unevenly distributed precipitation events [23]. The region comprises diverse land cover types, including primarily deserts, grasslands, and croplands. The distribution of vegetation is predominantly observed in mountainous and oases [24].

2.2. The 30-Meter Global Land Cover Products

Three widely used LC products were selected for evaluation: GlobeLand30, GLC_FCS30, and CLCD (Figure 2). Among them, GLC_FCS30 is a 30-m global land cover product developed using time-series Landsat data. The product covers the time span from 1985 to 2020, with update cycles occurring every 5 years. It incorporates 16 primary LCCS land cover types and 14 secondary land cover types (Table 2). The overall global accuracy of the GLC_FCS30 is reported as 82.5% [1]. CLCD is a 30-m-resolution annual land cover dataset for China. This product is derived from the utilization of Landsat data in conjunction with a random forest classifier. The dataset includes yearly land cover information for China from 1985 to 2019. Its classification system includes nine land cover types (Table 2). The overall accuracy of the CLCD reaches 79.31% [5]. GlobeLand30 is a high-precision global land cover product with a spatial resolution of 30 m. It was updated with three sets of data, in 2000, 2010, and 2020. The development of GlobeLand30 was based on the "Pixel-Object-Knowledge" (POK) method. Its classification system includes 10 land cover types (Table 2), and the overall accuracy is 83.5% [3]. By utilizing the vector boundaries of Xinjiang, the original LC products were clipped and reclassified to standardize their classification system, as shown in Figure 2.

Figure 1. Study area and the land cover type along with the corresponding examples of their appearance on Google Earth: (source: Google Earth).

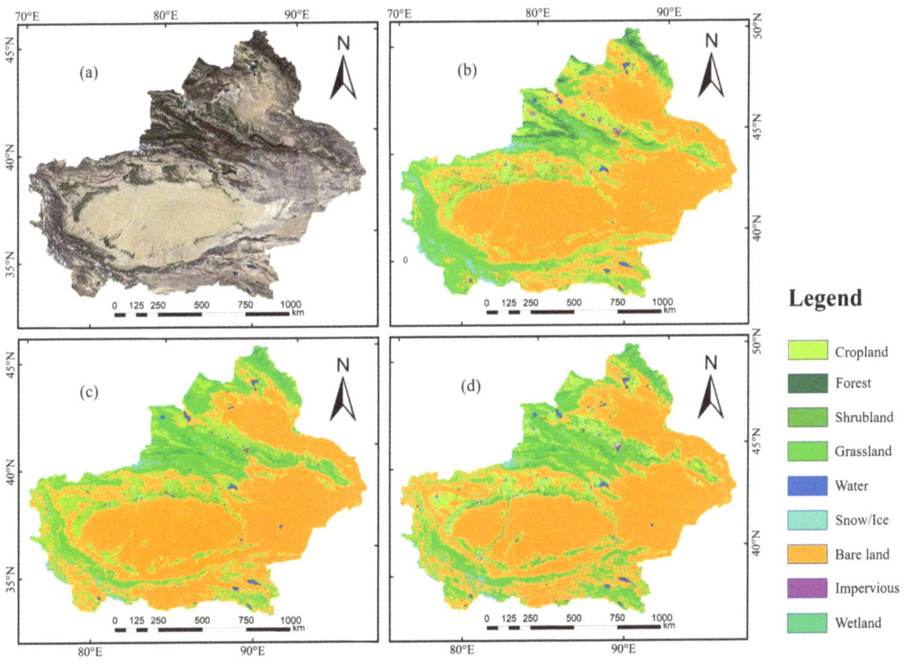

Figure 2. The 30-m global land cover products in harmonized classification systems: (**a**) remote sensing image of Xinjiang, (**b**) GLC_FCS30, (**c**) CLCD, and (**d**) GlobeLand30.

Table 2. The classification systems of the three LC products. The classification system for GLC_FCS30, CLCD, and GlobeLand30 were sourced from Zhang et al. (2021) [1], Yang et al. (2021) [5], and Chen et al. (2015) [3], respectively. Note: the land cover types we used in this study are in bold.

CLCD	Id	GlobeLand30	Id	GLC_FCS30	Id
Cropland	1	Cultivated land	10	Rain-fed cropland	10
				Herbaceous cover	11
				Tree or shrub cover (orchard)	12
				Irrigated cropland	20
Forest	2	Forest	20	Evergreen broadleaved forest	50
				Deciduous broadleaved forest	60
				Closed deciduous broadleaved forest	61
				Open deciduous broadleaved forest	62
				Evergreen needleleaved forest	70
				Closed evergreen needleleaved forest	71
				Open evergreen needleleaved forest	72
				Deciduous needleleaved forest	80
				Closed deciduous needleleaved forest	81
				Open deciduous needleleaved forest	82
				Mixed-leaf forest	90
Shrub	3	Shrubland	40	Shrubland	120
				Evergreen shrubland	121
				Deciduous shrubland	122
Grassland	4	Grassland	30	Grassland	130
Wetland	9	Wetland	50	Wetlands	180
Impervious	8	Artificial surfaces	80	Impervious surfaces	190
Bare land	7	Bare land	90	Lichens and mosses	140
				Sparse vegetation	150
				Sparse shrubland	152
				Sparse herbaceous cover	153
				Bare areas	200
				Consolidated bare areas	201
				Unconsolidated bare areas	202
Water	5	Water bodies	60	Water body	210
Snow/Ice	6	Permanent snow and ice	100	Permanent ice and snow	220
		Tundra	70		

2.3. *Climate Zone and Population Datasets*

In order to scientifically generate sufficient validation samples in different regions and to avoid spatial imbalance in sample distribution, climate zone and population datasets were also collected. The climate zone data used in this study were from the Köppen–Geiger climate spatial distribution dataset, which was derived based on temperature and precipitation observations from global climate models (http://koeppen-geiger.vu-wien.ac.at/shifts.htm (accessed on 16 June 2023)) [25]. The climate zones are defined as follows: primary climate groups include equatorial (A), arid (B), warm temperate (C), snow (D), and polar (E). Precipitation categories include desert (W), steppe (S), fully humid (f), summer dry (s), winter dry (w), and monsoonal (m). Temperature categories include hot arid (h), cold arid (k), hot summer (a), warm summer (b), cool summer (c), extremely continental (d), polar frost (F), and polar tundra (T). The Köppen–Geiger climate classification spatial distribution data are shown in Figure 3. We clipped the Köppen-Geiger climate spatial distribution dataset using the vector boundary data of Xinjiang (Figure 3b). Xinjiang includes 10 different climate types, namely BSk, BWk, Dfa, Dfb, Dfc, Dsa, Dsb, Dsc, Dwc, and ET. Due to the limited spatial extent and similar characteristics of certain climate classifications (Dsa, Dsb, Dfa) in Xinjiang, this study employed the merging method utilized by Olofsson et al. (2012) for collecting validation data on global land cover products [25–27]. Classes were manually merged into different classes. For example, Dsa: "Continental, dry and hot summer", Dsb: "Continental, dry and warm summer", and BSk: "Arid, desert, cold" were merged into one desert class in Xinjiang, and Dsa: "Continental, dry summer, hot summer" and Dsa: "Continental, dry summer, warm summer" were merged with the continental forest class. Table 3 summarizes the initial 10 classes into the final 5 climate classifications.

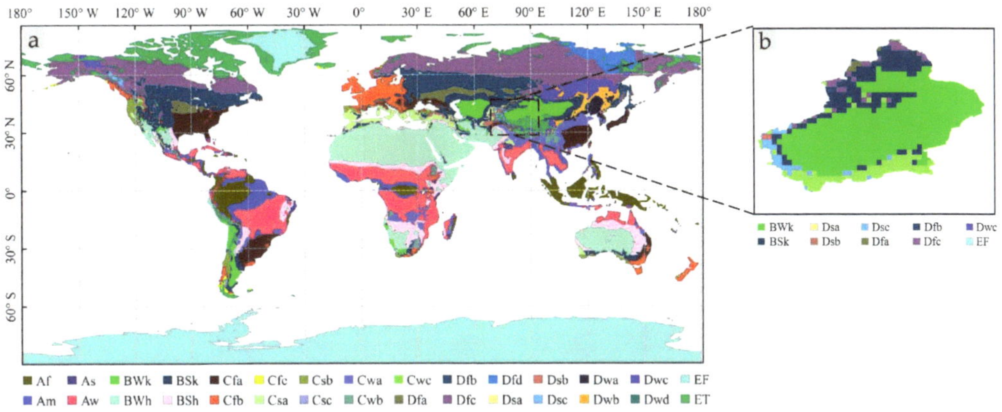

Figure 3. The Köppen-Geiger climate classification spatial distribution data for the world (**a**) and Xinjiang (**b**).

Table 3. Edited Köppen–Geiger climate classification in Xinjiang.

New Code	Climate	Climate Code
1	Desert	BWk
2	Grassland	BSk + Dsa + Dsb
3	Continental forest	Dfa + Dfb
4	Boreal forest	Dsc + Dwc + Dfc
5	Frost	EF

The Global Human Settlement Layer (GHSL) dataset provides spatial information on global population distribution and residential areas. It offers four levels of global geospatial population data with resolutions of 100 m, 1 km, and more, covering the period from 1975 to 2030 with a 5-year interval (https://ghsl.jrc.ec.europa.eu/download.php?ds=pop (accessed on 16 June 2023)). Due to the availability of the latest global geospatial population data, GHSL-POP2020 was used as a source for stratified population data in this study. Based on the population density data in GHS-POP2020, the population data of the Xinjiang were reclassified using a threshold of five people per km^2 [28], resulting in the creation of the population and un-population layers.

2.4. The Existing Land Cover Validation Dataset

Two published global land cover validation datasets were collected in our study to cross-compare with our developed HDLV-XJ and to further illustrate the accuracy of our validation dataset. The first was the global validation dataset (SRS_Val) released by Zhao et al. (2023) [16]. The SRS_Val adopted a stratified random sampling method to distribute 79,112 validation samples globally in 2020, of which 659 validation points were distributed in Xinjiang. The second were the global land cover validation samples (GLV_2015) produced by Zhang et al. (2019) in 2015 [1]. The GLV_2015 was generated through integrating the GLCNMO2008 training dataset, VIIRS reference dataset, STEP reference dataset, global cropland reference data, and high-resolution imagery in Google Earth. This dataset contains a total of 403 sample points in Xinjiang. Both SRS_Val and GLV_2015 adopted a standardized classification system derived from UN-LCCS, and

GLV_2015 is only distributed in six categories in Xinjiang. The quantities and spatial distributions of the validation points in these two datasets are illustrated in Figure 4a,b. The validation samples of SRS_val are scattered throughout Xinjiang. The most prevalent land cover in SRS_Val is bare area (with 296 samples), followed by grassland (with 143 samples) and sparse vegetation (with 83 samples). Conversely, the sample points of GLV_2015 in the Xinjiang region are predominantly concentrated in the south. Furthermore, bare land presented the highest counts in the GLV_2015 (with 222 samples), followed by permanent ice and snow (with 96 samples) and sparse vegetation (with 56 samples). The land types in the two datasets were modified according to the classification system used in this study (Table 2). The number of samples corresponding to each class after adjustment is shown in Figure 4c.

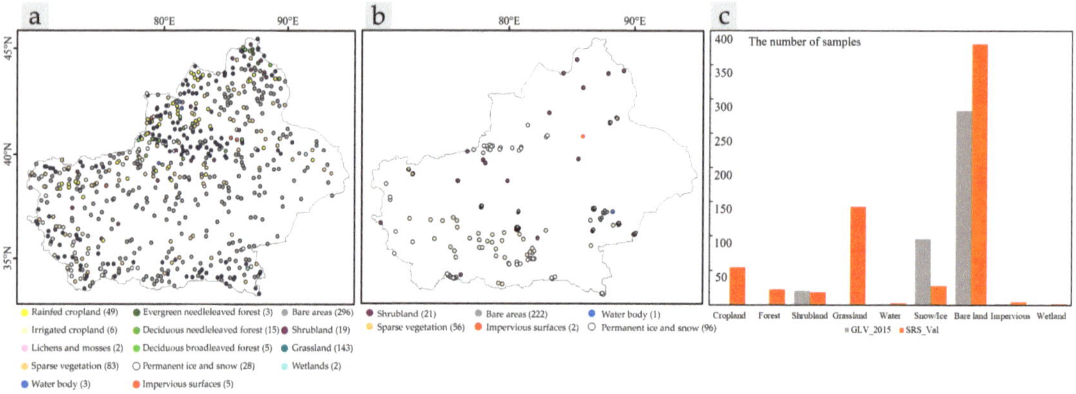

Figure 4. The quantities and spatial distributions of (**a**) SRS_Val and (**b**) GLV_2015, and (**c**) the number of samples corresponding to each class after LC classification system adjustment. Note: each number following the class represents the sample quantity for that category.

3. Methods

3.1. Harmonization of the Land Cover Classification Systems

This study encounters the challenge of dealing with the different classification systems employed by the LC products GLC_FCS30, CLCD, and GlobeLand30. When computing the consistency of LC products, the differences in classification systems used in LC products can affect the consistency and accuracy of those different products [8–10]. The classification systems of CLCD and GlobeLand30 solely encompass land cover's first-level classes, while GLC_FCS30 incorporates three-level classes; therefore, a classification system based on those of CLCD and GlobeLand30 was adopted (Table 2). It should be noted that the tundra in GlobeLand30 was not included in our classification system because this land cover class is almost nonexistent in the Xinjiang region. Meanwhile, the classification system of GLC_FCS30 was merged and adjusted to the categories used in this study (Table 2).

3.2. Construction of the Land Cover Validation Dataset in Xinjiang

The flowchart is illustrated in Figure 5. First, we performed a spatial overlay operation on the Köppen–Geiger climate spatial distribution data and GHSL-POP2020 data to obtain a climate–population density constraint layer. Furthermore, the Shannon diversity index (*SHDI*) was calculated to derive the landscape heterogeneity, and the hexagonal discrete global grid system (HDGGS) was used to generate basic sample-allocated hexagonal grids. Then, a spatial join operation was utilized to combine the above constraint layers to determine the number of samples distributed in each hexagonal grid. A total of 22,000 sample points were allocated within each grid based on the derived constraint layers. Finally, the category of each validation sample point was annotated based on visual interpretation to generate the HDLV-XJ.

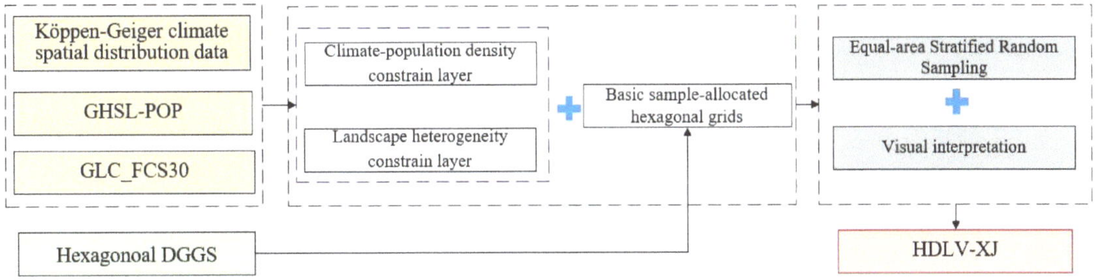

Figure 5. Flowchart for constructing the HDLV-XJ.

3.2.1. The Equal-Area Stratified Random Sampling Method Based on Multiple Indicator Constraints

To ensure a sufficient number of samples in complex areas and to position appropriate sampling points in homogeneous and heterogeneous areas, the equal-area stratified random sampling method was utilized, based on multiple indicators [16]. This method accomplishes two objectives: it augments the sample size for uncommon land cover categories and concurrently reduces the standard deviation in accuracy assessment [26]. The methodology primarily relies on two constraint indices, climate–population density data and landscape heterogeneity, to determine the number of validation points in each region. A hexagonal grid was constructed as the fundamental unit for assigning sampling points, in order to ensure adequate representation of each region and sufficient data for validation purposes.

First, since land cover information is intricately intertwined with climate change and human activities [29,30], climate and population density data were incorporated to generate a climate–population density layer, serving as the primary foundation for the layout of samples. Specifically, the population and un-population layers in Xinjiang were delineated based on the population density data. The climate data were reclassified into five categories (Table 3) and overlaid with the population and un-population layers to obtain the climate–population density layer.

Then, because landscape variables significantly influence the accuracy of LC products and classification results are generally more attainable in areas with homogeneous landscapes than those with heterogeneous landscapes [31], landscape heterogeneity was used as a quantitative measure of landscape fragmentation within the study area, enabling the development of a more robust land cover validation dataset. Therefore, in addition to establishing the climate–population density layer, this study also incorporated landscape heterogeneity as a criterion for the allocation of sample points. The objective was to allocate a greater number of sample points in highly heterogeneous areas, thereby establishing a more rigorous validation framework to evaluate the accuracy of the LC products [32]. The Shannon diversity index was used to describe the degree of landscape heterogeneity [33]. In a landscape ecosystem, the more diverse the land use types and the higher the fragmentation, the greater the amount of information contained in the patches, resulting in a higher calculated value of the Shannon diversity index (*SHDI*) [34]:

$$SHDI = -\sum_{i=1}^{m} (P_i)(\ln P_i) \quad (1)$$

where P_i is the proportion occupied by landscape patch type i and m is the total number of land cover types in the landscape patch; the GLC_FCS30 was utilized in this study to compute the *SHDI*. The calculated *SHDI* map was classified into different layers at intervals of 0.2 for the creation of a landscape constraint layer that represents landscape heterogeneity.

Finally, in order to ensure a representative sample distribution across the Xinjiang province, the hexagonal discrete global grid system (HDGGS) was employed to gather validation datasets [6,21]. The R programming language was used to generate HDGGS grids. These grids were then employed as the foundational units for allocating samples, while the generated climate–population density layer and the landscape heterogeneity layer were used to constrain the samples in each HDGGS grid. Specifically, all samples were first stratified and allocated to each constraint layer based on the proportional area of each constraint layer. Then, the number of HDGGS grids within each constraint layer was calculated, and samples within each constraint layer were equally distributed in each HDGGS grid.

3.2.2. Labeling HDLV-XJ Based on Visual Interpretation Method

Based on the aforementioned constraints, a total of 22,000 samples were allocated to the HDGGS grids based on the area of each separate region formed by the constraint layers. Since high-resolution imagery in Google Earth (version: 7.3.6.9345) is sufficient and often used to assist in the visual interpretation of datasets for validation purposes [35–37], Google Earth was used for the visual interpretation of the real land cover of each validation sample. It enables the careful examination of shapes, sizes, and patterns, thereby facilitating the assessment of human impacts on the Earth's surface [3,15,38]. A total of nine land cover types were classified, which encompassed cropland, forest, shrubland, grassland, water, permanent snow/ice, bare land, impervious, and wetland. Table 4 presents these land cover types, along with examples showcasing their appearance on Google Earth. In order to maintain the accuracy of the generated validation dataset, samples that were difficult to identify were directly excluded.

Table 4. The land cover types and examples of their appearance on Google Earth.

LC Type	Typical Imagery on Google Earth
Cropland	
Forest	
Shrubland	
Grassland	

Table 4. *Cont.*

LC Type	Typical Imagery on Google Earth
Water	
Snow/Ice	
Bare land	
Impervious	
Wetland	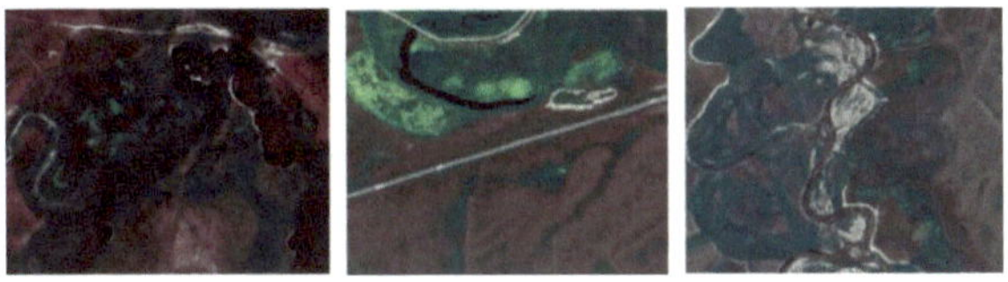

3.3. Accuracy and Consistency Assessment for Land Cover Products

3.3.1. Accuracy Assessment

The calculation of accuracy metrics, utilizing a confusion matrix, is a widely employed approach for quantitative remote-sensing accuracy assessment [31,39]. By utilizing the confusion matrix, it becomes possible to calculate the overall accuracy of the mapping results, as well as the user accuracy and producer accuracy for each land cover type. The confusion matrix offers a comprehensive evaluation of remote sensing classification outcomes, aiding in the comprehension of accuracy levels and the identification of error sources within the mapping results [40]. The calculation formulas for overall accuracy (*O.A.*), user's accuracy (*U.A.*), producer's accuracy (*P.A.*), and kappa coefficients are provided by

$$O.A. = \left(\sum\nolimits_{i=1}^{m} n_{ii}\right)/n \tag{2}$$

$$U.A. = \frac{n_{ii}}{n_{+i}} \tag{3}$$

$$P.A. = \frac{n_{ii}}{n_{i+}} \tag{4}$$

$$Kappa = \frac{n\sum_{i=1}^{m} n_{ii} - \sum_{i=1}^{m}(n_{+i} + n_{i+})}{n^2 - \sum_{i=1}^{m}(n_{+i} + n_{i+})} \tag{5}$$

respectively, where n_{ii} is the correctly classified pixel number of type i, m is the number of land cover types, n is the total pixel number in the study area, n_{i+} is the total pixel number of type i in the validation dataset, and n_{+i} is the total pixel number of type i in the ground truth data.

3.3.2. Consistency Analysis

The evaluation of LC product consistency can be performed considering both area and pixel levels. Area consistency examines the variations in the proportion of land cover types across different LC products. By comparing the respective areas of each land cover type among different products, it becomes possible to visually assess the consistency and diversity of these LC products [41–43]. The pixel consistency in LC products can reflect the spatial consistency of each land cover type. The pixel consistency is obtained using the spatial overlay method. The overall similarity coefficient (OS) and class similarity coefficient (CS) between different products can be calculated using [44]

$$OS = \frac{\sum_{1}^{n} XY_{ii}}{M} \times 100\% \tag{6}$$

$$CS_i = \frac{XY_{ii}}{(X_i+Y_i)/2} \times 100\% \tag{7}$$

where X_i is the number of pixels of the i-th LC type in LC product X; Y_i is the number of pixels of the i-th LC type in the comparison LC product Y; XY_{ii} is the number of consistent pixels between X and Y; n is the number of land cover types, which is 9 in this case; and M is the total number of pixels in the study area.

4. Results

4.1. The High-Density Land Cover Validation Dataset for Xinjiang

In order to generate a high-density and representative validation dataset for Xinjiang, the HDGGS was used to generate hexagonal grids as the basic unit for sample allocation. Figure 6a shows the results of the generated hexagonal grid units, including 1604 grids in the Xinjiang region. In addition, a total of 20,932 validation samples were generated within these grid units. The spatial distribution and distribution via the categories of these validation samples are illustrated in Figure 6b. The interpretation of land features relied on the comprehensive status observed in high-resolution imagery from Google Earth for 2020. It can be observed that the sample points are distributed across the study area. The equal-area stratified random sampling method employed in this study ensures the distribution of sample points in different areas.

Furthermore, in order to demonstrate the accuracy of the HDLV-XJ dataset that we developed, we merged the validation samples from SRS_Val and GLV_2015 to cross-compare with our HDLV-XJ product. Considering that the locations of the sample points in the three validation datasets do not correspond one-to-one, the nearest neighbor points were selected as the reference points for accuracy calculation [45]. The confusion matrix was computed using Equations (2)–(5), and the resulting values are presented in Table 5. When SRS_Val and GLV_2015 were used as reference benchmarks, the overall accuracy of our HDLV_XJ dataset exceeded 80% for both. Specifically, the producer's accuracy for bare land and snow/ice was around 90%. In the case of SRS_Val vs. HDLV-XJ, the producer's accuracy for grassland and cropland was above 90%. Although forest and shrubland showed a lower accuracy than the above land cover types, the agreement still reached 75%. Taking into account errors in the visual interpretation of the validation samples, as well as differences in the years represented by the validation datasets, the comparison results with the third-party dataset demonstrated the reliability of our HDLV-XJ dataset.

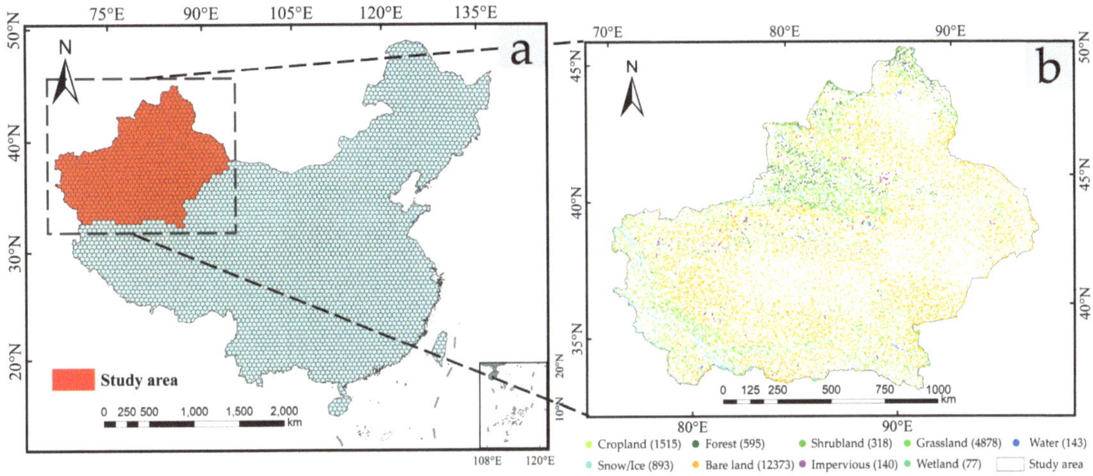

Figure 6. (**a**) HDGGS grids and (**b**) the spatial distribution of validation samples in HDLV-XJ.

Table 5. The accuracy of HDLV_XJ as compared with SRS_Val and GLV_2015.

Classes	SRS_Val vs. HDLV-XJ		GLV_2015 vs. HDLV-XJ	
	P.A.	U.A.	P.A.	U.A.
Cropland	100.00%	77.78%	0.00%	0.00%
Forest	75.00%	100.00%	0.00%	0.00%
Shrubland	75.00%	75.00%	0.00%	0.00%
Grassland	90.63%	82.86%	0.00%	0.00%
Water	0.00%	0.00%	0.00%	0.00%
Snow/Ice	75.00%	50.00%	94.44%	77.27%
Bare land	89.09%	100.00%	89.80%	100.00%
Impervious	0.00%	0.00%	0.00%	0.00%
Wetland	0.00%	0.00%	0.00%	0.00%
O.A.	88.68%		82.43%	

4.2. Accuracy Assessment of GLC_FCS30, GlobeLand30, and CLCD

The confusion matrices for the three LC products in Xinjiang were calculated using the validation dataset constructed for this study, utilizing Equations (2)–(5). Based on the obtained results, the GLC_FCS30 exhibits the highest overall accuracy of 88.10%. The GlobeLand30 follows with the second highest overall accuracy of 83.58%, while the CLCD demonstrates the lowest overall accuracy of 81.57%.

The GLC_FCS30 demonstrates an overall accuracy of 88.10%, along with a kappa coefficient of 0.799 (Table 6). Regarding the producer's accuracy, forest exhibits the highest accuracy, followed by bare land, shrubland, cropland, and water. However, impervious and wetland display lower accuracies. These findings suggest that regions with homogenous surface cover types occupying larger proportions in the study area generally exhibit higher accuracy levels. Conversely, complex surface cover types often exhibit confusion with other types. For instance, wetland, which possess particularly intricate spectra, are prone to confusion with vegetation [46]. According to Table 6, approximately 49.4% of wetland validation points were incorrectly classified as vegetation types, including cropland, forest, shrubland, and grassland. In terms of user's accuracy, cropland, water, permanent snow/ice, and bare land exhibit similar accuracies to the producer's accuracy. Water demonstrates the highest user's accuracy, reaching 96.24%. This indicates the product's remarkable capability to accurately assign samples to their respective water categories.

Table 6. The accuracy matrix for the GLC_FCS30.

		Classified									
		CRP	**FST**	**SHR**	**GRS**	**Wat**	**SI**	**BaL**	**IMP**	**WET**	**Total**
Reference	**CRP**	1369	4	59	13	0	0	61	7	2	1515
	FST	1	570	4	6	0	0	14	0	0	595
	SHR	1	1	294	7	0	0	15	0	0	318
	GRS	130	204	51	3679	0	20	788	1	5	4878
	Wat	1	0	3	1	128	2	6	1	1	143
	SI	0	1	2	89	2	781	18	0	0	893
	BaL	6	4	181	542	2	108	11,513	1	16	12,373
	IMP	18	0	16	3	0	0	17	86	0	140
	WET	12	3	9	14	1	1	16	0	21	77
	Total	1538	787	619	4354	133	912	12,448	96	45	20,932
	P.A.	90.36%	95.80%	92.45%	75.42%	89.51%	87.46%	93.05%	61.43%	27.27%	
	U.A.	89.01%	72.43%	47.50%	84.50%	96.24%	85.64%	92.49%	89.58%	46.67%	
	O.A.					88.10%					
	Kappa					0.798716					

Note: CRP: cropland; FST: forest; SHR: shrubland; GRS: grassland; Wat: water; SI: permanent snow/ice; BaL: bare land; IMP: impervious; WET: wetland.

The CLCD achieves an overall accuracy of 81.57%, with a kappa coefficient of 0.675 (Table 7). Considering producer's accuracy, bare land exhibits the highest accuracy at 91.01%, followed by water, cropland, and grassland. However, shrubland and wetland display relatively low producer's accuracies, possibly due to identification difficulties within the CLCD. In terms of user's accuracy, cropland and forest demonstrate higher accuracies at 93.51% and 87.84%, respectively, indicating strong classification abilities for these land cover types. Notably, the producer's accuracy and user's accuracy for the shrubland in the CLCD classifier are both zero, suggesting that no samples were correctly classified within this category in the training dataset for the CLCD classifier, as depicted in Figure 7c,d. This deficiency in training data representation for the shrubland potentially undermines the classification performance of this specific class.

Table 7. The accuracy matrix for the CLCD.

		Classified									
		CRP	**FST**	**SHR**	**GRS**	**Wat**	**SI**	**BaL**	**IMP**	**WET**	**Total**
Reference	**CRP**	1211	7	0	261	0	0	24	11	1	1515
	FST	3	289	0	295	0	1	4	3	0	595
	SHR	8	0	0	203	0	0	101	6	0	318
	GRS	26	29	0	3620	1	14	1173	14	1	4878
	Wat	0	1	0	11	115	1	11	4	0	143
	SI	0	0	0	52	11	551	279	0	0	893
	BaL	17	0	0	1040	10	26	11,261	19	0	12,373
	IMP	22	1	0	77	0	0	18	22	0	140
	WET	8	2	0	34	4	1	18	5	5	77
	Total	1295	329	0	5593	141	594	12,889	84	7	20,932
	P.A.	79.93%	48.57%	0.00%	74.21%	80.42%	61.70%	91.01%	15.71%	6.49%	
	U.A.	93.51%	87.84%	0.00%	64.72%	81.56%	92.76%	87.37%	26.19%	71.43%	
	O.A.					81.57%					
	Kappa					0.675249					

Note: CRP: cropland; FST: forest; SHR: shrubland; GRS: grassland; Wat: water; SI: permanent snow/ice; BaL: bare land; IMP: impervious; WET: wetland.

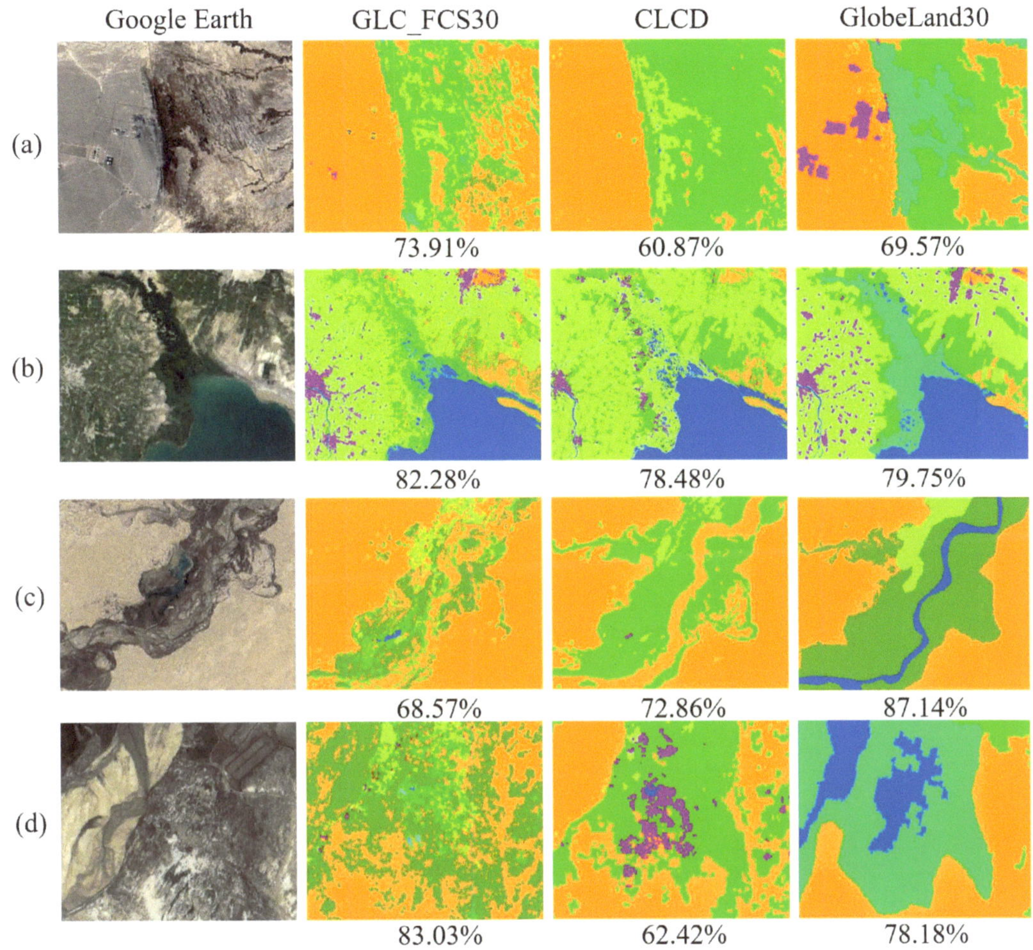

Figure 7. Misclassification errors of (**a**,**b**) wetland and (**c**,**d**) shrubland.

The overall accuracy of the GlobeLand30 is 83.58%, with a kappa coefficient of 0.717 (Table 8). Regarding the producer's accuracy, bare land obtains the highest accuracy at 90.28%, followed by cropland, water, and wetland. On the other hand, shrubland has a relatively low producer's accuracy of 18.24%, indicating significant challenges in classifying shrubland. As for the user's accuracy, cropland and bare land demonstrate higher accuracies of 90.13% and 89.60%, respectively, suggesting GlobeLand30's strong ability to correctly classify instances of cropland and bare land. However, the user's accuracy for shrubland is lower, at 32.77%. In summary, GlobeLand30 performs well in some land cover categories, such as cropland and bare land, but there is room for improvement in the classification of other categories.

In conclusion, it is evident that complex land cover types are more susceptible to misclassification. For example, based on Table 6, it is apparent that wetlands in GLC_FCS30 exhibit a higher confusion ratio, with over 50% of validation samples being misclassified as other types, as demonstrated in Figure 7a,b. There are also numerous instances of misclassification between similar land cover types. In particular, approximately 20% of shrubland was classified incorrectly as forest and grassland, as depicted in Figure 7c,d.

Table 8. The accuracy matrix for the GlobeLand30.

		Classified									
		CRP	**FST**	**SHR**	**GRS**	**Wat**	**SI**	**BaL**	**IMP**	**WET**	**Total**
Reference	**CRP**	1360	2	1	109	1	0	15	21	6	1515
	FST	5	274	6	297	1	1	9	2	0	595
	SHR	18	5	58	120	2	0	105	3	7	318
	GRS	48	58	46	3800	7	40	854	4	21	4878
	Wat	1	1	0	3	126	1	2	1	8	143
	SI	0	5	3	37	2	543	302	0	1	893
	BaL	59	9	61	988	25	33	11,170	8	20	12,373
	IMP	18	0	1	11	0	0	6	104	0	140
	WET	0	1	1	3	8	0	3	0	61	77
	Total	1509	355	177	5368	172	618	12,466	143	124	20,932
	P.A.	89.77%	46.05%	18.24%	77.90%	88.11%	60.81%	90.28%	74.29%	79.22%	
	U.A.	90.13%	77.18%	32.77%	70.79%	73.26%	87.86%	89.60%	72.73%	49.19%	
	O.A.					83.58%					
	Kappa					0.717466					

Note: CRP: cropland; FST: forest; SHR: shrubland; GRS: grassland; Wat: water; SI: permanent snow/ice; BaL: bare land; IMP: impervious; WET: wetland.

4.3. Consistency Analysis for Global Land Cover Products

Through the spatial overlay of the three LC products, a combination of visual mapping and quantitative expression approaches was employed to illustrate the overall pixel consistency between two or more products, as well as the pixel consistency for each land cover type.

To begin, the overall similarity coefficient (*OS*) and class similarity coefficient (*CS*) were computed for different products using Equations (6) and (7), as presented in Table 9. The overall similarity coefficient between CLCD and GlobeLand30 was 83.32%, whereas between GLC_FCS30 and CLCD, it was 78.46%. Notably, the lowest overall similarity coefficient of 75.16% was observed between GLC_FCS30 and GlobeLand30. The above analysis reveals that approximately 75% of pixels share the same land cover label between any two LC products. Additionally, through overlaying the three products, an overall similarity coefficient of 69.96% was obtained, indicating a lower coefficient compared to the pairwise similarities. Consequently, it can be inferred that the land cover types are completely identical in 69.96% of the study area.

Table 9. Overall similarity coefficient (*OS*) (%) and individual class similarity coefficient (*CS*) (%) among different products.

Similarity Coefficient	Three Maps	GLC_FCS30 vs. CLCD	CLCD vs. GlobeLamd30	GLC_FCS30 vs. GlobeLand30
Cropland	66.01%	69.76%	80.60%	64.47%
Forest	30.95%	42.48%	55.20%	35.27%
Shrubland	0.00%	0.00%	0.00%	2.10%
Grassland	37.76%	48.53%	67.80%	43.26%
Water	66.13%	71.81%	74.80%	61.89%
Snow/Ice	51.84%	55.52%	72.42%	47.02%
Bare land	74.14%	73.46%	90.41%	69.55%
Impervious	12.14%	17.64%	15.44%	40.84%
Wetland	0.78%	2.34%	8.96%	8.12%
OS	69.96%	78.46%	83.32%	75.16%

Subsequently, the class similarity coefficients were calculated among the three products. The analysis reveals that the class similarity coefficients for bare land, water, and cropland exceed 60%, indicating a spatial correspondence of over 60% between these land cover types (Figure 8). Remarkably, the average value of the class similarity coefficient for bare land stands is highest, at 77.14%. In contrast, wetland and shrubland exhibit the lowest class similarity coefficients, suggesting limited overlap among all three products. Additionally, the consistency for the impervious category is notably low, with an average class similarity coefficient of 21.52%.

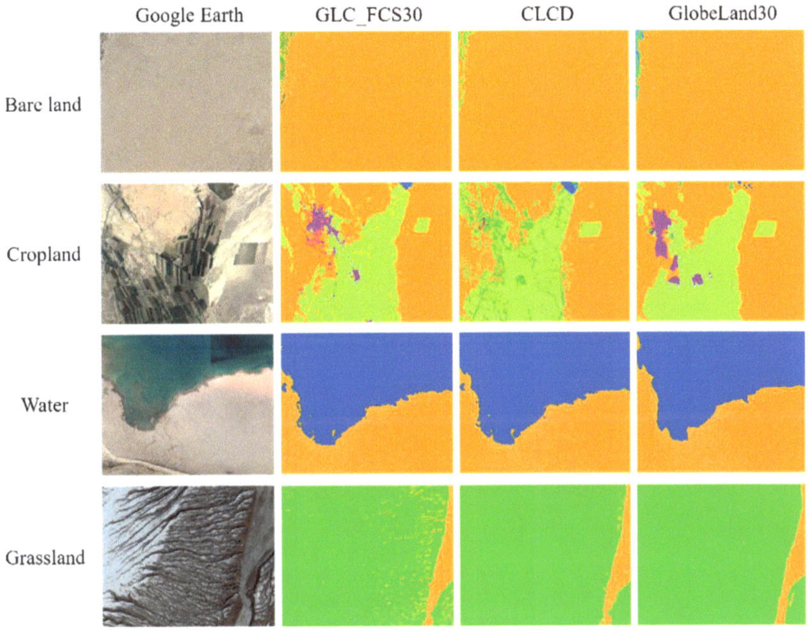

Figure 8. Examples of LC types with high consistency.

Simultaneously, the pixel consistency between the two products across different land cover types aligns with the aforementioned similarity of the three products. The highest consistency is observed in bare land, with class similarity coefficients of 73.46%, 90.41%, and 69.55% for GLC_FCS30 vs. CLCD, CLCD vs. GlobeLand30, and GLC_FCS30 vs. GlobeLand30, respectively. Following that, cropland exhibits the next highest consistency, with class similarity coefficients ranging from 64% to 80%. Conversely, the class similarity coefficients for shrubland and wetland are below 9%.

The average areas of the nine land classes for the three LC products, as presented in Table 10, can be sorted in descending order based on their average area: bare land, grassland, cropland, permanent snow/ice, forest, shrubland, water, impervious, and wetland. Notably, the average areas of bare land, grassland, and cropland surpass 90,000 km^2. Their respective average areas are 1,102,577.54 km^2; 350,050.94 km^2; and 98,989.83 km^2. Bare land covers 66.22% of the region, making it the predominant geographical landscape.

Table 11 presents the comparative area results for each land class in three LC products: GLC_FCS30, CLCD, and GlobeLand30. There is a relatively high consistency in the area measurements for bare land, cropland, and permanent snow/ice across the three products. The consistency is moderate for water and impervious. However, significant variations exist in the area measurements for some land classes, particularly for shrubland and wetland, indicating low consistency. The area of shrubland in GLC_FCS30 is nearly four times larger than in GlobeLand30, while CLCD only reports shrubland of 1.27 km^2, significantly smaller than both GLC_FCS30 and GlobeLand30. Moreover, GlobeLand30 indicates a larger wetland area of 9197.70 km^2 compared to the other two products. In contrast, CLCD reports a mere 500.83 km^2 of wetland area, accounting for a mere 0.03% of the total area.

Table 10. The average area of LC types of the LC products (unit: km^2).

LC Type	Mean Area	Percentage Mean
Cropland	98,989.83	5.95%
Forest	29,424.22	1.77%
Shrubland	18,850.94	1.13%
Grassland	350,050.94	21.03%
Water	11,369.57	0.68%
Snow/Ice	42,092.11	2.53%
Bare land	1,102,577.54	66.22%
Impervious	7339.36	0.44%
Wetland	4202.49	0.25%

Table 11. The areas of LC types of the LC products (unit: km^2).

LC Type	GLC_FCS30	Percentage in GLC_FCS30	CLCD	Percentage in CLCD	GlobeLand30	Percentage in GlobeLand30
Cropland	104,931.34	6.30%	87,744.63	5.27%	104,293.52	6.26%
Forest	48,311.59	2.90%	18,576.16	1.12%	21,384.91	1.28%
Shrubland	44,592.01	2.68%	1.27	0.00%	11,959.54	0.72%
Grassland	307,358.94	18.46%	383,017.16	23.01%	359,776.72	21.61%
Water	10,028.85	0.60%	10,792.64	0.65%	13,287.22	0.80%
Snow/Ice	55,086.92	3.31%	35,631.59	2.14%	35,557.83	2.14%
Bare land	1,084,205.57	65.12%	1,123,584.54	67.49%	1,099,942.51	66.07%
Impervious	7472.84	0.45%	5048.18	0.30%	9497.05	0.57%
Wetland	2908.94	0.17%	500.84	0.03%	9197.70	0.55%

An analysis of area consistency between GLC_FCS30 and GlobeLand30 exposes notable variations across different land cover types. The areas of bare land in both products exhibit high similarity, encompassing approximately 66% of the total study area. A similar pattern emerges in cropland, where GLC_FCS30 reports an area of 104,931.34 km^2 and GlobeLand30 reports an area of 104,293.52 km^2. However, for other categories, the consistency between the two products is relatively low.

5. Discussion

5.1. The Advantages of the HDLV-XJ Dataset

In this study, we combined multiple indicators and the equal-area stratified sampling method to generate over 20,000 validation samples in Xinjiang province. The high-density HDLV-XJ validation dataset ensured a sufficient quantity of validation data in both homogeneous and heterogeneous regions. In order to demonstrate the advantages of our dataset, the *SHDI* of Xinjiang was calculated and categorized into 10 levels to exhibit the degree of landscape fragmentation in this region (Figure 9a). A higher level indicates a greater diversity of land-use types and a higher degree of fragmentation. Figure 9b illustrates the sample proportions at different landscape heterogeneities. In addition, we also assigned the same number of samples into this region using random sampling as a comparison. As illustrated in Figure 9b, the sample proportion of HDLV-XJ is higher than the distribution proportion of random sampling results in the heterogeneous regions. The random sampling allocation results in an excessive number of samples in homogeneous regions, resulting in a lower representation in heterogeneous regions. As a comparison, our HDLV-XJ dataset takes into account the distribution of samples in heterogeneous regions while ensuring that homogeneous regions are allocated with sufficient samples.

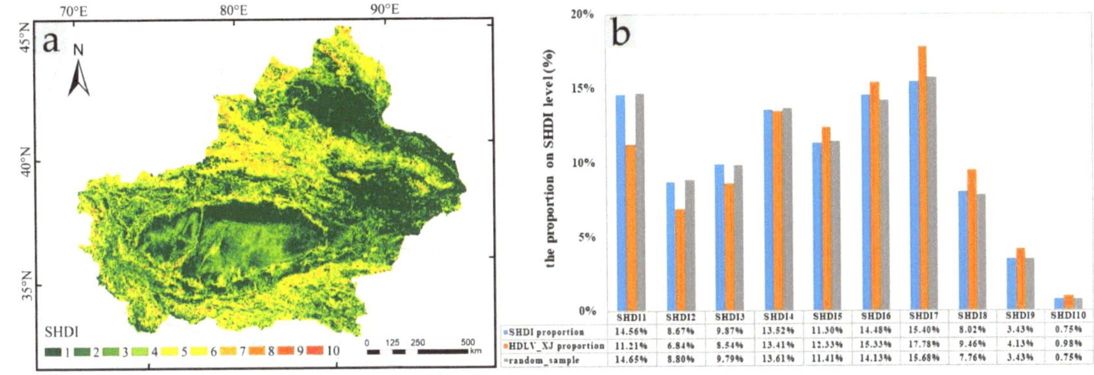

Figure 9. (**a**) The Shannon diversity index (*SHDI*) and (**b**) sample proportions of different landscape heterogeneity levels. The calculated *SHDI* map was classified into different layers at intervals of 0.2.

In addition, by comparing the validation point quantities of two global land cover validation datasets (SRS_Val, GLV_2015) and our HDLV-XJ dataset across different land cover types (Table 12), our HDLV-XJ dataset showed considerably higher sample numbers for each land cover type compared to the other datasets. This indicates that the global-scale validation data cannot accommodate regional-scale accuracy assessments. Our HDLV-XJ provides representative validation data with sufficient samples for rare categories (such as water (143 HDLV-XJ samples vs. 3 SRS_Val samples and 1 GLV_2015 samples) and impervious (140 HDLV-XJ samples vs. 5 SRS_Val samples and 2 GLV_2015 samples), enabling a more accurate assessment of the accuracy of land cover products in the Xinjiang area. Our HDLV-XJ product provides scientific data to support the use of land cover products in the Xinjiang region.

Table 12. The comparison of the validation point quantities for each LC type among the three datasets.

LC Type	CRP	FST	SHR	GRS	Wat	SI	BaL	IMP	WET	Total
HDLV-XJ	1515	595	318	4878	143	893	12,373	140	77	20,932
SRS_Val	55	23	19	143	3	28	381	5	2	659
GLV_2015	0	0	32	0	1	96	283	2	0	403

Note: CRP: cropland; FST: forest; SHR: shrubland; GRS: grassland; Wat: water; SI: permanent snow/ice; BaL: bare land; IMP: impervious; WET: wetland.

5.2. Analysis of the Relationship between Different Environment Conditions and the Performance of Land Cover Products

Due to the potential impact of landscape heterogeneity on LC mapping accuracy [26,47–49], we conducted an analysis to examine the relationship between landscape heterogeneity and the accuracy of LC products (Figure 10). The results showed that the accuracy of all three LC products significantly decreased (p-value: <0.05) with increasing landscape heterogeneity. Specifically, the LC product with the highest slope of decrease was CLCD (with a slope of −0.32), followed by GLC_FCS30 (with a slope of −0.22), and finally GlobeLand30 (with a slope of −0.17). Furthermore, the accuracy of all three LC products can generally reach 80% in homogeneous areas (low heterogeneity). In particular, the accuracy of GLC_FCS30 is close to 0.9 when the heterogeneity is below 0.1. This suggests that the degree of landscape heterogeneity indeed has a significant negative impact on LC mapping accuracy, and the results of the three products are more reliable in homogeneous areas. Additionally, GlobeLand30 generally exhibits the most robust performance in areas with high landscape heterogeneity. As can also be observed from Figure 10d–k, the consistency of classification results for the three LC products varied significantly in areas with high landscape heterogeneity, indicating a greater degree of uncertainty. The accuracy in these regions ranges from 60% to 80%. Therefore, in order to enhance the accuracy of LC products in complex regions, producers may need to pay more attention to transitional areas where landscape heterogeneity is high. These areas exhibit higher probabilities of misclassification compared to homogeneous regions.

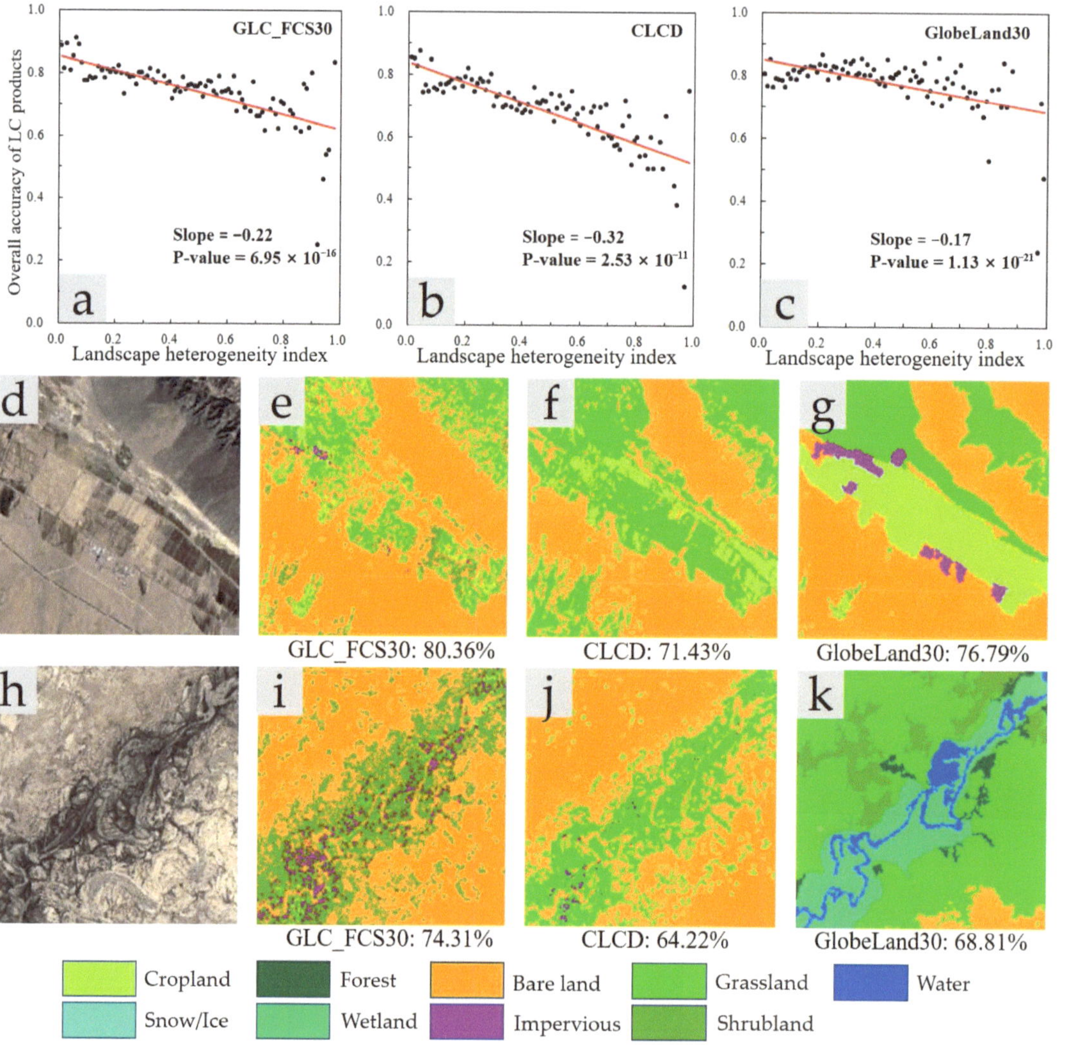

Figure 10. (**a–c**) The relationship between landscape heterogeneity and overall accuracy for LC products and (**d–k**) the comparison of accuracy among three products in heterogeneous regions.

Then, we analyzed the relationship between the area of land cover types and the mapping classification accuracy (Figure 11). It can be seen that the accuracy of most LC types in GLC_FCS30 is relatively stable, particularly performing well in cases with small areas. However, the classification accuracy for impervious surfaces and wetlands is still limited. Due to their strong spectral heterogeneity, impervious surfaces and wetlands are also recognized as land cover classes that are relatively difficult to classify accurately [3]. As a comparison, apart from water bodies, the mapping accuracy of each land cover class in CLCD decreases noticeably as the area decreases, with a correlation coefficient (r) of 0.56

between area and accuracy. Due to the significant differences in spectral characteristics between water bodies and other land cover classes [50], water bodies are relatively easy to identify and can maintain a high level of accuracy in CLCD. Furthermore, GlobeLand30 exhibits a similar pattern to CLCD, but its accuracy for impervious surfaces and wetlands, which are small-area land classes, is still higher than that of GLC_FCS and CLCD. This may be attributed to the incorporation of manual post-processing after classification. Therefore, GLC_FCS30 performs the best in handling small-area land classes, but its classification ability for impervious and wetland still needs improvement. On the other hand, GlobeLand30 demonstrates good classification performance in impervious and wetland classes.

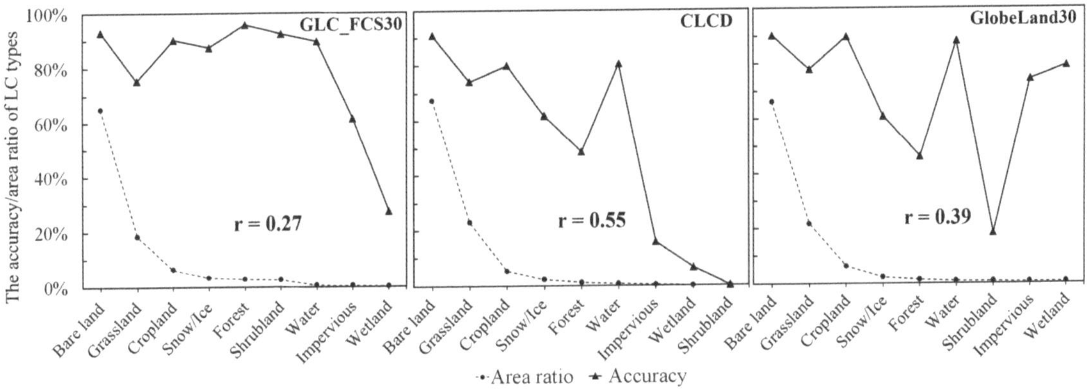

Figure 11. The relationship between area ratio and accuracy of LC types.

Based on the above, the performance of CLCD is more sensitive to landscape fragmentation compared to GlobeLand30 and GLC_FCS30 (Figure 10). The mean landscape heterogeneity index of Xinjiang is approximately 0.65, which is relatively high; this may be the reason for the relatively lower accuracy of CLCD compared to the other two LC products. Furthermore, it can be seen that GlobeLand30 and CLCD are more sensitive to the area ratio of land classes compared to GLC_FCS30 (Figure 11). Because our HDLV-XJ is a regionally intensive validation dataset with a high density of samples, there are still a significant number of samples (as shown in Table 12), even in land classes with relatively low coverage (such as cropland, snow and ice, forest, and shrubland) in Xinjiang. Therefore, this leads to lower assessment accuracies of CLCD and GlobeLand30 for these land classes, which in turn results in the highest validation accuracy for GLC_FCS30 in Xinjiang. The differences in mapping methods may be the main reason for the variations in accuracy among the three LC products. From Table 13, it can be seen that CLCD and GlobeLand30 are both based on global classification models, while GLC_FCS30 is based on a local adaptive classification model. Due to the requirement of ensuring overall accuracy across a large-scale area during training, there may be significant differences in the number of samples available in local areas when building global classification models [51,52].In contrast, the local adaptive classification strategy divides the large-scale area into different sub-regions and constructs training data within each sub-region [1]. Therefore, this mapping strategy can better balance the number of training data for each land cover in local areas [1]. However, GLC_FCS30 still faces challenges in accurately extracting land cover types such as wetlands and impervious surfaces. On the other hand, GlobeLand30 incorporates manual post-processing in its mapping process, which further improves the accuracy of these fragmented land cover types [3]. As a result, GlobeLand30 has a higher level of accuracy in these cases. Therefore, when conducting large-scale LC mapping, the local adaptive classification strategy may be more appropriate. Additionally, producers should also pay attention to individually addressing land cover types that are difficult to accurately identify, such as wetlands and impervious surfaces. This ensures that these specific land cover types receive appropriate handling and improves the overall accuracy of the mapping process.

Table 13. The mapping methods of 30-m GLC_FCS30, GlobeLand30, and CLCD.

LC Product	Method	Literature
GLC_FCS30	Local adaptive random forest models were trained for each $5^{\circ} \times 5'$ geographical grid element to generate the land-cover maps.	Zhang et al. (2021) [1]
CLCD	A global random forest classifier was trained to classify the whole of China.	Yang et al. (2021) [5]
GlobeLand30	A global pixel- and object-based classification model was applied to classify global land covers, and a knowledge-based interactive post-process was applied to improve the mapping accuracy.	Chen et al. (2015) [3]

5.3. Limitations

There are still some limitations in the present research, mainly in three aspects: (1) The constructed HDLV-XJ dataset introduced climatic factors, population density, and a hexagonal grid combined with landscape heterogeneity to design the sampling scheme, ensuring a representative and effective sample layout. However, this dataset is only applicable for land cover validation for Xinjiang in 2020, which hinders its use in long-term and large-scale accuracy assessment studies. To address this limitation, the study will explore extending the validation dataset through the combination of time-series remote sensing images and fitting algorithms and encourage public participation in the construction and interpretation of the validation dataset through open collaboration. (2) The generated validation dataset can only validate the nine primary land cover types and is not suitable for validating the secondary classification system. To overcome this limitation, the study will incorporate multiple sources of geospatial data and design a more detailed classification system. This will allow for the interpretation and validation of more detailed land cover categories. (3) Considering the importance of 30-m resolution land cover products in fine-resolution long-term monitoring, we only validated three mainstream 30-m land cover products in this study. However, with the recent availability and sharing of 10-m-resolution Sentinel data, 10-m land cover products are also gradually being generated. In the future, we will also include 10-m land cover products for further evaluation.

6. Conclusions

Fine-resolution LC products have been developed in recent years. However, the accuracy evaluation of the developed LC products is typically conducted at the global and national levels, with limited consideration for their accuracy and applicability in regional areas. Therefore, conducting a comprehensive accuracy assessment and consistency analysis of LC products in local areas is crucial for users to effectively compare the performance of different products. This study examined Xinjiang as the research area and constructed a high-density land cover validation dataset (HDLV-XJ, containing 20,932 validation samples) based on multi-source remote sensing data for Xinjiang. The accuracy and consistency of three mainstream land cover products in 2020 with a resolution of 30-m (GLC_FCS30, CLCD, GlobeLand30) were analyzed based on the constructed validation dataset.

The results indicated that the CLC_FCS30 exhibited the highest overall accuracy (88.10%) in Xinjiang, followed by GlobeLand30 (with an overall accuracy of 83.58%). By contrast, CLCD demonstrated the lowest overall accuracy of 81.57%. In terms of consistency, Xinjiang demonstrates high-consistency patterns for bare land, farmland, and water. Specifically, the average consistencies between the GLC_FCS30 and GlobeLand30, GLC_FCS30 and CLCD, and GlobeLand30 and CLCD are 77.14%, 70.21%, and 68.66%, respectively. However, this consistency drops significantly when it comes to wetland and shrubland, with a value of less than 1% for each. Furthermore, among the LC products, GlobeLand30 demonstrated the highest performance in regions characterized by high landscape fragmentation. On the other hand, GLC_FCS30 emerged as the superior product in areas with uneven proportions of land cover types. Additionally, the utilization of a local adaptive classification mapping strategy offers significant advantages in enhancing the accuracy of land cover mapping. The study provided important insights for the application of current land cover products in Xinjiang, uncovering both the accuracy and limitations associated with these products.

Author Contributions: Conceptualization, X.C. and J.L.; funding acquisition, X.C.; methodology, X.C., J.L. and Y.R.; software, J.L. and Y.R.; visualization, J.L. and Y.R.; writing—original draft preparation, J.L. and X.C.; writing—review and editing, J.L., Y.R. and X.C. All authors have read and agreed to the published version of the manuscript.

Funding: This research was funded by the National Natural Science Foundation of China (grant no. 42301336) and the National Key Research and Development Program of China (grant no. 2021YFE0117800).

Data Availability Statement: Our code and developed high-density land cover validation dataset for Xinjiang (HDLV-XJ) in 2020 are now available at https://doi.org/10.5281/zenodo.10029022.

Acknowledgments: The authors sincerely thank the production agencies that provided free land cover datasets and validation reference datasets.

Conflicts of Interest: The authors declare no conflicts of interest.

References

1. Zhang, X.; Liu, L.; Chen, X.; Gao, Y.; Xie, S.; Mi, J. GLC_FCS30: Global land-cover product with fine classification system at 30 m using time-series Landsat imagery. *Earth Syst. Sci. Data* **2021**, *13*, 2753–2776. [CrossRef]
2. Chen, X.; Liu, L.; Zhang, X.; Li, J.; Wang, S.; Gao, Y.; Mi, J. Long-term water clarity patterns of lakes across China using Landsat series imagery from 1985 to 2020. *Hydrol. Earth Syst. Sci.* **2022**, *26*, 3517–3536. [CrossRef]
3. Chen, J.; Chen, J.; Liao, A.; Cao, X.; Chen, L.; Chen, X.; He, C.; Han, G.; Peng, S.; Lu, M. Global land cover mapping at 30 m resolution: A POK-based operational approach. *ISPRS J. Photogramm. Remote Sens.* **2015**, *103*, 7–27. [CrossRef]
4. Giri, C.; Pengra, B.; Long, J.; Loveland, T.R. Next generation of global land cover characterization, mapping, and monitoring. *Int. J. Appl. Earth Obs. Geoinf.* **2013**, *25*, 30–37. [CrossRef]
5. Yang, J.; Huang, X. The 30 m annual land cover dataset and its dynamics in China from 1990 to 2019. *Earth Syst. Sci. Data* **2021**, *13*, 3907–3925. [CrossRef]
6. Gong, P.; Wang, J.; Yu, L.; Zhao, Y.; Zhao, Y.; Liang, L.; Niu, Z.; Huang, X.; Fu, H.; Liu, S. Finer resolution observation and monitoring of global land cover: First mapping results with Landsat TM and ETM+ data. *Int. J. Remote Sens.* **2013**, *34*, 2607–2654. [CrossRef]
7. Homer, C.; Dewitz, J.; Jin, S.; Xian, G.; Costello, C.; Danielson, P.; Gass, L.; Funk, M.; Wickham, J.; Stehman, S. Conterminous United States land cover change patterns 2001–2016 from the 2016 national land cover database. *ISPRS J. Photogramm. Remote Sens.* **2020**, *162*, 184–199. [CrossRef]
8. Bai, Y.; Feng, M.; Jiang, H.; Wang, J.; Zhu, Y.; Liu, Y. Assessing consistency of five global land cover data sets in China. *Remote Sens.* **2014**, *6*, 8739–8759. [CrossRef]
9. Kang, J.; Wang, Z.; Sui, L.; Yang, X.; Ma, Y.; Wang, J. Consistency analysis of remote sensing land cover products in the tropical rainforest climate region: A case study of Indonesia. *Remote Sens.* **2020**, *12*, 1410. [CrossRef]
10. Tsendbazar, N.-E.; de Bruin, S.; Mora, B.; Schouten, L.; Herold, M. Comparative assessment of thematic accuracy of GLC maps for specific applications using existing reference data. *Int. J. Appl. Earth Obs. Geoinf.* **2016**, *44*, 124–135. [CrossRef]
11. Liu, L.; Zhang, X.; Gao, Y.; Chen, X.; Shuai, X.; Mi, J. Finer-resolution mapping of global land cover: Recent developments, consistency analysis, and prospects. *J. Remote Sens.* **2021**, *2021*, 5289697. [CrossRef]
12. Tsendbazar, N.-E.; De Bruin, S.; Fritz, S.; Herold, M. Spatial accuracy assessment and integration of global land cover datasets. *Remote Sens.* **2015**, *7*, 15804–15821. [CrossRef]
13. Sun, B.; Chen, X.; Zhou, Q. Uncertainty assessment of GlobeLand30 land cover data set over central Asia. *Int. Arch. Photogramm. Remote Sens. Spat. Inf. Sci.* **2016**, *41*, 1313–1317. [CrossRef]
14. Cui, P.; Chen, T.; Li, Y.; Liu, K.; Zhang, D.; Song, C. Comparison and Assessment of Different Land Cover Datasets on the Cropland in Northeast China. *Remote Sens.* **2023**, *15*, 5134. [CrossRef]
15. Zhao, Y.; Gong, P.; Yu, L.; Hu, L.; Li, X.; Li, C.; Zhang, H.; Zheng, Y.; Wang, J.; Zhao, Y. Towards a common validation sample set for global land-cover mapping. *Int. J. Remote Sens.* **2014**, *35*, 4795–4814. [CrossRef]
16. Zhao, T.; Zhang, X.; Gao, Y.; Mi, J.; Liu, W.; Wang, J.; Jiang, M.; Liu, L. Assessing the Accuracy and Consistency of Six Fine-Resolution Global Land Cover Products Using a Novel Stratified Random Sampling Validation Dataset. *Remote Sens.* **2023**, *15*, 2285. [CrossRef]
17. Stehman, S.V.; Foody, G.M. Key issues in rigorous accuracy assessment of land cover products. *Remote Sens. Environ.* **2019**, *231*, 111199. [CrossRef]

18. Tsendbazar, N.; De Bruin, S.; Herold, M. Assessing global land cover reference datasets for different user communities. *ISPRS J. Photogramm. Remote Sens.* **2015**, *103*, 93–114. [CrossRef]
19. Olofsson, P.; Foody, G.M.; Herold, M.; Stehman, S.V.; Woodcock, C.E.; Wulder, M.A. Good practices for estimating area and assessing accuracy of land change. *Remote Sens. Environ.* **2014**, *148*, 42–57. [CrossRef]
20. Wickham, J.D.; Stehman, S.V.; Gass, L.; Dewitz, J.; Fry, J.A.; Wade, T.G. Accuracy assessment of NLCD 2006 land cover and impervious surface. *Remote Sens. Environ.* **2013**, *130*, 294–304. [CrossRef]
21. Sahr, K. Hexagonal discrete global grid systems for geospatial computing. *Arch. Fotogram. Kartogr. I Teledetekcji* **2011**, *22*, 363–376.
22. Liang, S.; Yi, Q.; Liu, J. Vegetation dynamics and responses to recent climate change in Xinjiang using leaf area index as an indicator. *Ecol. Indic.* **2015**, *58*, 64–76.
23. Yu, H.; Bian, Z.; Mu, S.; Yuan, J.; Chen, F. Effects of climate change on land cover change and vegetation dynamics in Xinjiang, China. *Int. J. Environ. Res. Public Health* **2020**, *17*, 4865. [CrossRef] [PubMed]
24. Xu, Y.; Yang, J.; Chen, Y. NDVI-based vegetation responses to climate change in an arid area of China. *Theor. Appl. Climatol.* **2016**, *126*, 213–222. [CrossRef]
25. Rubel, F.; Kottek, M. Observed and projected climate shifts 1901-2100 depicted by world maps of the Köppen-Geiger climate classification. *Meteorol. Z.* **2010**, *19*, 135. [CrossRef]
26. Olofsson, P.; Stehman, S.V.; Woodcock, C.E.; Sulla-Menashe, D.; Sibley, A.M.; Newell, J.D.; Friedl, M.A.; Herold, M. A global land-cover validation data set, part I: Fundamental design principles. *Int. J. Remote Sens.* **2012**, *33*, 5768–5788. [CrossRef]
27. Peel, M.C.; Finlayson, B.L.; McMahon, T.A. Updated world map of the Köppen-Geiger climate classification. *Hydrol. Earth Syst. Sci.* **2007**, *11*, 1633–1644. [CrossRef]
28. Yang, Z.; Lei, J.; Duan, Z.; Dong, J.; Su, C. Spatial distribution of population in Xinjiang. *Geogr. Res* **2016**, *35*, 2333–2346.
29. Dale, V.H. The relationship between land-use change and climate change. *Ecol. Appl.* **1997**, *7*, 753–769.
30. Valjarević, A.; Popovici, C.; Štilić, A.; Radojković, M. Cloudiness and water from cloud seeding in connection with plants distribution in the Republic of Moldova. *Appl. Water Sci.* **2022**, *12*, 262. [CrossRef]
31. Friedl, M.A.; Sulla-Menashe, D.; Tan, B.; Schneider, A.; Ramankutty, N.; Sibley, A.; Huang, X. MODIS Collection 5 global land cover: Algorithm refinements and characterization of new datasets. *Remote Sens. Environ.* **2010**, *114*, 168–182. [CrossRef]
32. Chen, F.; Chen, J.; Wu, H.; Hou, D.; Zhang, W.; Zhang, J.; Zhou, X.; Chen, L. A landscape shape index-based sampling approach for land cover accuracy assessment. *Sci. China Earth Sci.* **2016**, *59*, 2263–2274. [CrossRef]
33. Morris, E.K.; Caruso, T.; Buscot, F.; Fischer, M.; Hancock, C.; Maier, T.S.; Meiners, T.; Müller, C.; Obermaier, E.; Prati, D. Choosing and using diversity indices: Insights for ecological applications from the German Biodiversity Exploratories. *Ecol. Evol.* **2014**, *4*, 3514–3524. [CrossRef] [PubMed]
34. Spellerberg, I.F.; Fedor, P.J. A tribute to Claude Shannon (1916–2001) and a plea for more rigorous use of species richness, species diversity and the 'Shannon–Wiener'Index. *Glob. Ecol. Biogeogr.* **2003**, *12*, 177–179. [CrossRef]
35. Karydas, C.; Panagos, P. Towards an assessment of the ephemeral gully erosion potential in Greece using Google Earth. *Water* **2020**, *12*, 603. [CrossRef]
36. Sato, H.; Harp, E. Interpretation of earthquake-induced landslides triggered by the 12 May 2008, M7. 9 Wenchuan earthquake in the Beichuan area, Sichuan Province, China using satellite imagery and Google Earth. *Landslides* **2009**, *6*, 153–159. [CrossRef]
37. Taylor, J.R.; Lovell, S.T. Mapping public and private spaces of urban agriculture in Chicago through the analysis of high-resolution aerial images in Google Earth. *Landsc. Urban Plan.* **2012**, *108*, 57–70. [CrossRef]
38. Potere, D.; Schneider, A.; Angel, S.; Civco, D.L. Mapping urban areas on a global scale: Which of the eight maps now available is more accurate? *Int. J. Remote Sens.* **2009**, *30*, 6531–6558. [CrossRef]
39. Foody, G. Local characterization of thematic classification accuracy through spatially constrained confusion matrices. *Int. J. Remote Sens.* **2005**, *26*, 1217–1228. [CrossRef]
40. Stehman, S.V. Estimating area and map accuracy for stratified random sampling when the strata are different from the map classes. *Int. J. Remote Sens.* **2014**, *35*, 4923–4939. [CrossRef]
41. Bie, Q.; Shi, Y.; Li, X.; Wang, Y. Contrastive Analysis and Accuracy Assessment of Three Global 30 m Land Cover Maps Circa 2020 in Arid Land. *Sustainability* **2023**, *15*, 741. [CrossRef]
42. Fritz, S.; See, L. Identifying and quantifying uncertainty and spatial disagreement in the comparison of Global Land Cover for different applications. *Glob. Change Biol.* **2008**, *14*, 1057–1075. [CrossRef]
43. See, L.M.; Fritz, S. A method to compare and improve land cover datasets: Application to the GLC-2000 and MODIS land cover products. *IEEE Trans. Geosci. Remote Sens.* **2006**, *44*, 1740–1746. [CrossRef]
44. Wang, J.; Yang, X.; Wang, Z.; Cheng, H.; Kang, J.; Tang, H.; Li, Y.; Bian, Z.; Bai, Z. Consistency Analysis and Accuracy Assessment of Three Global Ten-Meter Land Cover Products in Rocky Desertification Region—A Case Study of Southwest China. *ISPRS Int. J. Geo Inf.* **2022**, *11*, 202. [CrossRef]
45. Pérez-Hoyos, A.; García-Haro, F.; San-Miguel-Ayanz, J. Conventional and fuzzy comparisons of large scale land cover products: Application to CORINE, GLC2000, MODIS and GlobCover in Europe. *ISPRS J. Photogramm. Remote Sens.* **2012**, *74*, 185–201. [CrossRef]
46. Ludwig, C.; Walli, A.; Schleicher, C.; Weichselbaum, J.; Riffler, M. A highly automated algorithm for wetland detection using multi-temporal optical satellite data. *Remote Sens. Environ.* **2019**, *224*, 333–351. [CrossRef]

47. Tsendbazar, N.; Herold, M.; De Bruin, S.; Lesiv, M.; Fritz, S.; Van De Kerchove, R.; Buchhorn, M.; Duerauer, M.; Szantoi, Z.; Pekel, J.-F. Developing and applying a multi-purpose land cover validation dataset for Africa. *Remote Sens. Environ.* **2018**, *219*, 298–309. [CrossRef]
48. Herold, M.; Mayaux, P.; Woodcock, C.; Baccini, A.; Schmullius, C. Some challenges in global land cover mapping: An assessment of agreement and accuracy in existing 1 km datasets. *Remote Sens. Environ.* **2008**, *112*, 2538–2556. [CrossRef]
49. Fitzpatrick-Lins, K. Comparison of sampling procedures and data analysis for a land-use and land-cover map. *Photogramm. Eng. Remote Sens.* **1981**, *47*, 343–351.
50. Lv, Z.; Wang, P.; Yan, C.; Nie, M.; Xiong, X.; Ding, M. Spectral characteristic of the waters with different sizes of particles: Impact of water quality and land-use type. *Environ. Sci. Pollut. Res.* **2023**, *30*, 9543–9557. [CrossRef]
51. Kang, J.; Yang, X.; Wang, Z.; Cheng, H.; Wang, J.; Tang, H.; Li, Y.; Bian, Z.; Bai, Z. Comparison of three ten meter land cover products in a Drought region: A case study in Northwestern China. *Land* **2022**, *11*, 427. [CrossRef]
52. Zhu, L.; Jin, G.; Gao, D. Integrating Land-Cover Products Based on Ontologies and Local Accuracy. *Information* **2021**, *12*, 236. [CrossRef]

remote sensing

Article

Impacts of Green Fraction Changes on Surface Temperature and Carbon Emissions: Comparison under Forestation and Urbanization Reshaping Scenarios

Faisal Mumtaz [1,2], Jing Li [1,2,*], Qinhuo Liu [1,2], Aqil Tariq [3,4], Arfan Arshad [5], Yadong Dong [1], Jing Zhao [1,2], Barjeece Bashir [1,2], Hu Zhang [1,2], Chenpeng Gu [1,2] and Chang Liu [1,2]

1 State Key Laboratory of Remote Sensing Sciences, Aerospace Information Research Institute, Chinese Academy of Sciences (AIRCAS), Beijing 100094, China
2 University of Chinese Academy of Sciences, Beijing 100049, China
3 Department of Wildlife, Fisheries and Aquaculture, College of Forest Resources, Mississippi State University, 775 Stone Boulevard, Starkville, MS 39762, USA
4 State Key Laboratory of Information Engineering in Surveying, Mapping and Remote Sensing, Wuhan University, Wuhan 430072, China
5 Department of Biosystems and Agricultural Engineering, Oklahoma State University, Stillwater, OK 74075, USA
* Correspondence: lijing01@radi.ac.cn or lijing200531@aircas.ac.cn

Abstract: Global land cover dynamics alter energy, water, and greenhouse gas exchange between land and atmosphere, affecting local to global weather and climate change. Although reforestation can provide localized cooling, ongoing land use land cover (LULC) shifts are expected to exacerbate urban heat island impacts. In this study, we monitored spatiotemporal changes in green cover in response to land use transformation associated with the Khyber Pakhtunkhwa (KPK) provincial government's Billion Tree Tsunami Project (BTTP) and the Ravi Urban Development Plan (RUDP) initiated by the provincial government of Punjab, both in Pakistan. The land change modeler (LCM) was used to assess the land cover changes and transformations between 2000 and 2020 across Punjab and KPK. Furthermore, a curve fit linear regression model (CFLRM) and sensitivity analysis were employed to analyze the impacts of land cover dynamics on land surface temperature (LST) and carbon emissions (CE). Results indicated a significant increase in green fraction of +5.35% under the BTTP, achieved by utilizing the bare land with an effective transition of 4375.87 km^2. However, across the Punjab province, an alarming reduction in green fraction cover by −1.77% and increase in artificial surfaces by +1.26% was noted. A significant decrease in mean monthly LST by −4.3 °C was noted in response to the BTTP policy, while an increase of 5.3 °C was observed associated with the RUDP. A substantial increase in LST by 0.17 °C was observed associated with transformation of vegetation to artificial surfaces. An effective decrease in LST by −0.21 °C was observed over the opposite transition. Furthermore, sensitivity analysis suggested that LST fluctuations are affecting the % of CO_2 emission. The current findings can assist policymakers in revisiting their policies to promote ecological conservation and sustainability in urban planning.

Keywords: Google Earth Engine (GEE); LULC transitions; vegetation dynamics; CFLR model; land use policies; Billion Tree Tsunami Project (BTTP); Ravi Urban Development Plan (RUDP); Master Plan 2050

Citation: Mumtaz, F.; Li, J.; Liu, Q.; Tariq, A.; Arshad, A.; Dong, Y.; Zhao, J.; Bashir, B.; Zhang, H.; Gu, C.; et al. Impacts of Green Fraction Changes on Surface Temperature and Carbon Emissions: Comparison under Forestation and Urbanization Reshaping Scenarios. *Remote Sens.* **2023**, *15*, 859. https://doi.org/10.3390/rs15030859

Academic Editor: Dino Ienco

Received: 13 December 2022
Revised: 23 January 2023
Accepted: 29 January 2023
Published: 3 February 2023

1. Introduction

In the 21st century, increased urbanization is intrinsically linked to the phenomenon of global warming (GW) [1,2], which is primarily associated with rapid urbanization [3,4]. With further urbanization, the percentage of the global population residing in cities is projected to grow from 56% to 66% by the year 2050 [5–7]. Rapid urbanization produces significant land use change from forest cover and agricultural landscapes to impermeable surfaces [8–11] to meet the urban population's needs, a crucial variable affecting local to global climate [12]. These transformations have led to ecological and environmental challenges, such as compromising urban ecosystem sustainability [13–15], influencing land surface temperature (LST) [16–18], and contributing to anthropogenic-induced carbon emissions (CE) [19,20]. For example, evapotranspiration, albedo, energy radiation, and other biochemical and biophysical parameters are affected by the surface reflectance and roughness of various LULC types in distinct ways [21,22]. Due to increased urbanization, forest cover is rapidly decreasing and this is boosting global warming trends [19,23]. Forests are a billion-dollar economic treasure that absorbs about 30% of the world's yearly carbon dioxide emissions [6,23,24]. In addition, this transition is responsible for about a third of the CE produced by human activities [25,26], mainly due to the shift from forested to developed land [25–27].

According to the Intergovernmental Panel on Climate Change's (IPCC) latest report (2019), human shifts in LULC lead to a depletion of natural resources as well as an increase in atmospheric CO_2 concentrations and nitrogen deposition [28–30]. From 1750 to the present, CO_2 emissions associated with deforestation have contributed ~180 ± 80 PgC to the cumulative anthropogenic CO_2 emission [31]. Moreover, several studies at both the national [32,33] and regional [32,34,35] scales have already demonstrated the phenomenon that LULC transitions affect the carbon balance [36–38]. It is estimated that global forest cover decreased by 7.8 million hectares per year between 1990 and 2000, 5.2 million hectares per year between 2000 and 2010, and 4.7 million hectares per year between 2010 and 2020 [39], and led to a rise in CE by 12.5% [40]. According to Statista [41], the atmospheric level of carbon dioxide had increased from 354.45 ppm in 1990 to 414.24 ppm by 2020 (1.99 ppm per year) [42,43].

The 2017 Atlas of Sustainable Development Goals (SDGs) reports that the growth in CO_2 emissions and other greenhouse gases has caused an increase of around 0.8 degrees Celsius in the mean global temperature over pre-industrial times [44]. According to the Intergovernmental Panel on Climate Change (IPCC) Assessment Report 6 (AR6) A.1.2, the global average temperature in the first two decades of the 21st century (2001–2020) was 0.99 (0.84–1.10) °C higher than in the period 1850–1900. The increase rate across land surfaces (1.59 (1.34 to 1.83) °C) was higher than the temperature increase over the ocean (0.88 (0.68 to 1.01) °C) [45]. The Paris Agreement, adopted in 2015 under the United Nations Framework Convention on Climate Change (UNFCCC), was signed by 195 countries to reduce global warming to below 2 °C and, preferably, 1.5 °C [46].

In the context of national and global warming, one of the most pressing challenges is determining the impact of urban land cover transition on climate change [17]. Several studies [32–38] have mainly focused on carbon fluctuation over a specific ecosystem by applying the bookkeeping [47–49] and processing models [50,51] to analyze the dynamics of carbon emissions in response to land cover transition over a specific ecosystem. However, a comparison of different ecosystem interactions was lacking. Zhu E. et al. [52] indicated that it would be difficult to precisely determine the spatiotemporal scales of LULC and the corresponding induced carbon emissions. Nevertheless, carbon levels can be reduced through artificial adjustments of land cover patterns [53,54]. The development of conservation management plans and strategies to ensure the long-term sustainability of ecosystems can be significantly assisted by evidence collected through observing changes in land use and land cover (LULC) [18].

Since 1972, Landsat satellites have continually provided remote sensing (RS) data for diverse applications regarding metropolitan areas' social, economic, and environmental aspects [55]. RS technology has tremendously facilitated data collection by eliminating the previous hassles and limitations of traditional field surveys and methods [6]. Remote sensing for forest biomass and carbon stock estimation can be implemented with acceptable accuracy [56]. RS combined with geographic information systems (GISs) is a valuable tool for investigating LULC classification changes, ecological monitoring, and management [21,57]. Developing a baseline and reference levels for Reducing Emissions from Deforestation and Forest Degradation (REDD+) implementation highly depends on accurate carbon stock assessment, which is possible using the combined data of RS and the National Forest Inventory (NFI) [58].

According to the Global Climate Risk Index (2021) published by Germanwatch, Pakistan is eighth on the list of countries most affected by climate change in the past two decades [59]. Pakistan (a country of roughly 220 million people) is one of the world's least wooded countries, with forest covering only 5% of its total land area [60–62], mostly in its northern region [14,16]. According to the forest department of Pakistan, the forest cover, which was 4,981,163 (ha) in 2004, had decreased to 4,786,831 (ha) by 2012 [63]. The BTT project was initiated in 2013 by the interim administration of KPK Province (northern Pakistan) to combat deforestation in the region [64], and one billion trees had been planted in the province by 2017 [65]. This was an excellent effort toward mitigating deforestation's adverse impacts [6]. In contrast, over the past two decades, the capital city of Punjab province in Pakistan (Lahore) has experienced substantial urban expansion and a decrease in its green proportion [66]. The local developers and government authorities (e.g., the Lahore Development Authority (LDA)) are continuously expanding the urban settlements through various housing communities and other projects [6,67,68] which require clearing part of the green fraction [6]. In addition, in 2014, the Government of Punjab (Pakistan) initiated an urban project called the "Ravi Urban Development Project" (RUDP) close to the interim capital for up to 35 million residents on an area of 40,000 hectares, most of which was farmland. The local farmers protested against such plans by visiting the Lahore High Court (LHC). Later the rulings of the LHC found "gross irregularities" in the development process, along with concerns about the environmental impact, and ordered work on this project to stop [69]. However, the government still intends to complete this project in the future (Master Plan 2050; included in the Supplementary Materials). According to the population census of 2017, 36.4% of the population of Pakistan lives in urban areas. The UN Population Division estimates that, by 2025, nearly half the country's population will live in cities.

Considering the above developments and the literature on rapid urbanization and its associated problems, this study aimed to (1) monitor and compare the land cover dynamics in scenarios from 2000–2020, (2) determine the contribution rates of LULC transitions to LST change by utilizing the cutting-edge CFLRM method, (3) determine changes in the carbon stock in response to green fraction transitions, and (4) determine the sensitivity between CO_2 variation (%) and LST under the Khyber Pakhtunkhwa (KPK) government's Billion Tree Tsunami Project (BTTP) (forestation) and for urbanization reshaping scenarios in Punjab as a consequence of the RUDP.

2. Materials and Methods

2.1. Study Area

To achieve the research objectives, we choose two provinces of Pakistan, (i) KPK, where the provisional government started a forestation project, the BTTP, in 2013 to reduce the impact of deforestation, and (ii) Punjab, which is experiencing urban growth reshaping scenarios and losing its vegetation cover. The three sample sites included (a) Peshawar (capital of KPK Province and located in northern Pakistan), which belonged to the "Central-Southern Forest Region-I Peshawar" in the BTTP. It is geographically located at 44°15′N, 71°42′E at 330 m (amsl) and covers an area of 1264 km^2. In terms of climatic characteristics, the region is classified as "dry-summer subtropical" ("Csa" in Koppen climate classification). The average temperature is about 23.9 °C, with a minimum of 17.4 °C and a maximum of 30.5 °C, but which can drop to 12 °C on cold nights (https://en.climate-data.org/asia/pakistan) (accessed on 18 November 2022). According to the 2017 census, Peshawar had 1.97 million people (http://www.pbs.gov.pk/) (accessed on 2 June 2022). The second site was (b) the Swat District, which lies in "Northern Forest Region-III Malakand" in the BTTP. Geographically, Swat is located at 35°22°N′, 72°42′E, with a total population of 2.31 million (2017) and an area of 5337 km^2. The regional climate passes through four seasons: a cool winter from November through February; a mild spring from March through May; the summer season from June through September; and the retreating monsoon period of October and November in autumn (https://aboutkp.kp.gov.pk/page/climate) (accessed on 12 July 2021). Finally, the third site was (c) Lahore, located at 31°15′N, 74°10′E, and covering 1842 km^2. The 2017 census showed 11.13 million people living in Lahore (http://www.pbs.gov.pk/) (accessed on 24 July 2021). In the last few decades, the city has been experiencing massive urbanization at the cost of losing green fraction. It contains a semi-arid climate ("BSh" in the Köppen climate classification). The city's maximum temperature was 48.3 °C, recorded on 30 May 1944, with 48 °C recorded on 9 June 2007. Spatial location of the study area, including the sample sites, is presented in Figure 1.

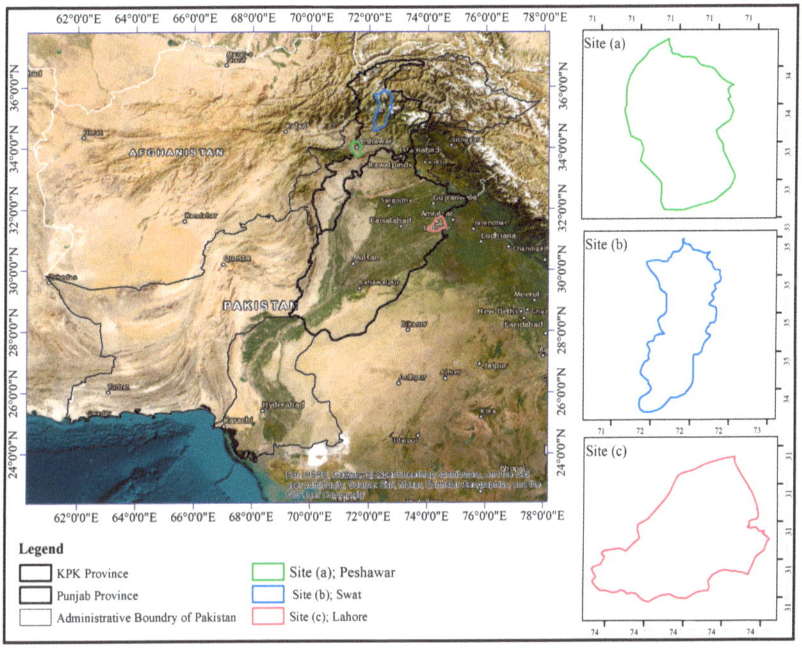

Figure 1. Maps showing the locations of study sites in the KPK and Punjab provinces, Pakistan.

2.2. Data Sets and Pre-Processing

2.2.1. Land Use Land Cover Maps

To achieve the set objectives, this study utilized high-resolution datasets, including "Global land cover—GLC30" (http://www.globeland30.org/) (accessed on 2 June 2022), which provides 30 m multispectral images based on TM, ETM+, OLI Landsat (USA), HJ-1 (China Environment and Disaster Reduction Satellite), and 16 m resolution Gaofen (GF)-1 (China High-Resolution Satellite). The product provided data for the years 2000, 2010, and 2020. GlobeLand30 product data has already been submitted to the United Nations [70–72]. The data of GlobeLand30 were reclassified and divided into six types according to the first-level classification standard of Land Use Status Classification (GB/T2010-2017). Details regarding the LULC class divisions are included in the Supplementary Materials for reference. The GlobeLand30 provided data with an overall accuracy of 83.50–85.72%, and a Kappa coefficient of 0.78–0.82. Furthermore, for cross-validation, the accuracy of the LULC classifications (obtained with the LCM30m product) was reconfirmed using ground truth data obtained from Google Earth, and this indicated an overall accuracy of >80% for all LULC classes.

2.2.2. LULC Transition Analysis

After acquiring the LULC maps, the cross-tabulation was performed using Terreset 2020 (https://clarklabs.org/terrset/land-change-modeler/) (accessed on 4 August 2022) to quantify the net change by category in km^2, patterns of aerial exchanges, and the gain and loss for each LULC class between 2000 and 2020. The Markov Chain model (MCM) was utilized to determine the transition probabilities. MCM can be mathematically expressed as follows [6,73]:

$$S\ (t+1) = Pij * S(t) \tag{1}$$

where S represents the land-use status at time t, and S (t + 1) is the land-use status at time t + 1, while Pij is the transition probability matrix in a state which is calculated as follows [6,74]

$$\|Pij\| = \left\| \begin{matrix} P_{1,1} & P_{1,2} & P_{1,\,N} \\ P_{2,1} & P_{2,2} & P_{2,\,N} \\ P_{N,1} & P_{N,2} & P_{N,\,N} \end{matrix} \right\| \tag{2}$$

$$(0 \leq Pij\ \leq 1) \tag{3}$$

P is the transition probability, Pij stands for the likelihood of converting from current state i to another state j next time, and PN is the state probability of any time. The low transition will have a chance near 0, and high growth has possibilities near 1 [6,74].

2.2.3. Land Surface Temperature (LST) Maps

Mean LST (for the March-June months) was retrieved from all possible cloud-free imagery of Landsat TM and OLI. Considering the operational time of Landsat, we used Landsat-5 TM for 2000 and 2010 and Landsat-8 TIRS for 2020 (imagery details are mentioned in Table S1—Supplementary Materials). Band6 (thermal band) of Landsat-5 TM was utilized to retrieve LST by using a mono-window algorithm [75]. On the other hand, Landsat-8 TIRS contains two thermal bands (10 and 11), but USGS (United States Geological Survey) discourages the use of the split-window algorithm to obtain the LST because of the instabilities with the participation of band 11 (https://www.usgs.gov/faqs) (accessed on 23 July 2021). Therefore, considering the accuracy and feasibility, we decided on the single-channel algorithm [76], for which all the necessary atmospheric corrections were performed [77,78]. Additionally, it should be noted that there is a 15 min time lag between Equatorial Crossing Time for Landsat OLI and Landsat TM, because of which it is necessary to be sure that time lag is not an obstacle to the analysis.

2.2.4. Retrieval of Normalized Difference Vegetation Index (NDVI)

NDVI facilitates identification of the relationship between land use type and estimated surface temperature quantitatively (Schmidt, et al., 2000). NDVI is derived using following mathematical expressions:

$$NDVI = \frac{NIR - RED}{NIR + RED} \tag{4}$$

The NIR band represents band 4 with wavelength (λ = 0.76–0.90 μm) of Landsat5 (T.M.), while RED corresponds to band 3 with wavelength (λ = 0.63–0.69 μm). On the other hand, in Landsat8 (OLI/TIRS), the NIR band depicts band5 with (λ = 0.85–0.87 μm), and RED represents band 4 with wavelength (λ = 0.63–0.67 μm) (Landsat Project Science Office 2016).

2.2.5. Correlation Analysis between LST and LULC Transitions

The relationships between LST and LULC transitions were calculated by using the land change modeler (LCM) and the curve fit linear regression model—CFLRM (an external extension for GIS) [79], which allows users to run regression analysis on a series of raster datasets. The equation of linear regression is described as follows:

$$Y = aX + b \tag{5}$$

where Y is the dependent variable, X is the independent variable, b is the slope of the line, and a is the y-intercept. Furthermore, zonal statistics as a table were applied using ArcGIS 10.8 to get zonal values of the LST trend for each land transition category.

2.2.6. Carbon Emission (CE) Estimation

The carbon emission from different LULC classes over the study period was estimated by biomass carbon stock assessment (BCSA) using the forestry inventory (FI) data collected by the Ministry of Climate Change of Pakistan (http://www.mocc.gov.pk/) (accessed on 2 June 2022) and the Pakistan Forest Institute Peshawar (https://www.pfi.gov.pk/) (accessed on 2 June 2022). In addition, we collected the biomass and carbon stock estimation data under the Reducing Emissions from Deforestation and Forest Degradation REDD+ project [63] using the tools presented in Table 1 (details about the REDD+ Project are included in the Supplementary Materials). Furthermore, for the estimation of carbon emission from different LULC classes, zonal statistical analysis was carried out in Arc GIS10.4 [29].

Table 1. Tools used in field sampling.

Materials	Purpose
Diameter Caliper (120 cm)	Diameter Measurement
Haga Altimeter	Height Measurement
Sunnto Clinometer	Slope and Height Measurement
Measuring Tape (100 m)	Measuring Distance
Ranging Rods	Plot Center and Other Location
Garmin GPS	Navigation
Landsat Images	Computation of Vegetation Indices and Carbon Stock Assessment

Tree height and DBH $\geq$ 5 cm were measured in all plots to calculate aboveground biomass [80,81] by using a diameter caliper of 100 cm in length. As per UNFCCC guidelines regarding tree measurement, forked and defective trees were also subjected to size measurement as they are believed to have a quantity of carbon stock. The Haga Altimeter used for height measurement is a gravity-oriented pointer device for height measurement within a fixed range of distances (20, 25, and 30 m). As height measurement is difficult and time-consuming for variable distances and viewpoints, height was only measured for dominant trees in the plots. A Sunnto Clinometer was used for slope estimation, the distance was measured with the help of a measuring tape (100 m), and GPS was used for location identification or navigation purposes. In contrast, a compass was used for the proper direction of plot selection in the area for aspect identification. Aboveground biomass (AGB) was estimated using Equation (6)

$$\text{AGB (kg)} = 0.0673 \times \left(\text{pDBH}^2\text{H}\right)^{0.976} \tag{6}$$

where DBH is the diameter in millimeters at breast height, H is the height of a tree, and the standard wood specific gravity was 1.54 g cm^{-3} [29,82]. Furthermore, the carbon stock was estimated by data conversion from aboveground biomass to carbon stock [83] using Equation (7).

$$\text{Carbon (kg)} = 0.5 \times \text{biomass} \tag{7}$$

The carbon stock data was superimposed on the LULC and LST data to determine the carbon concentration for each LULC and LST and to calculate the carbon emission according to zonal statistics [29]. Furthermore, the relationship between CE and LST was explored using the sensitivity analysis [29] in R.

2.2.7. Carbon Stock Data Validity

This study's estimated carbon stock maps were compared with ORNL CBD data [84,85] and WCMC CBD data (https://data-gis.unep-wcmc.org/portal/home/item.html?id=8a8d4e24683a46e6b039aea78c8af20f (accessed on 18 November 2022)). The overall Kappa Statistic was 0.92, indicating that the carbon stock maps are reasonable and can be used for further applications. A detailed methodology flowchart is attached as Figure 2.

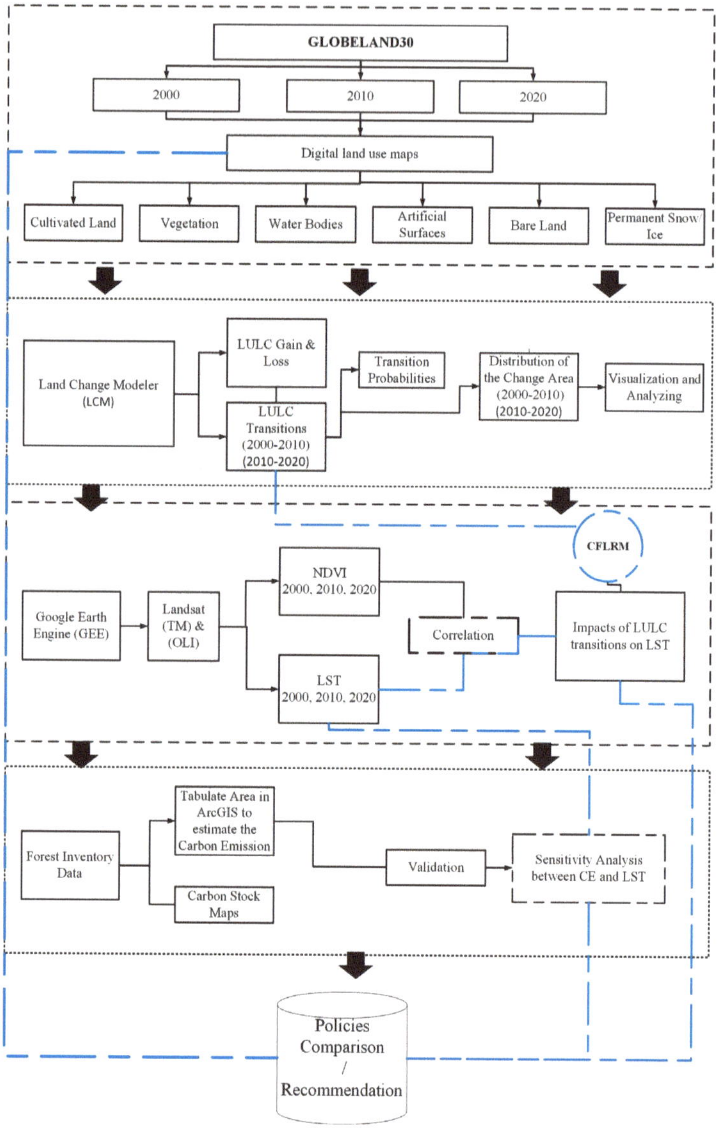

Figure 2. Flowchart indicating the detailed procedures for data collection, preprocessing, and evaluation of results.

3. Results

3.1. Spatial Patterns of LULC under BTTP and Urbanization in Punjab

The patterns of the LULC scenarios from 2000–2020, obtained from GLC30, were divided into six classes (the details are presented in the Supplementary Materials); the scenarios over KPK Province are shown in Figure 3A, while Figure 3B illustrates the LULC scenarios over Punjab Province. Furthermore, Table 2 elaborates on the dynamics of LULC from 2000–2020 for both Punjab and KPK provinces.

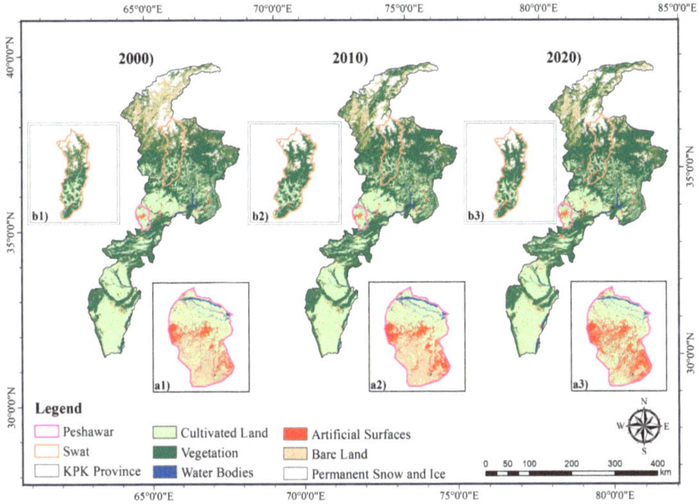

(A)

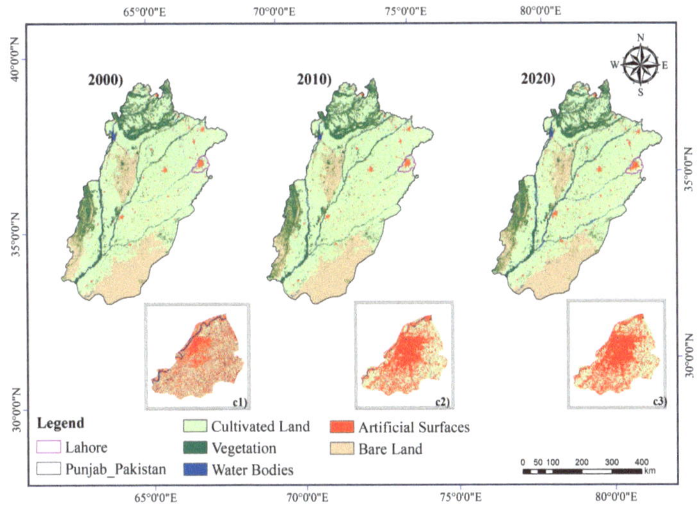

(B)

Figure 3. (**A**). Spatial patterns of land use land cover classes in KPK Province, including the sample sites (a) Peshawar (a1) 2000, (a2) 2010, and (a3) 2020; and (b) Swat (b1) 2000, (b2) 2010, and (b3) 2020. (**B**). Spatial patterns of land use land cover classes in Punjab Province, including sample site (c) Lahore (c1) 2000, (c2) 2010, and (c3) 2020.

Table 2. Trends of LULC in KPK (top) and Punjab (bottom) provinces between 2000 and 2020 (km^2).

		2000	%	2010	%	2020	%
KPK	Cultivated land	25,956.87	33.71	25,714.37	33.39	25,781.15	33.48
	Vegetation	29,337.59	38.10	30,257.76	39.29	33,458.05	43.45
	Water bodies	727.82	0.95	842.2	1.09	695.58	0.90
	Artificial surface	664.18	0.86	1196.17	1.55	1375.82	1.79
	Bare land	14,598.18	18.96	13,285.62	17.25	10,346.19	13.44
	Ice and snow	5724.44	7.43	5712.96	7.42	5352.29	6.95
	Total	77,009.08	100	77,009.08	100	77,009.08	100
Punjab	Cultivated land	136,014.21	64.91	137,125.65	65.44	137,244.7	65.50
	Vegetation	21,250.54	10.14	21,303.2	10.17	19,634.2	9.37
	Water surface	2459.14	1.17	2325.64	1.11	2240.43	1.07
	Artificial surface	3846.3	1.84	4065.38	1.94	6497.6	3.10
	Bare land	45,960.09	21.93	44,710.43	21.34	43,913.37	20.96
	Total	209,530.30	100	209,530.30	100	209,530.30	100

By combining the findings from Figure 3A and Table 2, it can be seen that over the last two decades, in the KPK province, a significant increase in artificial surfaces has taken place, increasing from 0.86% in 2000 to 1.79% in 2020. On the other hand, the percentage of vegetated land appears to be on the rise, growing from 38.10 percent in 2000 to 43.45 percent in 2020; this increase was relatively modest until 2010 (38.10 percent to 39.29 percent), but it was boosted during the period 2010–2020 when the provisional government initiated the BTTP. That increase in artificial surfaces and vegetation mainly came from bare land, which decreased from 18.96% (in 2000) to 17.25% in 2010 and 13.44% in 2020.

Figure 3B shows an opposite trend for the Punjab province, where the percentage of vegetated land decreased from 10.17% in 2000 to 10.10% in 2010, and then to 9.37% in 2020, while the percentage of bare land was reduced from 21.93% in 2000 to 21.34% by 2010, and then to 20.96% in 2020. Figure 4 further reveals this loss of vegetation and that artificial surfaces occupied the former bare land, with an increase from 1.84% in 2000 to 1.94% by 2010, and an apparent rise to 3.10% by 2020. Similar findings were observed over the sample sites a, b, and c, with an evident increase in vegetation land from 25.6% in 2000 to 37.02% in 2010 and 52.07% in 2020, and a decrease in bare land of −17% by 2010 and −25.3% by 2020 being noted. In contrast, for site (c), the trends were opposite, with a significant loss of vegetation land of −1.50% by 2010 being noted, which then accelerated from 2010–2020 with a loss of −2.90%, and a similar decrease in bare land (2000–2010 (−3.6%) and 2010–2020 (−5.2%)). Artificial surfaces mostly replaced the lost vegetation and barren land.

3.1.1. LULC Transitions over the Period 2000–2020

Furthermore, LCM was utilized to analyze the transitions of LULC from 2000–2020 to understand the transitions among the land cover classes. The results in Table 3 and Figure 4 present the KPK and Punjab Province scenarios over two decades, along with sample sites a, b, and c (Supplementary Materials, Tables S2 and S3). It was observed that a significant portion of the bare land in KPK province, along with sites (a) and (b), was transformed into vegetation and artificial surfaces, with a significant transition from bare land to vegetation (4375.87 km^2) from 2010–2020, which supported the BTTP's main objective and indicated a better use of bare land by the KPK government. On the other hand, over the Punjab province, along with site (c), the scenarios were alarming, with the transition of 1619.9 km^2 of cultivated land to artificial surfaces and 2263.5 km^2 of vegetation land to bare land. This indicates clearly contradictory policies for bare land use in the two provinces.

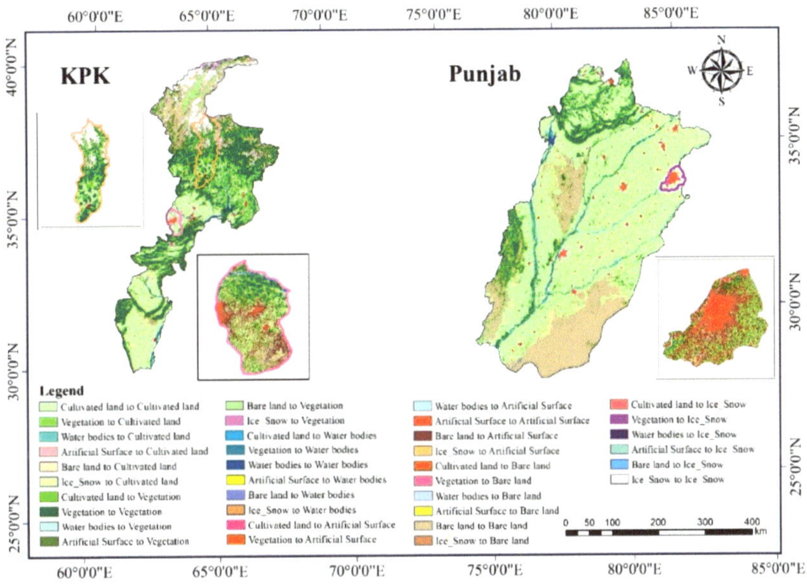

Figure 4. The LULC transition categories for KPK (**Left**) and Punjab (**right**) from 2000 to 2020.

Table 3. LULC transitions in KPK and Punjab province between 2000 and 2020 (area in sq. km).

		Cultivated Land	Vegetation	Water Bodies	Artificial Surfaces	Bare Land	Permanent Snow and Ice
KPK	Cultivated land	24,674.29	705	72.94	466.51	38.07	NA
	Vegetation	685.04	26,541.1	198.82	33.79	1711.54	167.24
	Water bodies	57.02	108.02	508.49	1.42	51.35	1.5
	Artificial surfaces	66.92	6.71	0.72	589.22	0.59	NA
	Bare land	231.05	4375.87	60.97	5.21	9750.15	174.9
	Permanent snow and ice	NA	520.97	0.22	NA	233.89	4969.31
Punjab	Cultivated land	13,2359.4	698.7	1072.7	1619.9	256.0	NA
	Vegetation	1067.0	16,655.6	1187.0	1074.9	2263.5	NA
	Water bodies	422.7	626.0	1332.3	4.2	73.6	NA
	Artificial surfaces	584.1	14.1	5.7	3227.0	15.3	NA
	Bare Land	1211.3	58.7	242.3	1840.3	41,605.0	NA

NA (Not Applicable) was utilized for the LULC classes where "Permanent Snow and Ice" was not present/applicable. Additionally, LULC transitions for all 3 sample sites (a), (b), and (c) are presented in the Supplementary Materials (Tables S2 and S3).

The Markov Chain model was used to explore the transition probabilities of LULC classes from 2000–2020. The results are presented in Table S3 (Supplementary Materials). The highest transition probability was found to be for the artificial surfaces in Punjab province and for vegetation in KPK province.

3.1.2. Gain and Loss between LULC Classes over the Period 2000–2020

We employed a "Land Change modeler (LCM) for ecological sustainability" after assessing LULC trends, transitions, and probabilities to pinpoint precisely where the LULC shifts will occur (TerrSet 2020). Results are presented in Figure 5A,B (Provisional Scenarios), and Figure 5C (sample sites) for the last two decades.

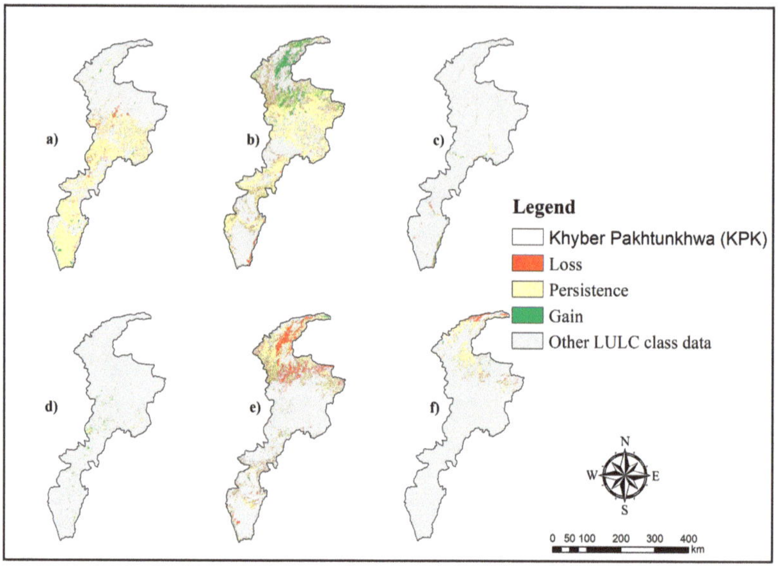

(A)

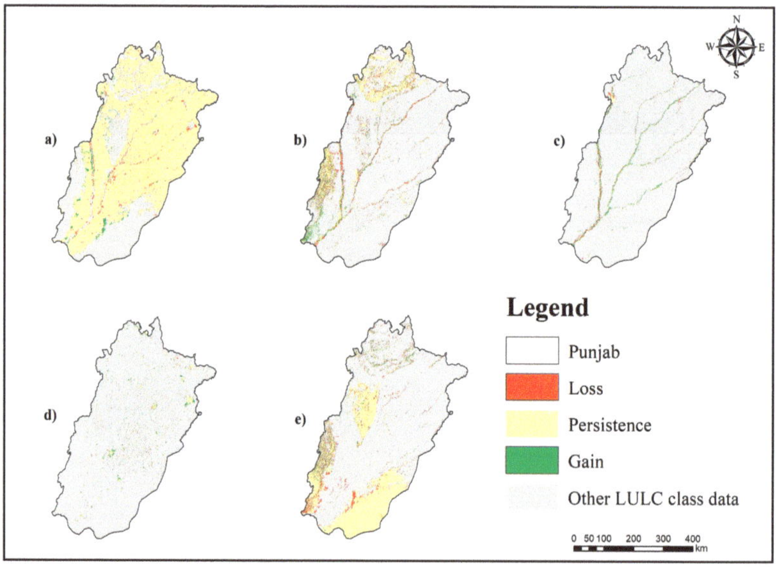

(B)

Figure 5. *Cont.*

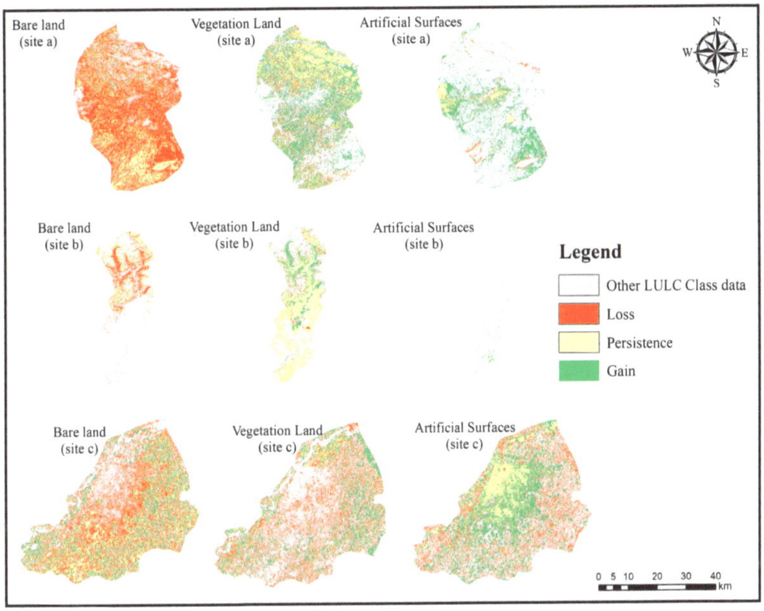

(C)

Figure 5. (**A**) Gain and loss for all LULC classes over KPK province from 2000–2020: (a) cultivated land, (b) vegetation, (c) water bodies, (d) artificial surfaces, (e) bare land, and (f) permanent snow and ice. (**B**) Gain and loss for all LULC classes over Punjab province from 2000–2020: (a) cultivated land, (b) vegetation, (c) water bodies, (d) artificial surfaces, and (e) bare land. (**C**) Gain and loss for all LULC classes over 2000–2020: sites (a) Peshawar (top), (b) Swat (middle), and (c) Lahore (bottom).

Analysis indicated a significant change in the vegetation and barren land for KPK province, where a total of 4375.87 sq. km area was changed from bare land to vegetation (Table 3). Further results for the sample sites (a, b, and c) in Figure 5C indicate significant gains for vegetation cover of 5716.6 km^2 and 506.9 km^2 for artificial surfaces, mainly coming from the apparent loss of bare land of 2035 km^2 (Table S4, Supplementary Materials). This also supports and validates our above results about the spatial patterns of LULC and transitions over KPK province.

In contrast, for the Punjab province, a significant loss was observed in both vegetation (−4592.4 km^2) and bare land (−4352.6 km^2), with a gain in artificial surfaces (1757.6 km^2). This also validated our transitions and probabilities analysis; similar trends were observed over site (c) Lahore, with a significant loss of bare land and vegetation land and substantial gain in artificial surfaces (Table S5 in the Supplementary Materials). Meanwhile, the scenarios (Figure 5C) highlighted apparent improvements in vegetation area over the sample sites (a) and (b). In contrast, site (c) indicated a significant loss of bare land and green fraction cover with a dominant gain in artificial surfaces.

3.2. Land Surface Temperature (LST) Variations

Spatial patterns of the monthly (March-June) mean LST (Figure 6) indicated a significant decrease of −4.3 °C from 2000–2020 over sites (a) and (b). The fluctuation in the LST (increase/decrease) was dominant over the areas of green fraction transitions. Several studies have already indicated the vegetation fraction in this phenomenon [86–92]. Over sites (a) and (b), apparent changes can be observed, especially after 2010 in (a3) and (b3) when they increased their vegetation cover under the BTTP, as compared to (a1) and (b1) for both high and low LST. On the other hand, site (c) indicated an opposite trend, with an increase of 5.3 °C from 2000–2020, with a significant fluctuation over the central and southern parts, where it lost the green fraction cover (Figure 5C).

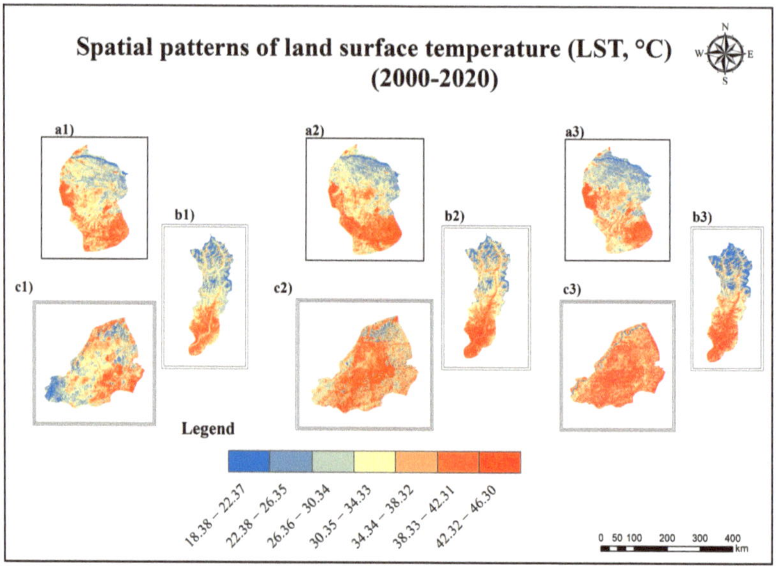

Figure 6. Spatial patterns of LST over sites a, b, and c from 2000–2020: (**a1**) Peshawar 2000, (**a2**) Peshawar 2010, (**a3**) Peshawar 2020; (**b1**) Swat 2000, (**b2**) Swat 2010, (**b3**) Swat 2020; (**c1**) Lahore 2000, (**c2**) Lahore 2010, and (**c3**) Lahore 2020.Furthermore, we calculated the NDVI to highlight the relation of the green fraction with LST; the results are presented in Figure 7.

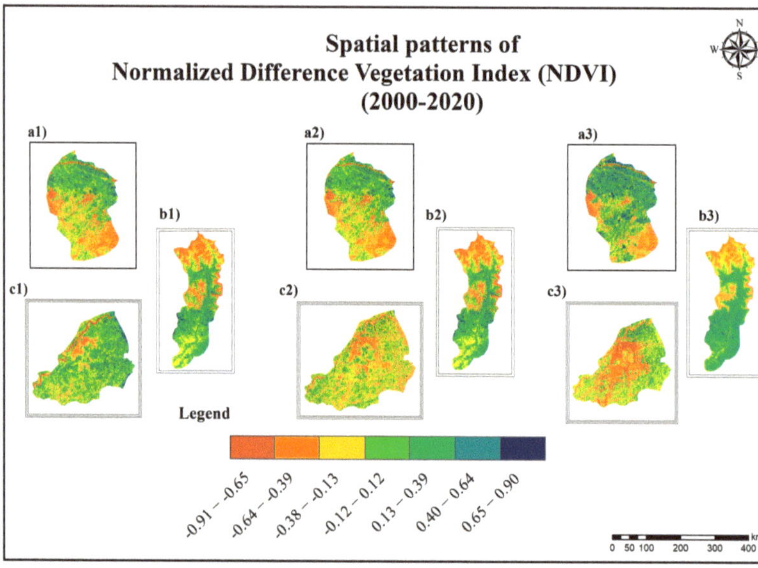

Figure 7. Spatial patterns of normalized difference vegetation index (NDVI) over sites a, b, and c from 2000–2020. Note: (**a1**) Peshawar 2000, (**a2**) Peshawar 2010, (**a3**) Peshawar 2020; (**b1**) Swat 2000, (**b2**) Swat 2010, (**b3**) Swat 2020; (**c1**) Lahore 2000, (**c2**) Lahore 2010, and (**c3**) Lahore 2020.

3.3. Spatial Patterns of Normalized Difference Vegetation Index (NDVI)

To further analyze the relationship between LST and NDVI, we calculated the spatial patterns of NDVI (Figure 7) to build a linear regression between LST and NDVI (Figure 8; Table 4) over the sites (a, b, and c).

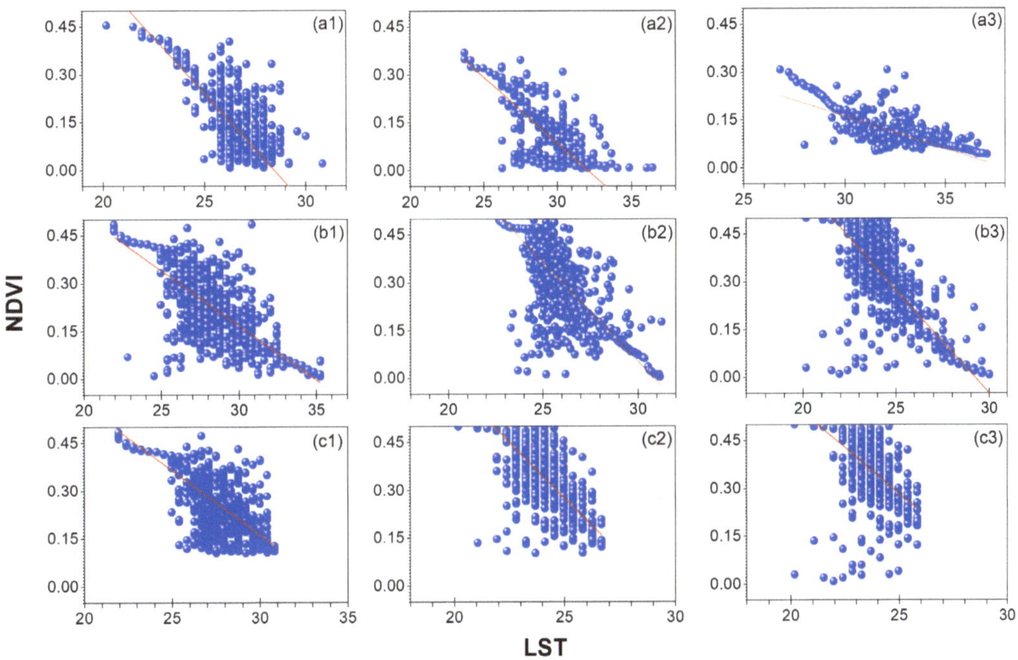

Figure 8. Correlation between LST and NDVI over the sample sites (a, b, and c). Note: (**a1**) Peshawar 2000, (**a2**) Peshawar 2010, (**a3**) Peshawar 2020; (**b1**) Swat 2000, (**b2**) Swat 2010, (**b3**) Swat 2020; (**c1**) Lahore 2000, (**c2**) Lahore 2010, and (**c3**) Lahore 2020.

Correlation statistics revealed a strong relationship for sites (a) and (c) with overall values of 0.70 or above for R^2 (Table 4). For site (c) Lahore, a decreasing trend for NDVI produced R^2 values of (c1) 0.83, (c2) 0.72, and (c3) 0.59, with a significant decrease over the central and southern parts of the area. Similarly, for sites (a) Peshawar and (b) Swat, an increasing trend and strong correlation was found for the R^2 at (a1) 0.78, (a2) 0.70, (a3) 0.84, (b1) 0.58, (b2) 0.61, and (b3) 0.69.

Table 4. Linear regression analysis over the sites (a, b, and c).

	Equation	**y = a + b × x**	**y = a + b × x**	**y = a + b × x**
Peshawar	Plot	NDVI 2000	NDVI 2010	NDVI 2020
	Weight	No Weighting	No Weighting	No Weighting
	Intercept	1.20 ± 0.01	1.92479 ± 0.02	1.89 ± 0.02
	Slope	−0.03 ± 4.24	−0.06 ± 0.00	−0.06 ± 8.20
	Residual sum of squares	4.74	5.56	6.27
	Pearson's r	−0.88	−0.83	−0.89
	R^2 (COD)/	0.78	0.70	0.80

Table 4. *Cont.*

	Equation	**y = a + b × x**	**y = a + b × x**	**y = a + b × x**
Swat	Plot	NDVI 2000	NDVI 2010	NDVI 2020
	Weight	No Weighting	No Weighting	No Weighting
	Intercept	1.38 ± 0.02	2.02 ± 0.03	1.70 ± 0.04
	Slope	−0.04 ± 9.85	−0.06 ± 0.00	−0.05 ± 0.00
	Residual sum of squares	2.88	3.80	4.43
	Pearson's r	−0.78	−0.83	−0.68
	R^2 (COD)	0.58	0.61	0.69
Lahore	**Equation**	**y = a + b × x**	**y = a + b × x**	**y = a + b × x**
	Plot	NDVI 2000	NDVI 2010	NDVI 2020
	Weight	No Weighting	No Weighting	No Weighting
	Intercept	1.99 ± 0.02	1.32 ± 0.02	0.77 ± 0.01
	Slope	−0.07 ± 8.53	−0.04 ± 8.42	−0.02 ± 3.32
	Residual sum of squares	3.35	1.66	0.75
	Pearson's r	−0.89	−0.84	−0.84
	R^2 (COD)	0.83	0.72	0.59

3.4. Relationship between LULC Transitions and LST

After evaluating the relationship between NDVI and LST, we further analyzed the impacts of LULC transitions on LST (Table 5) using CFLRM. Results from Table 5 indicated a decreasing trend for LST over the LULC classes associated with land transformation from bare or built-up land to vegetation and cultivated land. Over site (c), a significant gain of LST (0.17 °C) was noted in response to the transformation of VG to BU. In contrast, substantial reductions were seen in LST (−0.21 °C) after transition from built-up land/artificial surfaces (BU) to vegetation (VG), (−0.19 °C) after transition from BU to CL (cultivated land), and (−0.22 °C) after transition from bare land (BR) to CL.

Table 5. CFLRM-based average contribution rate of LULC class transition to LST (°C) from 2000 to 2020.

		Cultivated Land	**Vegetation**	**Water Bodies**	**Artificial Surfaces**	**Bare Land**	**Permanent Snow and Ice**
Peshawar	Cultivated land	−0.01	−0.05	0.07	0.11	0.16	NA
	Vegetation	0.03	−0.01	0.06	0.18	0.13	−0.1
	Water bodies	−0.2	−0.13	0.01	0.14	0.17	0.1
	Artificial surfaces	−0.18	−0.15	−0.1	0.01	0.09	NA
	Bare land	−0.08	−0.24	−0.1	0.12	0.01	−0.14
Swat	Cultivated land	−0.02	−0.04	0.05	0.09	0.14	−0.01
	Vegetation	0.02	−0.02	0.04	0.13	0.16	−0.03
	Water bodies	0.1	0.06	0.02	0.1	0.15	−0.1
	Artificial surfaces	−0.12	0.16	−0.2	0	0.1	−0.09
	Bare land	−0.05	−0.19	−0.02	0.14	0	−0.17

Table 5. *Cont.*

		Cultivated Land	Vegetation	Water Bodies	Artificial Surfaces	Bare Land	Permanent Snow and Ice
Lahore	Cultivated land	−0.01	−0.03	0.1	0.13	0.14	NA
	Vegetation	0.02	−0.01	−0.09	0.17	0.16	NA
	Water	−0.02	0.06	0.1	0.13	0.14	NA
	Artificial surfaces	−0.19	−0.21	−0.11	0.19	0.17	NA
	Bare land	−0.22	−0.18	−0.01	0.16	0.06	NA

NA = Not applicable.

3.5. Carbon Stock Maps and Analysis between Carbon Emission and LST

Using the CFLRM, we analyzed the scenarios of land cover dynamics and LULC transitions, including the geographical locations of gain and loss, variations in LST throughout the research period, and the contribution rates of LULC transitions to LST.

The results shown in Figure 9 indicated low carbon stock over bare land/water, medium over built-up land, and high stock over vegetation-covered areas, which were in line with previous studies in this domain [93–96]. Furthermore, the sensitivity analysis [29] of the relationship between LST and CO_2 variation (%), shown in Figure 10, indicated that CO_2 emission variation (%) was highly sensitive to LST. Results suggested that fluctuations in LST are affecting the % of emission. Over Lahore, we noticed a growing LST and a corresponding increase in the percent of CO_2 flux. In contrast, the study indicated a decreasing LST and a definite drop in the percentage of CO_2 variation over Peshawar and Swat. Studies have highlighted the fact that green fraction cover fluctuation affects the LST and the CE, as a decrease in vegetation fraction results in a massive reduction of carbon sink from the environment and traps CO_2 in the atmosphere, which also contributes to fluctuations in the LST [97–104].

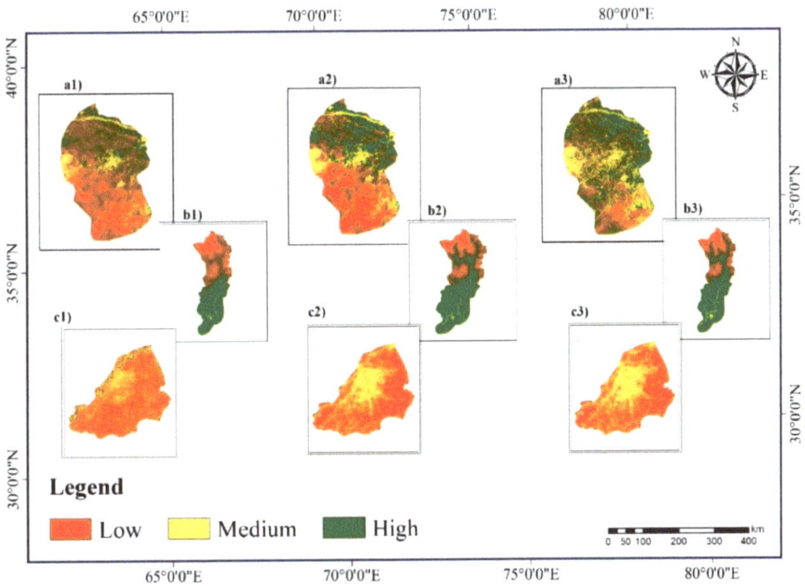

Figure 9. Maps of carbon stock over sites (**a1**) Peshawar 2000, (**a2**) Peshawar 2010, (**a3**) Peshawar 2020; (**b1**) Swat 2000, (**b2**) Swat 2010, (**b3**) Swat 2020; (**c1**) Lahore 2000, (**c2**) Lahore 2010, and (**c3**) Lahore 2020.

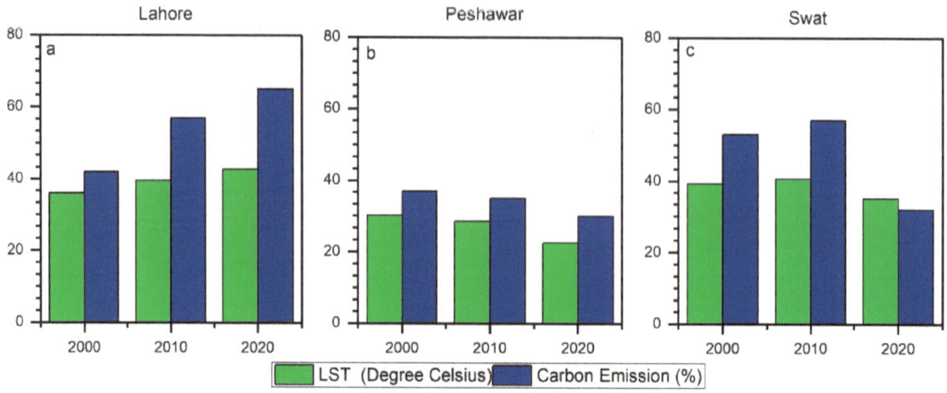

Figure 10. Sensitivity analysis between LST and carbon emission.

4. Discussion

This research was carried out to monitor and compare scenarios of land cover dynamics from 2000–2020 associated with the different land-use policies of local governments and to analyze their impacts on LST in the Punjab and KPK provinces of Pakistan. Furthermore, carbon stock maps were prepared, and the linkage between CE and LST was explored. Findings indicated a significant increase in artificial surfaces over KPK from 664.18 km^2 in 2000 to 1196.17 km^2 in 2010 and 1375.82 km^2 in 2020. A significant part of the increase in artificial surfaces was due to the mass migration of people from Afghanistan toward the KPK [105]. In addition, the same trends were observed over Punjab province and site "c", where a significant increase in artificial surfaces from 3846.3 km^2 in 2000 to 4065.38 km^2 in 2010 and 6497.6 km^2 by 2020 were noted. This was mainly because of the development of new housing communities in response to a growing urban population [50,51], mega-development projects (including Metro Bus, Orange Train) [50,51,106], and the RUDP (designed for up to approximately 35 million inhabitants and using an area of 40,000 ha). Furthermore, our results indicated that the increase in artificial surfaces was mainly at the cost of losing green fraction, which is essential for equilibrium between the land surface and atmospheric parameters [66,107]. An increase in population required new commercial and residential areas, as well as public utilities and transportation infrastructure, which also caused a decrease in green vegetation [108,109].

Previous studies [53,64,110–113] also reported that transitions in LULC, especially an increase in artificial surfaces, affect the LST and carbon emission and profoundly impact local and regional climates. As a result of variations in surface roughness and reflectance among the LULC categories, the amount of energy radiated and absorbed by these objects can vary significantly [21,22]. Our study explored these conclusions by evaluating the spatio-temporal changes in LST associated with the historical transitions of each LULC category. We concluded that there was a significant decrease in LST by 4.3 °C from 2000–2020 over sites (a) and (b), while an increase of 5.3 °C over site (c) was observed. To further understand the relationship of LST with vegetation cover, we utilized the linear regression between LST and NDVI [87,92] and found a strong positive correlation. Results indicated a significant gain in LST (0.17 °C) in response to the transformation of VG to BU. In contrast, there was a significant decrease in LST (-0.21 °C) over the transition from built-up land/artificial surfaces (BU) to vegetation (VG), (-0.19 °C) over the transition from BU to CL (cultivated land), and (-0.22 °C) over the transition from bare land (BR) to CL was noted.

The decreasing trend of LST can be seen after 2010, when the provisional government initialized the BTTP project to minimize the effects of green fraction loss [64], and then achieved its goals in 2017 [65]. Several pieces of research have already indicated the importance of the vegetation fraction in this phenomenon; e.g., [86] reported that safeguarding biodiversity and vegetation such as forests, agricultural land, and dense or sparse vegetation, is necessary to maintain the soil cover and balance in the ecosystem, hydrological cycle, surface temperatures, and air because each land-use category has a different magnitude of surface reflectance and roughness [114,115]. Research findings indicated that in response to the increase in vegetation land under the BTTP, the LST of the area saw a decreasing trend. In contrast, Lahore, in the Punjab province of Pakistan, has shown a decreasing trend for vegetation and an increasing trend for artificial surfaces since 2000, which significantly alter the LST and carbon emission [98,101–104].

We next examined the carbon stock and carbon emission over the research region after determining the LULC dynamics, LST pattern, and the contribution rate of each LULC transition on LST. Several studies [97–100] have reported that transitions in LULC, especially the loss/gain of green fraction, impact the LST, carbon stocks, and CE [93–96]. Our results indicated low carbon stock over bare land/water, medium over built-up land, and high stock over vegetation-covered areas. Kafy et al. [29] also reported the same results in Bangladesh. Green portion conversion to high-rise infrastructure boosts carbon emissions [27]. The primary driver of carbon emissions is the expansion of artificial surfaces [116], which has been a primary concern since the industrial revolution [26]. Updated LULC information is essential for understanding environmental pressures such as carbon emissions, biosafety, resource depletion, and climate change [9,117,118], and it is crucial for mesoscale dynamic understanding [119–121]. For example, climate models specify that the Paris Agreement's purpose can be achieved by efficient and sustainable policies relating to low carbon emissions [46,52]. Depleting carbon levels can be reduced through artificial adjustments of land cover patterns [53,54].

5. Conclusions

The study findings confirmed that over the study period, the KPK province, along with sites (a) Peshawar and (b) Swat, was predominantly gaining vegetation cover by +5.35%. A 1.13% increase was noted between 2000–2010, but the ratio accelerated to +4.16% from 2010–2020 under the BTTP project with an effective transition of bare land totalling 4375.87 km^2 to vegetation. On the other hand, the scenarios were alarming for the Punjab province, along with site (c) Lahore, where there was an increase in artificial surfaces from 1.83% in 2000 to 3.10% in 2020 at the cost of vegetation and bare land loss, amounting to a significant transition of 1074.9 km^2 of vegetation land and 1840.3 km^2 of bare ground to artificial surfaces. Shifts in LULC can alter LST. This study reported an increase of 5.3 °C LST over Lahore, Punjab, whereas a significant drop in LST by −4.3 °C was detected over KPK province. Furthermore, a significant association was found between the LST and NDVI with $R^2 > 0.70$.

The contribution rate of LULC transitions to LST suggested a considerable influence of land cover transitions on LST, with a 0.17 °C increase seen across regions transformed from vegetation to artificial surfaces, whereas there was an effective decrease of −0.21 °C in LST over artificial surfaces transformed to vegetation, −0.19 °C over artificial surfaces transformed to cultivated land, and −0.22 °C over bare land transformed to cultivated land. The results for carbon stock indicated a low carbon stock over bare land/water, medium over built-up land, and high stock over the vegetation-covered areas for both Punjab and KPK provinces. The sensitivity analysis between LST and carbon emission indicated that CO_2 emission variation (%) was highly sensitive to LST. The results suggested that fluctuations in LST are affecting the % of emission.

The research indicated that the boost in vegetation cover at the cost of bare land in response to the BTTP project produced a positive and robust environmental impact. On the other hand, the site (c) scenarios indicated dangerous trends which not only assisted in increasing LST but also caused a high carbon emission rate. The findings of this research and several previous studies have already highlighted the importance of vegetation fraction for the urban environment and the consequences of rapid urban development. Therefore, the government of Punjab should revisit its LULC policies, especially, in the context of the Master Plan 2050, the plans to build a new city (RUDP) adjacent to site (c) by acquiring farming land and transforming it into artificial surfaces, which would accelerate the negative consequences for the environment and local communities. Instead of converting farmland to artificial surfaces, authorities might follow KPK's policies (e.g., the BTTP) to boost the green proportion.

Supplementary Materials: The following supporting information can be downloaded at: https://www.mdpi.com/article/10.3390/rs15030859/s1, Table S1: Satellite imagery details; REDD+ Project; LULC class divisions [122–126]; Table S2: Land use transitions 2000–2010 (Provinces); Table S3: Land use transitions 2000–2020 (Sites); Table S4: LULC transition probabilities KPK and Punjab 2000–2020; Table S5: LULC gain and loss over KPK and Punjab 2000–2020; Figure S1: (a). Lahore metropolitan area Master Plan 2021; (b). Lahore metropolitan area Master Plan 2050.

Author Contributions: Conceptualization, F.M., J.L. and Q.L.; data curation, F.M., J.L. and J.Z.; formal analysis, F.M., J.L., Y.D. and C.G.; funding acquisition, J.L. and Q.L.; investigation, F.M. and J.L.; methodology, F.M., J.L., A.A. and B.B.; project administration, J.L., and Q.L.; resources, J.L., and Q.L.; software, F.M., C.L; supervision, J.L. and Q.L.; validation, F.M., J.L., Q.L. and J.Z.; visualization, F.M.; writing—original draft, F.M., A.T., C.G., H.Z., C.L. and B.B.; writing—review and editing, F.M., Q.L., J.L., J.Z., and A.A. All authors have read and agreed to the published version of the manuscript.

Funding: This work was supported by the National Key Research and Development Program (Grants # 2019YFE0126700).

Data Availability Statement: The datasets used and/or analyzed during the current study are available from the corresponding author on request.

Conflicts of Interest: The authors declare no conflict of interest.

References

1. Al Rakib, A.; Akter, K.S.; Rahman, M.N.; Arpi, S.; Kafy, A.A. Analyzing the pattern of land use land cover change and its impact on land surface temperature: A remote sensing approach in Mymensingh, Bangladesh. In Proceedings of the 1st International Student Research Conference-2020, Dhaka, Bangladesh, 1 April 2020.
2. Nurwanda, A.; Honjo, T. The prediction of city expansion and land surface temperature in Bogor City, Indonesia. *Sustain. Cities Soc.* **2019**, *52*, 101772. [CrossRef]
3. Omar, N.Q.; Sanusi, S.A.M.; Hussin, W.M.W.; Samat, N.; Mohammed, K.S. Markov-CA model using analytical hierarchy process and multiregression technique. In Proceedings of the IOP Conference Series: Earth and Environmental Science, Kuala Lumpur, Malaysia, 22–23 April 2014.
4. Chen, X.-L.; Zhao, H.-M.; Li, P.-X.; Yin, Z.-Y. Remote sensing image-based analysis of the relationship between urban heat island and land use/cover changes. *Remote Sens. Environ.* **2006**, *104*, 133–146. [CrossRef]
5. Mitchell, L.; Moss, H.O.N. *Urban Mobility in the 1st Century*; NYU Rudin Center for Transportation Policy: New York, NY, USA, 2012.
6. Mumtaz, F.; Tao, Y.; de Leeuw, G.; Zhao, L.; Fan, C.; Elnashar, A.; Bashir, B.; Wang, G.; Li, L.; Naeem, S.; et al. Modeling Spatio-Temporal Land Transformation and Its Associated Impacts on land Surface Temperature (LST). *Remote Sens.* **2020**, *12*, 2987. [CrossRef]
7. Bank, T.W. Urban Population. 2018. Available online: https://data.worldbank.org/indicator/SP.URB.TOTL.IN.ZS/ (accessed on 2 June 2022).
8. De Sherbinin, A. *A CIESIN Thematic Guide to Land-Use and Land-Cover Change (LUCC)*; Center for International Earth Science Information Network, Columbia University: New York, NY, USA, 2002.
9. Eastman, J.; Van Fossen, M.; Solarzano, L. *Transition Potential Modeling for Land Cover Change*; GIS, Spatial Analysis and Modeling; ESRI Press: Redlands, CA, USA, 2005; pp. 357–386.
10. Rai, R.; Zhang, Y.; Paudel, B.; Li, S.; Khanal, N.R. A Synthesis of Studies on Land Use and Land Cover Dynamics during 1930–2015 in Bangladesh. *Sustainability* **2017**, *9*, 1866. [CrossRef]
11. Briassoulis, H. *Analysis of Land Use Change: Theoretical and Modeling Approaches*; University of the Aegean: Mitilini, Greece, 2020.
12. Lin, Y.; Qiu, R.; Yao, J.; Hu, X.; Lin, J. The effects of urbanization on China's forest loss from 2000 to 2012: Evidence from a panel analysis. *J. Clean. Prod.* **2019**, *214*, 270–278. [CrossRef]
13. Wang, Z.-H. Reconceptualizing urban heat island: Beyond the urban-rural dichotomy. *Sustain. Cities Soc.* **2021**, *77*, 103581. [CrossRef]

14. Dubovyk, O.; Sliuzas, R.; Flacke, J. Spatio-temporal modelling of informal settlement development in Sancaktepe district, Istanbul, Turkey. *ISPRS J. Photogramm. Remote Sens.* **2011**, *66*, 235–246. [CrossRef]
15. Poelmans, L.; Van Rompaey, A. Detecting and modelling spatial patterns of urban sprawl in highly fragmented areas: A case study in the Flanders–Brussels region. *Landsc. Urban Plan.* **2009**, *93*, 10–19. [CrossRef]
16. Song, X.; Zeng, X. Evaluating the responses of forest ecosystems to climate change and CO_2 using dynamic global vegetation models. *Ecol. Evol.* **2017**, *7*, 997–1008. [CrossRef] [PubMed]
17. Bowman, D.M.J.S.; Kolden, C.A.; Abatzoglou, J.T.; Johnston, F.H.; van der Werf, G.R.; Flannigan, M. Vegetation fires in the Anthropocene. *Nat. Rev. Earth Environ.* **2020**, *1*, 500–515. [CrossRef]
18. Gidado, K.A.; Kamarudin, M.K.A.; Firdausaq, N.A.; Nalado, A.M.; Saudi, A.S.M.; Saad, M.H.; Ibrahim, S. Analysis of Spatiotemporal Land Use and Land Cover Changes using Remote Sensing and GIS: A Review. *Int. J. Eng. Technol.* **2018**, *7*, 159–162. [CrossRef]
19. Waseem, S.; Khayyam, U. Loss of vegetative cover and increased land surface temperature: A case study of Islamabad, Pakistan. *J. Clean. Prod.* **2019**, *234*, 972–983. [CrossRef]
20. Tang, X.; Woodcock, C.E.; Olofsson, P.; Hutyra, L.R. Spatiotemporal assessment of land use/land cover change and associated carbon emissions and uptake in the Mekong River Basin. *Remote Sens. Environ.* **2021**, *256*, 112336. [CrossRef]
21. Ahmad, A.; Nizami, S.M. Carbon stocks of different land uses in the Kumrat valley, Hindu Kush Region of Pakistan. *J. For. Res.* **2015**, *26*, 57–64. [CrossRef]
22. Abutaleb, K.; Ngie, A.; Darwish, A.; Ahmed, M.; Arafat, S.; Ahmed, F. Assessment of Urban Heat Island Using Remotely Sensed Imagery over Greater Cairo, Egypt. *Adv. Remote Sens.* **2015**, *04*, 35–47. [CrossRef]
23. Canadell, J.G.; Raupach, M.R. Managing Forests for Climate Change Mitigation. *Science* **2008**, *320*, 1456–1457. [CrossRef]
24. Mackey, B. Counting trees, carbon and climate change. *Significance* **2014**, *11*, 19–23. [CrossRef]
25. Penman, J.; Gytarsky, M.; Hiraishi, T.; Krug, T.; Kruger, D.; Pipatti, R.; Buendia, L.; Miwa, K.; Ngara, T.; Tanabe, K.; et al. Good Practice Guidance for Land Use, Land-Use Change and Forestry. 2003. Available online: https://www.ipcc-nggip.iges.or.jp/public/gpglulucf/gpglulucf_files/GPG_LULUCF_FULL.pdf (accessed on 2 June 2022).
26. Houghton, R.A.; House, J.I.; Pongratz, J.; Van Der Werf, G.R.; DeFries, R.S.; Hansen, M.C.; Le Quéré, C.; Ramankutty, N. Carbon emissions from land use and land-cover change. *Biogeosciences* **2012**, *9*, 5125–5142. [CrossRef]
27. Chuai, X.; Huang, X.; Wang, W.; Zhao, R.; Zhang, M.; Wu, C. Land use, total carbon emission's change and low carbon land management in Coastal Jiangsu, China. *J. Clean. Prod.* **2015**, *103*, 77–86. [CrossRef]
28. Djalante, R. Key assessments from the IPCC special report on global warming of 1.5 C and the implications for the Sendai framework for disaster risk reduction. *Prog. Disaster Sci.* **2019**, *1*, 100001. [CrossRef]
29. Kafy, A.A.; Al Rakib, A.; Fattah, M.A.; Rahaman, Z.A.; Sattar, G.S. Impact of vegetation cover loss on surface temperature and carbon emission in a fastest-growing city, Cumilla, Bangladesh. *Build. Environ.* **2022**, *208*, 108573. [CrossRef]
30. Mumtaz, F.; Arshad, A.; Mirchi, A.; Tariq, A.; Dilawar, A.; Hussain, S.; Shi, S.; Noor, R.; Noor, R.; Daccache, A.; et al. Impacts of reduced deposition of atmospheric nitrogen on coastal marine eco-system during substantial shift in human activities in the twenty-first century. *Geomatics, Nat. Hazards Risk* **2021**, *12*, 2023–2047. [CrossRef]
31. Jayakrishnan, K.U.; Bala, G.; Cao, L.; Caldeira, K. Contrasting climate and carbon-cycle consequences of fossil-fuel use versus deforestation disturbance. *Environ. Res. Lett.* **2022**, *17*, 064020. [CrossRef]
32. Fang, J.; Guo, Z.; Piao, S.; Chen, A. Terrestrial vegetation carbon sinks in China, 1981–2000. *Sci. China Ser. D Earth Sci.* **2007**, *50*, 1341–1350. [CrossRef]
33. Pilli, R.; Grassi, G. Bilancio del carbonio negli ecosistemi terrestri cinesi. *For.-J. Silvic. For. Ecol.* **2009**, *6*, 137.
34. Pan, G.; Li, L.; Wu, L.; Zhang, X. Storage and sequestration potential of topsoil organic carbon in China's paddy soils. *Glob. Chang. Biol.* **2004**, *10*, 79–92. [CrossRef]
35. Zhang, G.; Kang, Y.; Han, G.; Mei, H.; Sakurai, K. Grassland degradation reduces the carbon sequestration capacity of the vegetation and enhances the soil carbon and nitrogen loss. *Acta Agric. Scand. Sect. B—Soil Plant Sci.* **2011**, *61*, 356–364. [CrossRef]
36. DeFries, R.S.; Houghton, R.A.; Hansen, M.C.; Field, C.B.; Skole, D.; Townshend, J. Carbon emissions from tropical deforestation and regrowth based on satellite observations for the 1980s and 1990s. *Proc. Natl. Acad. Sci. USA* **2002**, *99*, 14256–14261. [CrossRef]
37. Houghton, R. Magnitude, distribution and causes of terrestrial carbon sinks and some implications for policy. *Clim. Policy* **2002**, *2*, 71–88. [CrossRef]
38. Leite, C.C.; Costa, M.H.; Soares-Filho, B.S.; Hissa, L.D.B.V. Historical land use change and associated carbon emissions in Brazil from 1940 to 1995. *Glob. Biogeochem. Cycles* **2012**, *26*. [CrossRef]
39. FAO; UN. *Global Forest Resources Assessment 2020: Key Findings*; FAO: Rome, Italy, 2020.
40. NOAA. NOAA's Greenhouse Gas Index Up 41 Percent Since 1990. 2018. Available online: https://research.noaa.gov/article/ArtMID/587/ArticleID/2359/NOAA%E2%80%99s-greenhouse-gas-index-up-41-percent-since-1990 (accessed on 2 June 2022).
41. Tiseo, I. Historic Average Carbon Dioxide (CO2) Levels in the Atmosphere Worldwide from 1959 to 2021 (in Parts Per Million). 2021. Available online: https://www.statista.com/statistics/1091926/atmospheric-concentration-of-co2-historic/ (accessed on 18 November 2022).

42. Wang, G.; Zhang, L.; Zhuang, Q.; Yu, D.; Shi, X.; Xing, S.; Xiong, D.; Liu, Y. Quantification of the soil organic carbon balance in the Tai-Lake paddy soils of China. *Soil Tillage Res.* **2016**, *155*, 95–106. [CrossRef]
43. IPCC (Intergovernmental Panel on Climate Change). *Synthesis Report. Summary for Policymakers*; Cambridge University Press: Cambridge, UK, 2007.
44. World Bank Data Team. The 2017 Atlas of Sustainable Development Goals: A New Visual Guide to Data and Development. 2017. Available online: https://blogs.worldbank.org/opendata/2017-atlas-sustainable-development-goals-new-visual-guide-data-and-development (accessed on 18 November 2022).
45. The Intergovernmental Panel on Climate Change (IPCC). Climate Change 2021—The Physical Science Basis in AR6. 2021. Available online: https://www.ipcc.ch/report/ar6/wg1/ (accessed on 18 November 2022).
46. Tollefson, J. The 2 C Dream. *Nature* **2015**, *527*, 436. [CrossRef] [PubMed]
47. Dixon, R.K.; Solomon, A.M.; Brown, S.; Houghton, R.A.; Trexier, M.C.; Wisniewski, J. Carbon Pools and Flux of Global Forest Ecosystems. *Science* **1994**, *263*, 185–190. [CrossRef] [PubMed]
48. Houghton, R.A.; Hackler, J.L. Emissions of carbon from forestry and land-use change in tropical Asia. *Glob. Chang. Biol.* **1999**, *5*, 481–492. [CrossRef]
49. Houghton, R.A.; Nassikas, A.A. Global and regional fluxes of carbon from land use and land cover change 1850–2015. *Glob. Biogeochem. Cycles* **2017**, *31*, 456–472. [CrossRef]
50. Han, J.; Meng, X.; Zhou, X.; Yi, B.; Liu, M.; Xiang, W.N. A long-term analysis of urbanization process, landscape change, and carbon sources and sinks: A case study in China's Yangtze River Delta region. *J. Clean. Prod.* **2017**, *141*, 1040–1050. [CrossRef]
51. Shevliakova, E.; Pacala, S.W.; Malyshev, S.; Hurtt, G.C.; Milly, P.C.D.; Caspersen, J.P.; Sentman, L.T.; Fisk, J.P.; Wirth, C.; Crevoisier, C. Carbon cycling under 300 years of land use change: Importance of the secondary vegetation sink. *Glob. Biogeochem. Cycles* **2009**, *23*. [CrossRef]
52. Zhu, E.; Deng, J.; Zhou, M.; Gan, M.; Jiang, R.; Wang, K.; Shahtahmassebi, A. Carbon emissions induced by land-use and land-cover change from 1970 to 2010 in Zhejiang, China. *Sci. Total. Environ.* **2019**, *646*, 930–939. [CrossRef]
53. Lal, R. Forest soils and carbon sequestration. *For. Ecol. Manag.* **2005**, *220*, 242–2588. [CrossRef]
54. Guadalupe, V.; Sotta, E.D.; Santos, V.F.; Aguiar, L.J.G.; Vieira, M.; de Oliveira, C.P.; Siqueira, J.V.N. REDD+ implementation in a high forest low deforestation area: Constraints on monitoring forest carbon emissions. *Land Use Policy* **2018**, *76*, 414–421. [CrossRef]
55. Wulder, M.A.; White, J.C.; Loveland, T.R.; Woodcock, C.E.; Belward, A.S.; Cohen, W.B.; Fosnight, E.A.; Shaw, J.; Masek, J.G.; Roy, D.P. The global Landsat archive: Status, consolidation, and direction. *Remote Sens. Environ.* **2016**, *185*, 271–283. [CrossRef]
56. Du, L.; Zhou, T.; Zou, Z.; Zhao, X.; Huang, K.; Wu, H. Mapping Forest Biomass Using Remote Sensing and National Forest Inventory in China. *Forests* **2014**, *5*, 1267–1283. [CrossRef]
57. Balcik, F.B.; Kuzucu, A.K. Determination of Land Cover/Land Use Using SPOT 7 Data With Supervised Classification Methods. In Proceedings of the 3rd International GeoAdvances Workshop, Istanbul, Turkey, 16–17 October 2016; pp. 143–146.
58. Mitchell, A.L.; Rosenqvist, A.; Mora, B. Current remote sensing approaches to monitoring forest degradation in support of countries measurement, reporting and verification (MRV) systems for REDD+. *Carbon Balance Manag.* **2017**, *12*, 1–22. [CrossRef] [PubMed]
59. Eckstein, A.; Künzel, V.; Schäfer, L. Global Climate Risk Index 2021. In *Who Suffers Most from Extreme Weather Events? Weather-Related Loss Events in 2019 and 2000–2019*; Germanwatch e.V.: Bonn, Germany, 2021; ISBN 978-3-943704-84-6.
60. Khan, W.R.; Khokhar, M.F.; Sana, M.; Naila, Y.; Qurban, A.P.; Muhammad, N.R. Assessing the context of REDD+ in Murree hills forest of Pakistan. *Adv. Environ. Biol.* **2015**, *9*, 15–20.
61. Munawar, S.; Khokhar, M.F.; Atif, S. Reducing emissions from deforestation and forest degradation implementation in northern Pakistan. *Int. Biodeterior. Biodegrad.* **2015**, *102*, 316–323. [CrossRef]
62. Yasmin, N. Dynamical Assessment of Vegetation Trends over Margalla Hills National Park by Using Modis Vegetation Indices. *Pak. J. Agric. Sci.* **2016**, *53*, 777–786.
63. ICIMOD. The REDD+ Project in Pakistan. 2013. Available online: https://redd.unfccc.int/files/pakistan_redd__strategy_final-oct_2_2021_edited_by_iucn-new_minister.pdf (accessed on 18 November 2022).
64. UNESCO. Pakistan: Green Again. 2019. Available online: https://en.unesco.org/courier/2019-3/pakistan-green-again (accessed on 18 November 2022).
65. WEF. Pakistan Has Planted over a Billion Trees. 2018. Available online: https://www.weforum.org/agenda/2018/07/pakistan-s-billion-tree-tsunami-is-astonishing/ (accessed on 18 November 2022).
66. Rana, I.A.; Bhatti, S.S. Lahore, Pakistan–Urbanization challenges and opportunities. *Cities* **2018**, *72*, 348–355. [CrossRef]
67. Shah, B.; Ghauri, B. Mapping urban heat island effect in comparison with the land use, land cover of Lahore district. *Pak. J. Meteorol. Vol* **2015**, *11*, 37–48.
68. Nespak, L. *Integrated Master Plan for Lahore-2021*; Lahore Development Authority: Lahore, Pakistan, 2004.
69. Baloch, S.M. We Will be Homeless': Lahore Farmers Accuse 'Mafia' of Land Grab for New City. 2021. Available online: https://www.theguardian.com/global-development/2021/nov/02/we-will-be-homeless-lahore-farmers-accuse-mafia-of-land-grab-for-new-city (accessed on 18 November 2022).

70. Xu, Z.; Luo, Q.; Xu, Z. Consistency of land cover data derived from remote sensing in Xinjiang. *J. Geo-Inf. Sci.* **2019**, *21*, 427–436.
71. Hou, W.; Hou, X. Consistency of the multiple remote sensing-based land use and land cover classification products in the global coastal zones. *J. Geo-Inf. Sci.* **2019**, *21*, 1061–1073.
72. Shi, X.; Nie, S.; Ju, W.; Yu, L. Climate effects of the GlobeLand30 land cover dataset on the Beijing Climate Center climate model simulations. *Sci. China Earth Sci.* **2016**, *59*, 1754–1764. [CrossRef]
73. Tariq, A.; Shu, H. CA-Markov Chain Analysis of Seasonal Land Surface Temperature and Land Use Land Cover Change Using Optical Multi-Temporal Satellite Data of Faisalabad, Pakistan. *Remote Sens.* **2020**, *12*, 3402. [CrossRef]
74. Kumar, S.; Radhakrishnan, N.; Mathew, S. Land use change modelling using a Markov model and remote sensing. *Geomat. Nat. Hazards Risk* **2013**, *5*, 145–156. [CrossRef]
75. Qin, Z.; Dall'Olmo, G.; Karnieli, A.; Berliner, P. Derivation of split window algorithm and its sensitivity analysis for retrieving land surface temperature from NOAA-advanced very high resolution radiometer data. *J. Geophys. Res. Atmos.* **2001**, *106*, 22655–22670. [CrossRef]
76. Jimenez-Munoz, J.C.; Cristobal, J.; Sobrino, J.A.; Soria, G.; Ninyerola, M.; Pons, X. Revision of the Single-Channel Algorithm for Land Surface Temperature Retrieval From Landsat Thermal-Infrared Data. *IEEE Trans. Geosci. Remote Sens.* **2008**, *47*, 339–349. [CrossRef]
77. Abuelgasim, A.; Ammad, R. Mapping soil salinity in arid and semi-arid regions using Landsat 8 OLI satellite data. *Remote Sens. Appl. Soc. Environ.* **2019**, *13*, 415–425. [CrossRef]
78. Onačillová, K.; Gallay, M.; Paluba, D.; Péliová, A.; Tokarčík, O.; Laubertová, D. Combining Landsat 8 and Sentinel-2 Data in Google Earth Engine to Derive Higher Resolution Land Surface Temperature Maps in Urban Environment. *Remote Sens.* **2022**, *14*, 4076. [CrossRef]
79. USGS. Curve Fit Linear Regression. 2017. Available online: https://www.usgs.gov/centers/upper-midwest-environmental-sciences-center/science/curve-fit-pixel-level-raster-regression# (accessed on 18 November 2022).
80. Nizami, S.M. The inventory of the carbon stocks in sub tropical forests of Pakistan for reporting under Kyoto Protocol. *J. For. Res.* **2012**, *23*, 377–384. [CrossRef]
81. Afzal, M.; Akhter, A. Estimation of biomass and carbon stock: Chichawatni Irrigated Plantation in Punjab, Pakistan. In Proceedings of the SDPI's Fourteenth Sustainable Development Conference, Islamabad, Pakistan, 13–15 December 2011.
82. Pearson, T.; Walker, S.; Brown, S. *Sourcebook for Land Use, Land-Use Change and Forestry Projects*; World Bank: Washington, DC, USA, 2013.
83. Chave, J.; Réjou-Méchain, M.; Búrquez, A.; Chidumayo, E.; Colgan, M.S.; Delitti, W.B.; Duque, A.; Eid, T.; Fearnside, P.M.; Goodman, R.C.; et al. Improved allometric models to estimate the aboveground biomass of tropical trees. *Glob. Chang. Biol.* **2014**, *20*, 3177–3190. [CrossRef] [PubMed]
84. Spawn, S.; Gibbs, H. *Global Aboveground and Belowground Biomass Carbon Density Maps for the Year 2010*; ORNL DAAC: Oak Ridge, TN, USA, 2020.
85. Spawn, S.A.; Sullivan, C.C.; Lark, T.J.; Gibbs, H.K. Harmonized global maps of above and belowground biomass carbon density in the year 2010. *Sci. Data* **2020**, *7*, 1–22. [CrossRef]
86. Aljerf, L. Biodiversity is key for More Variety for Better Society. *Biodivers. Int. J.* **2017**, *1*, 00002. [CrossRef]
87. Guha, S.; Govil, H.; Diwan, P. Monitoring LST-NDVI Relationship Using Premonsoon Landsat Datasets. *Adv. Meteorol.* **2020**, *2020*, 1–15. [CrossRef]
88. Sun, D.; Kafatos, M. Note on the NDVI-LST relationship and the use of temperature-related drought indices over North America. *Geophys. Res. Lett.* **2007**, *34*. [CrossRef]
89. Karnieli, A.; Agam, N.; Pinker, R.T.; Anderson, M.; Imhoff, M.L.; Gutman, G.G.; Panov, N.; Goldberg, A. Use of NDVI and Land Surface Temperature for Drought Assessment: Merits and Limitations. *J. Clim.* **2010**, *23*, 618–633. [CrossRef]
90. Yue, W.; Xu, J.; Tan, W.; Xu, L. The relationship between land surface temperature and NDVI with remote sensing: Application to Shanghai Landsat 7 ETM+ data. *Int. J. Remote Sens.* **2007**, *28*, 3205–3226. [CrossRef]
91. Marzban, F.; Sodoudi, S.; Preusker, R. The influence of land-cover type on the relationship between NDVI–LST and LST-T air. *Int. J. Remote Sens.* **2018**, *39*, 1377–1398. [CrossRef]
92. Govil, H.; Guha, S.; Diwan, P.; Gill, N.; Dey, A. Analyzing linear relationships of LST with NDVI and MNDISI using various resolution levels of Landsat 8 OLI and TIRS data. In *Data Management, Analytics and Innovation*; Springer: Berlin/Heidelberg, Germany, 2020; pp. 171–184.
93. Fernandes, M.M.; de Moura Fernandes, M.R.; Garcia, J.R.; Matricardi, E.A.; de Souza Lima, A.H.; de Araújo Filho, R.N.; Gomes Filho, R.R.; Piscoya, V.C.; Piscoya, T.O.; Cunha Filho, M. Land use and land cover changes and carbon stock valuation in the São Francisco river basin, Brazil. *Environ. Chall.* **2021**, *5*, 100247. [CrossRef]
94. Sohl, T.L.; Sleeter, B.M.; Zhu, Z.; Sayler, K.L.; Bennett, S.; Bouchard, M.; Reker, R.; Hawbaker, T.; Wein, A.; Liu, S.; et al. A land-use and land-cover modeling strategy to support a national assessment of carbon stocks and fluxes. *Appl. Geogr.* **2012**, *34*, 111–124. [CrossRef]
95. Barakat, A.; Khellouk, R.; Touhami, F. Detection of urban LULC changes and its effect on soil organic carbon stocks: A case study of Béni Mellal City (Morocco). *J. Sediment. Environ.* **2021**, *6*, 287–299. [CrossRef]

96. Rajbanshi, J.; Das, S. Changes in carbon stocks and its economic valuation under a changing land use pattern—A multitemporal study in Konar catchment, India. *Land Degrad. Dev.* **2021**, *32*, 3573–3587. [CrossRef]
97. Fattah; Morshed, S.R.; Morshed, S.Y. Multi-layer perceptron-Markov chain-based artificial neural network for modelling future land-specific carbon emission pattern and its influences on surface temperature. *SN Appl. Sci.* **2021**, *3*, 1–22. [CrossRef]
98. Fattah, A.; Morshed, S.R. Impacts of land use-based carbon emission pattern on surface temperature dynamics: Experience from the urban and suburban areas of Khulna, Bangladesh. *Remote Sens. Appl. Soc. Environ.* **2021**, *22*, 100508. [CrossRef]
99. Williams, R.G.; Roussenov, V.; Goodwin, P.; Resplandy, L.; Bopp, L. Sensitivity of Global Warming to Carbon Emissions: Effects of Heat and Carbon Uptake in a Suite of Earth System Models. *J. Clim.* **2017**, *30*, 9343–9363. [CrossRef]
100. Allen, M.R.; Frame, D.J.; Huntingford, C.; Jones, C.D.; Lowe, J.A.; Meinshausen, M.; Meinshausen, N. Warming caused by cumulative carbon emissions towards the trillionth tonne. *Nature* **2009**, *458*, 1163–1166. [CrossRef] [PubMed]
101. Sleeter, B.M.; Liu, J.; Daniel, C.; Rayfield, B.; Sherba, J.; Hawbaker, T.J.; Zhu, Z.; Selmants, P.C.; Loveland, T.R. Effects of contemporary land-use and land-cover change on the carbon balance of terrestrial ecosystems in the United States. *Environ. Res. Lett.* **2018**, *13*, 045006. [CrossRef]
102. Sarathchandra, C.; Abebe, Y.A.; Worthy, F.R.; Wijerathne, I.L.; Ma, H.; Yingfeng, B.; Jiayu, G.; Chen, H.; Yan, Q.; Geng, Y.; et al. Impact of land use and land cover changes on carbon storage in rubber dominated tropical Xishuangbanna, South West China. *Ecosyst. Heal. Sustain.* **2021**, *7*, 1915183. [CrossRef]
103. Yin, K.; Lu, D.; Tian, Y.; Zhao, Q.; Yuan, C. Evaluation of Carbon and Oxygen Balances in Urban Ecosystems Using Land Use/Land Cover and Statistical Data. *Sustainability* **2014**, *7*, 195–221. [CrossRef]
104. Zhou, J.; Zhao, Y.; Huang, P.; Zhao, X.; Feng, W.; Li, Q.; Xue, D.; Dou, J.; Shi, W.; Wei, W.; et al. Impacts of ecological restoration projects on the ecosystem carbon storage of inland river basin in arid area, China. *Ecol. Indic.* **2020**, *118*, 106803. [CrossRef]
105. Raziq, A.; Xu, A.; Li, Y. Monitoring of Land Use/Land Cover Changes and Urban Sprawl in Peshawar City in Khyber Pakhtunkhwa: An Application of Geo- Information Techniques Using of Multi-Temporal Satellite Data. *J. Remote Sens. GIS* **2016**, *5*, 174. [CrossRef]
106. Ullah, S.; Ahmad, K.; Sajjad, R.U.; Abbasi, A.M.; Nazeer, A.; Tahir, A.A. Analysis and simulation of land cover changes and their impacts on land surface temperature in a lower Himalayan region. *J. Environ. Manag.* **2019**, *245*, 348–357. [CrossRef]
107. Li, J.; Zheng, X.; Zhang, C. Retrospective research on the interactions between land-cover change and global warming using bibliometrics during 1991–2018. *Environ. Earth Sci.* **2021**, *80*, 1–17. [CrossRef]
108. Tariq, A.; Mumtaz, F. Modeling spatio-temporal assessment of land use land cover of Lahore and its impact on land surface temperature using multi-spectral remote sensing data. *Environ. Sci. Pollut. Res.* **2022**. [CrossRef] [PubMed]
109. Tariq, A.; Riaz, I.; Ahmad, Z.; Yang, B.; Amin, M.; Kausar, R.; Andleeb, S.; Farooqi, M.A.; Rafiq, M. Land surface temperature relation with normalized satellite indices for the estimation of spatio-temporal trends in temperature among various land use land cover classes of an arid Potohar region using Landsat data. *Environ. Earth Sci.* **2019**, *79*, 1–15. [CrossRef]
110. Hallegatte, S.; Corfee-Morlot, J. Understanding climate change impacts, vulnerability and adaptation at city scale: An introduction. *Clim. Chang.* **2010**, *104*, 1–12. [CrossRef]
111. Kant, Y.; Bharath, B.D.; Mallick, J.; Atzberger, C.; Kerle, N. Satellite-based analysis of the role of land use/land cover and vegetation density on surface temperature regime of Delhi, India. *J. Indian Soc. Remote Sens.* **2009**, *37*, 201–214. [CrossRef]
112. Khan, I.; Zhao, M. Water resource management and public preferences for water ecosystem services: A choice experiment approach for inland river basin management. *Sci. Total. Environ.* **2018**, *646*, 821–831. [CrossRef]
113. Jiang, J.; Tian, G. Analysis of the impact of Land use/Land cover change on Land Surface Temperature with Remote Sensing. *Procedia Environ. Sci.* **2010**, *2*, 571–575. [CrossRef]
114. Hou, G.L.; Zhang, H.Y.; Wang, Y.Q.; Qiao, Z.H.; Zhang, Z.X. Retrieval and Spatial Distribution of Land Surface Temperature in the Middle Part of Jilin Province Based on MODIS Data. *Sci. Geogr. Sin.* **2010**, *30*, 421–427.
115. Patz, J.A.; Campbell-Lendrum, D.; Holloway, T.; Foley, J.A. Impact of Regional Climate Change on Human Health. *Nature* **2005**, *438*, 310. [CrossRef]
116. Change, I. *2006 IPCC Guidelines for National Greenhouse Gas Inventories*; Institute for Global Environmental Strategies: Hayama, Japan, 2006.
117. Batty, M.; Xie, Y. *Urban Growth Using Cellular Automata Models*; ESRI Press: Redlands, CA, USA, 2005.
118. Mas, J.-F.; Kolb, M.; Paegelow, M.; Olmedo, M.T.C.; Houet, T. Inductive pattern-based land use/cover change models: A comparison of four software packages. *Environ. Model. Softw.* **2014**, *51*, 94–111. [CrossRef]
119. Cheng, F.-Y.; Byun, D.W. Application of high resolution land use and land cover data for atmospheric modeling in the Houston–Galveston metropolitan area, Part I: Meteorological simulation results. *Atmos. Environ.* **2008**, *42*, 7795–7811. [CrossRef]
120. De Meij, A.; Bossioli, E.; Penard, C.; Vinuesa, J.; Price, I. The effect of SRTM and Corine Land Cover data on calculated gas and PM10 concentrations in WRF-Chem. *Atmospheric Environ.* **2015**, *101*, 177–193. [CrossRef]
121. Ran, L.; Pleim, J.; Gilliam, R.; Binkowski, F.S.; Hogrefe, C.; Band, L. Improved meteorology from an updated WRF/CMAQ modeling system with MODIS vegetation and albedo. *J. Geophys. Res. Atmos.* **2016**, *121*, 2393–2415. [CrossRef]

122. Fry, I. Reducing emissions from deforestation and forest degradation: Opportunities and pitfalls in developing a new legal regime. *Rev. Eur. Community Int. Environ. Law* **2008**, *17*, 166–182. [CrossRef]
123. UNFCCC. Factsheet: Reducing Emissions from Deforestation in Developing Countries: Approaches to Stimulate Action. 2011. Available online: http://unfccc.int/files/press/backgrounders/application/pdf/fact_sheet_reducing_emissions_from_deforestation.pdf (accessed on 18 November 2022).
124. Page, P. *Report of the Ad Hoc Working Group on Long-Term Cooperative Action under the Convention on the First Part of Its Fifteenth Session, Held in Bonn from 15 to 24 May 2012*; United Nations Digital Library System: Geneva, Switzerland, 2012.
125. Herold, M.; Skutsch, M. Monitoring, reporting and verification for national REDD+ programmes: Two proposals. *Environ. Res. Lett.* **2011**, *6*, 014002. [CrossRef]
126. Skutsch, M.M.; Torres, A.B.; Mwampamba, T.H.; Ghilardi, A.; Herold, M. Dealing with locally-driven degradation: A quick start option under REDD+. *Carbon Balance Manag.* **2011**, *6*, 1–7. [CrossRef]

remote sensing

MDPI

Article

Data-Driven Approaches for Wildfire Mapping and Prediction Assessment Using a Convolutional Neural Network (CNN)

Rida Kanwal [1], Warda Rafaqat [1], Mansoor Iqbal [2] and Song Weiguo [1,*]

1 State Key Laboratory of Fire Science, University of Science and Technology of China (USTC), Hefei 230027, China; ridakanwal@mail.ustc.edu.cn (R.K.); warda@mail.ustc.edu.cn (W.R.)

2 Department of Electronic Engineering & Information Science, University of Science and Technology of China (USTC), Hefei 230027, China; man2017@mail.ustc.edu.cn

* Correspondence: wgsong@ustc.edu.cn

Abstract: As wildfires become increasingly perilous amidst Pakistan's expanding population and evolving environmental conditions, their global significance necessitates urgent attention and concerted efforts toward proactive measures and international cooperation. This research strives to comprehensively enhance wildfire prediction and management by implementing various measures to contribute to proactive mitigation in Pakistan. Additionally, the objective of this research was to acquire an extensive understanding of the factors that influence fire patterns in the country. For this purpose, we looked at the spatiotemporal patterns and causes of wildfires between 2000 and 2023 using descriptive analysis. The data analysis included a discussion on density-based clustering as well as the distribution of the data across four seasons over a period of six years. Factors that could indicate the probability of a fire occurrence such as weather conditions, terrain characteristics, and fuel availability encompass details about the soil, economy, and vegetation. We used a convolutional neural network (CNN) to extract features, and different machine learning (ML) techniques were implemented to obtain the best model for wildfire prediction. The majority of fires in the past six years have primarily occurred during the winter months in coastal locations. The occurrence of fires was accurately predicted by ML models such as random forest (RF), which outperformed competing models. Meanwhile, a CNN with 1D and 2D was used for more improvement in prediction by ML models. The accuracy increased from an 86.48 to 91.34 accuracy score by just using a CNN 1D. For more feature extraction, a CNN 2D was used on the same dataset, which led to state-of-the-art prediction results. A 96.91 accuracy score was achieved by further tuning the RF model on the total data. Data division by spatial and temporal changes was also used for the better prediction of fire, which can further be helpful for understanding the different prospects of wildfire. This research aims to advance wildfire prediction methodologies by leveraging ML techniques to explore the benefits and limitations of capturing complex patterns and relationships in large datasets. Policymakers, environmentalists, and scholars studying climate change can benefit greatly from the study's analytical approach, which may assist Pakistan in better managing and reducing wildfires.

Keywords: machine learning; wildfire assessment; CNN; random forest; fire occurrence

check for updates

Citation: Kanwal, R.; Rafaqat, W.; Iqbal, M.; Weiguo, S. Data-Driven Approaches for Wildfire Mapping and Prediction Assessment Using a Convolutional Neural Network (CNN). *Remote Sens.* **2023**, *15*, 5099. https://doi.org/10.3390/rs15215099

Academic Editors: Sawaid Abbas, Janet E. Nichol, Faisal M. Qamer and Jianchu Xu

Received: 6 September 2023
Revised: 20 October 2023
Accepted: 21 October 2023
Published: 25 October 2023

1. Introduction

The global risk of wildfire disasters has surged due to the effects of climate change. These wildfires pose a significant threat to wildland ecosystems, affecting their formation and stability on a global scale [1,2]. While fires have historically played a role in renewing and reshaping landscapes, the intensification of wildfires has led to devastating consequences for people, the environment, and economies [3]. Moreover, the impacts of wildfires extend beyond plants and animals, affecting the air, soil, and water quality [4,5]. The economic losses and threats to human lives and health are substantial [6–9].

The escalation in the frequency and severity of wildfires has been particularly noticeable across several Asian countries [10–12]. Research strongly suggests that with the ongoing impact of climate change, wildfires are projected to become more frequent and intense worldwide [13–17]. This phenomenon is also evident in Pakistan, where forest cover is notably low, accounting for just approximately 4.5 percent of the land. Consequently, Pakistan grapples with significant forest degradation issues [16].

During the latter half of the 20th century, there was a noticeable increase in fire occurrences in Pakistan and other nations, surpassing historical records. This period also witnessed a rise in large-scale wildfires and an expansion of their affected areas [2].

Understanding the factors that influence wildfire behavior such as ignition ease, rate of spread, management challenges, and overall impact is often referred to as "wildfire hazard" assessment [17]. Several components play a role in the occurrence and progression of wildfires including favorable weather conditions, the presence of combustible materials, the continuity of these materials, and potential ignition sources [18]. These factors can be categorized into topography, weather, and fuel-related factors. Each factor's spatial and temporal variability contributes differently to the overall fire risk.

Among these factors, climatic conditions have gained substantial attention in the context of wildfire changes [19,20]. The influence of topographic elements such as slope, aspect ratio, and elevation on the burnt area and ignition density has also been extensively studied [21]. Notably, fuel—comprising various plant species—plays a critical role in fire initiation and propagation [22,23]. While natural causes are important, human activities significantly contribute to fire incidents, impacting both ignition likelihood and subsequent fire behavior [24–26].

Human factors including population dynamics, socioeconomic elements, land use changes, and activities such as agriculture have a substantial impact on fuel availability and thus fire variation [27–29]. However, these human-related variables exhibit complex temporal and spatial dynamics.

Considering the availability of thermal sensors and RADAR systems, this research utilized space-borne remote sensing to derive relevant predictor variables (e.g., temperature, precipitation, population) from open-source satellite imagery, addressing the challenges associated with modeling numerous environmental and socioeconomic factors. Satellite remote sensing has emerged as a crucial tool for monitoring ecosystems and identifying potential hazards such as wildfires.

Recent advancements in ML algorithms offer significant potential for various data science applications [30–33]. These algorithms require high-quality training datasets to excel. To address this issue, remote sensing data are enhanced using a CNN. Previously, researchers utilized machine learning techniques, along with remotely sensed fire data and GIS, to develop a wildfire susceptibility map for the Adana and Mersin provinces in Turkey, aiming to identify and predict areas at high risk of wildfires by analyzing multiple factors [34]. Another researcher explored the use of remote sensing data and machine learning algorithms to predict and map wildfire occurrences over multiple time periods, providing valuable insights into the temporal patterns of wildfires through the generation of grid maps [35]. One further study introduced a machine-learning approach to identify and assess dense-fire events and their atmospheric emissions on the Indochina peninsula, utilizing data from 2010 to 2020 and providing insights into the impact of these fires on the atmosphere [36].

This study delved into ML techniques such as decision trees (DTs), RFs, K-nearest neighbors (KNNs), and support vector machines (SVMs) to map wildfires, with the goal to identify the most effective approach for model training and validation. Furthermore, to augment the features used in modeling, CNNs were employed—both 1D and 2D—to demonstrate their capabilities in wildfire assessment [37–42].

Compared to physics-based models, ML models train on satellite data and climate features from monitoring stations, covering different span periods. By incorporating historical data, ML models can learn from past patterns and trends, enabling them to make predictions based on historical patterns of wildfire occurrence and behavior. Physics-based models may require detailed knowledge of physical processes, which may not be readily available or practical for all regions. ML models offer improved prediction power by extracting patterns and relationships from complex datasets, resulting in more accurate predictions of wildfire occurrence. Being relatively novel, ML models provide fresh approaches to wildfire prediction, potentially uncovering new insights. Additionally, ML models can handle imbalanced datasets, which is crucial for obtaining reliable predictions in wildfire analysis [43]. The previous study emphasized the computational efficiency of ML models for fire forecasting, offering significant advantages over physics-based simulations. The integration of reduced-order modeling and ML ensures efficiency while maintaining accurate predictions. Data assimilation and error covariance tuning improve forecasting accuracy by incorporating real-time data. ML models accelerate fire forecasting, enabling near real-time predictions for effective decision-making and emergency response [44]. The application of existing models in other regions can be impracticable or very difficult due to distinct geographic characteristics and data availability. ML models, trained using a dataset specific to the Brazilian Federal District and enriched with various features, offer adaptability and outperform physics-based models, achieving an accuracy of 91% in predicting wildfire impacts, making them valuable for fire management agencies [45].

Despite the existing research, there remains a gap in thoroughly assessing fire susceptibility using remote sensing data. Previous research has highlighted the application of artificial neural networks and mapping techniques for fire risk assessment and vegetation analysis [46–48]. This study addresses this gap by focusing on a specific region with limited datasets. The research aims to build improved prediction models that can provide tailored recommendations for fire management and prevention, considering geographic, temporal, and seasonal aspects.

This research aimed to comprehensively assess wildfire prediction and management by evaluating model durability and reliability over an extended period, exploring the impact of seasonality on accuracy, capturing patterns linking precipitation and wildfire events, analyzing spatial distribution, developing targeted strategies, assessing advanced machine learning models, integrating real-time data, constructing a comprehensive management framework, implementing preemptive measures, and improving forecasting accuracy through data refinement. These objectives contribute to proactive wildfire mitigation in Pakistan. This study seeks to holistically address the wildfire conundrum in Pakistan by shedding light on its multifaceted dimensions. The overarching objective is to delve into the temporal and spatial facets of wildfire occurrences, aiming to advance our comprehension of their patterns and behaviors. At the heart of this pursuit is the integration of cutting-edge machine learning techniques including a CNN 2D and RF to bolster the accuracy of wildfire prediction models. The driving force behind this research is the aspiration to not only contribute to the academic discourse, but also to offer practical insights that have tangible implications for policymaking, disaster management frameworks, and sustainable community development in the face of escalating wildfire risks.

The primary objectives of this study encompass four key aspects. First, the study aims to undertake comprehensive data collection including historical fire data, and subsequently preprocesses the data by rectifying errors and standardizing formats. This process will involve the acquisition of pertinent details such as location, date, and pertinent factors such as temperature. Second, the study intends to perform model training and evaluation. This involves the identification of influential features, the division of data for training and testing purposes, and the training of a model—potentially employing techniques such as RF. Model performance evaluation will rely on metrics such as accuracy and recall. Third, the study seeks to acquire and refine data from the FIRMS dataset, which is utilized for fire detection. This refined dataset will be subjected to spatial analysis, allowing for a comparison with historical fire data and an assessment of alignment. Finally, the study will focus on identifying causal factors through an analysis of feature importance within the model. Insights garnered from this analysis will be leveraged to propose effective strategies aimed at fire management, prevention, and early detection. Collaboration with experts in the field will further enhance the applicability and efficacy of these recommendations.

The structure of the study comprises various sections: Section 2 gives an overview of the research area and past fire incidents; Section 3 details the techniques used and presents results; Section 4 covers data preparation, a CNN design, accuracy assessment, and key findings; Section 5 discusses the statistics and findings. Finally, Section 6 concludes the study with a brief summary.

2. Study Area

The research region is Pakistan, and the study period was from November 2000 to December 2020. Pakistan is situated in the Arabian Sea in South Asia. A total of 875,175 km^2 of surface area in Pakistan is made up of 25,683 km^2 of wildland, and 33.3% is agriculture. There are 292,882 km^2 of grassland and shrub land overall. The percentage of wilderness in Pakistan is 36.4%. There were 175,000 fires reported in Pakistan, of which 22,311 were wildfires. According to land cover categorization, Figure 1 represents the wilderness area of Pakistan, which is categorized as wildness. The area of interest is shown on the map, along with historical fire occurrences that occurred between 2001 and 2020. One of the most vulnerable regions of Pakistan is its eastern region. The least rainy and most prone to desertification weather, dry weather, is featured. The maximum elevation reaches 28,251 m, whereas the minimum altitude is at sea level, which is 0 m [49]. The northern region of the country is characterized by mountainous terrain, while the southern areas gradually transition into flatter landscapes. Notably, the northeast region exhibits the highest population density in the nation. The population is very evenly dispersed over the nation, with population clusters also being visible close to the largest cities. Southwest Pakistan is the least inhabited part of the country [50].

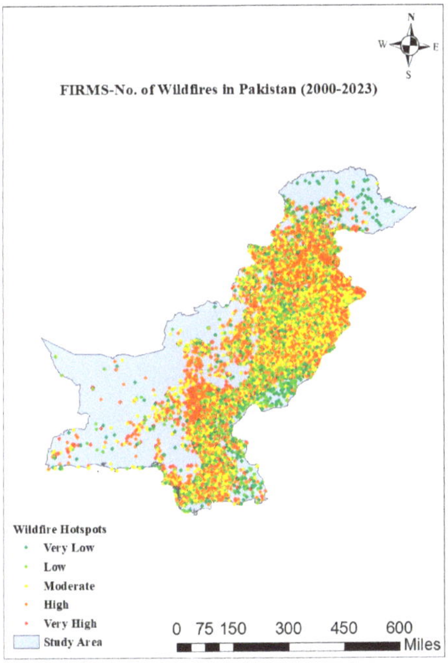

Figure 1. Map depicting the wildfire study area in Pakistan from November 2000 to October 2023.

Pakistan experiences four distinct seasons throughout the year. From December to February, the country experiences a cool and dry winter. The months of March to May mark the arrival of a hot and dry spring season. Then, from June to August, Pakistan experiences a wet monsoon season characterized by rainfall. Finally, the monsoon season gradually subsides from September to November, known as the waning monsoon season. Previous wildfire studies were not undertaken on a national scale due to the absence of a continuous dataset that spanned the entire country. Due to the excellent remotely sensed datasets, the best model for fire prediction may be studied using a publicly available worldwide dataset collected from remotely sensed data.

After defining the essential terms, we conducted an exploratory investigation into Pakistani wildfires. Our research approach involved using diverse remotely sensed datasets that are publicly accessible. Furthermore, we integrated datasets from other industries that also relied on satellite images as essential features (dependent variables) and influential factors (independent variables). The primary objective of this study was to gain a deeper understanding of Pakistani wildfires. By analyzing information from previous fire incidents, along with utilizing remote sensing data and incorporating datasets from various sources, we aimed to identify underlying patterns and factors contributing to wildfire occurrences in the region. This comprehensive methodology allowed us to explore potential relationships among different variables and contribute to the development of effective strategies for wildfire management and prevention.

To determine the total number of fires, data from the Fire Information for Resource Management System (FIRMS) were utilized. FIRMS provides data that are almost real-time and available approximately three hours after satellite observation. These data are transferred via FIRMS to the Terra and Aqua Earth Observing System (EOS) as part of NASA's Land, Atmosphere, Near Real-Time Capability (LANCE) for EOS. The data originate from the Visible Infrared Imaging Radiometer Suite (VIIRS) and the Moderate Resolution Imaging Spectroradiometer (MODIS). The wildfires currently occurring are highlighted in Figure 1, which displays pixels covering a square kilometer of the landscape. Within each 1 sq. km pixel, the publicly accessible dataset includes records for temperature, fire confidence, and fire type. From November 2000 to October 2023, users can obtain FIRMS data with a one-day temporal resolution and a one-kilometer square geographic precision.

In this study, we applied filtering criteria to remotely sensed fire data. Specifically, we categorized the fire data based on two criteria: confidence level and fire type. For the confidence level, we set a threshold of 60 percent confidence. This means that only fire observations with a confidence level of 60 percent or higher were chosen for further analysis. This filtering step helped ensure that we focused on more reliable fire detections and reduced the inclusion of potential false positives. Regarding fire type, we utilized the MODIS landcover vegetation data to filter the fire observations. We selected only those fire pixels that corresponded to vegetation-covered areas, as indicated by the land cover data. This helped us narrow down the analysis to fires occurring within vegetated regions and excluded non-vegetated areas or other land cover types. By applying these filtering criteria, we aimed to improve the quality and accuracy of the remotely sensed fire data used in our study.

Figure 2 illustrates the overall count of pixels indicating active wildfires in Pakistan during specific seasons from 2000 to 2023. In terms of fire activity, the data suggest that the year 2020 did not stand out significantly compared to the previous two decades. The most severe years in terms of fire activity were 2008, 2016, and 2017. Notably, the number of active fires in 2016 surpassed the combined count of 2015 and 2019. To identify irregularities across seasons, it is necessary to have a comprehensive understanding of fire activity throughout the year. The number of active fires in 2008 was significantly higher than that in earlier years. Satellite-based fire data, however, revealed that until 2023, every year, in December and February of each year for the previous 24 years, MODIS recorded 10,357 active wildfire indications throughout Pakistan.

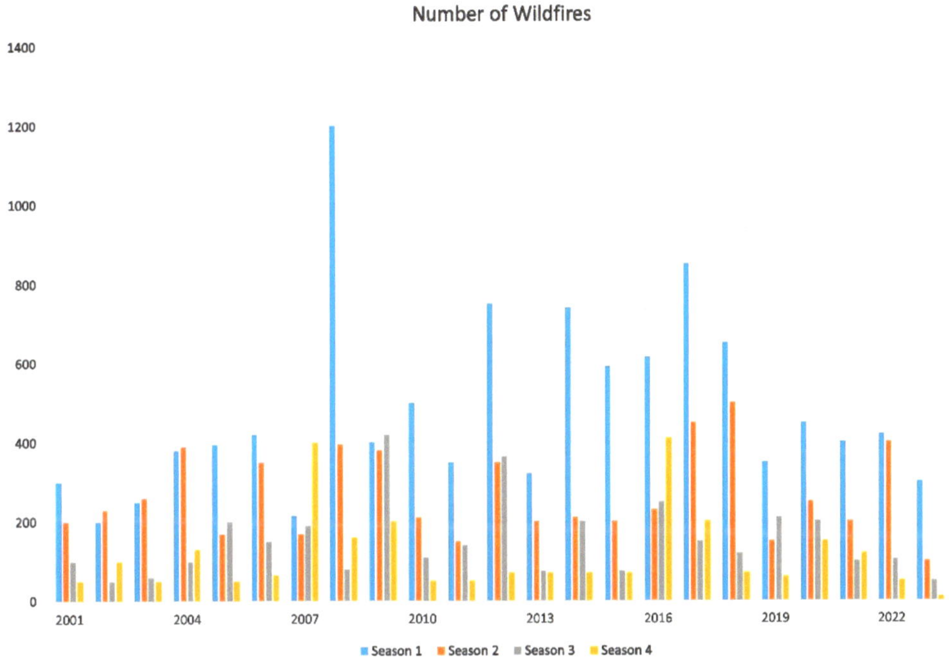

Figure 2. The distribution of the MODIS annual wildfire season is as follows: Seasons 1 and 2 cover December to February and March to May, respectively. Season 3 spans from June to August, and Season 4 covers the period from September to November.

Figure 2 presents the seasonal distribution of active fire sites between January 2001 and October 2023. It can be observed that the majority of fires tended to occur during the winter season, spanning from December to February. The seasons are categorized as follows: Seasons 1 and 2 encompass December to February and March to May, respectively; Season 3 encompasses June to August; Season 4 encompasses September to November.

A diverse range of studies utilizing machine learning techniques in the context of wildfire management and prediction [51] focused on the comprehensive application of deep learning methodologies using satellite remote sensing data for detecting, mapping, and predicting wildland fires. The authors in [52] explored the creation of wildfire susceptibility maps through the integration of interferometric synthetic aperture radar (InSAR) coherence, deep learning, and metaheuristic optimization techniques. The study by [53] contributed to the field by presenting a novel approach to forest fire prediction that leveraged long- and short-term time-series networks [54] and concentrated on active fire mapping in the Brazilian Pantanal region by employing deep learning and China-Brazil Earth Resources Satellite-4A (CBERS-4A) satellite imagery. These studies collectively showcase the increasing significance of machine learning approaches in advancing our understanding of wildfire behavior, enhancing prediction accuracy, and enabling effective fire management strategies using remote sensing data.

3. Methodology

In this section, we discuss how we went about achieving our main goals. The first stage was to find the optimum fire susceptibility model, and the second was to use a CNN to improve the characteristics of the features. The structure was divided into three sections: (a) data preparation and preprocessing; (b) a CNN; (c) model assessment (Figure 3). On a dataset with varied perspectives, the entire procedure was run four times. For the complete dataset, the data were split by year (six-year gap), the data were sorted regionally based on density, and the data were separated by season. To construct the training dataset, the three main driving factors of topography, weather, and fuel were considered. This information is gathered from various sources such as soil, socioeconomics, vegetation, and past fire incidents. The historical fire occurrences were divided into separate subgroups for training and testing purposes. Machine learning models were then trained using the training dataset, and the performance of the model was evaluated using the test dataset. The most effective ML model was employed to predict wildfire susceptibility.

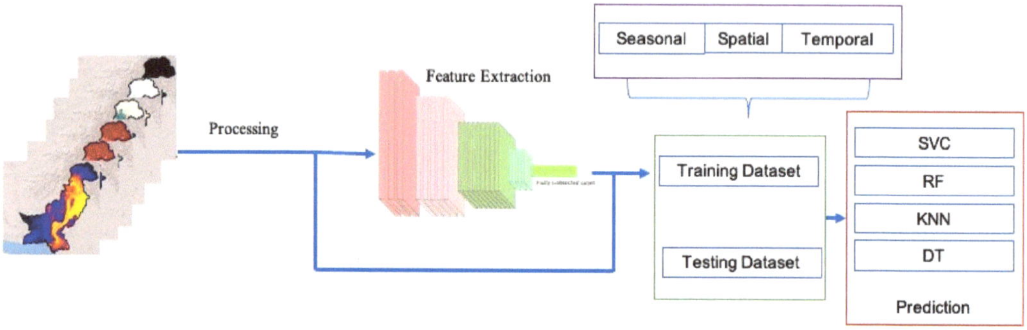

Figure 3. Workflow diagram for wildfire mapping and prediction using the CNN and ML models.

3.1. Feature Extraction Layer

The feature extraction layer in the workflow diagram for wildfire mapping and prediction patterns was extracted from the input data. In the context of this particular workflow, the feature extraction layer was responsible for identifying and extracting meaningful features from the wildfire data such as satellite imagery or other relevant sources. This layer played a crucial role in capturing the distinctive characteristics and patterns associated with wildfire events. Typically, techniques such as CNN or other machine learning algorithms are employed in this layer to automatically learn and extract these features from the input data. The extracted features are then used as inputs to subsequent layers or models for the further analysis, prediction, or mapping of wildfires. The goal of the feature extraction layer is to transform the raw input data into a more compact and representative feature representation that can be effectively utilized by the subsequent stages of the workflow such as classification or prediction models. By properly extracting and selecting relevant features, the feature extraction layer helped to improve the accuracy and effectiveness of the overall wildfire mapping and prediction process.

3.2. Data Extraction and Preprocessing

The passage describes the training dataset used for a specific analysis. The training dataset consists of two types of variables: independent variables (also known as predictors) and dependent variables (also known as response variables). Independent variables were the factors that were used to predict or explain the outcome, while dependent variables were the outcomes or responses that we were interested in understanding or predicting. The entire dataset was quite large and contained a substantial amount of data. It spans a 24-year period and covers information at the country level. Handling such a vast dataset manually would be impractical and time-consuming, so the creation of training datasets was automated. Automation allowed for a more efficient and streamlined process of preparing the necessary data to train the models. By providing the chosen models with more training data samples, their performance improved. This means that having a larger and more diverse set of data to train the models resulted in more accurate and robust predictions or classifications. The models were able to learn from a more extensive range of examples, leading to better generalization and higher accuracy in their predictions or classifications when applied to new, unseen data.

Fire and no-fire detail: The dependent variable was a binary variable that contrasted with areas where fire incidence occurred and those where it did not. Hence, mapping fire susceptibility can be viewed as a binary classification task in machine learning, with two distinct classes: fire and no-fire. This research served as the foundation for the development of an automated method for identifying where fires and no-fire incidents occurred.

To perform additional research on the driving mechanisms, no-fire instances must also be gathered. Pakistan's wildland land cover collection created a historical database of no-fire occurrences. Random locations in a wilderness area were selected and separated by a distance of one kilometer. To be classified as a no-fire occurrence, sites had to fall outside the firing zones within the 1 km buffer zone. From November 2000 to October 2023, six points were assigned to each day as no-fire points, corresponding to the same time period as the fire data. Additionally, latitude and longitude values were assigned to each month, resulting in a total of 184 points. Both fire and no-fire data had the same spatial resolution. Pakistan's dataset consisted of 22,311 fire points and an equal number of no-fire points.

Contributing factors: Choosing the independent variables, also referred to as predictors or conditioning factors, is a crucial step in predictive modeling. Based on field observations from several research projects as well as open-source and international satellite data, 34 driving elements were selected for this investigation [55–57]. While there are no strict guidelines regarding the inclusion of variables in a model, geomorphological, climatic, and human-related factors are commonly considered conditional components in many studies [58].

Wildfires are mostly caused by three factors: topography, weather, and fuel. The two primary applied wildfire conditioning variables related to fuel conditions are vegetation type and socioeconomic issues. A summary of each dataset used in this study can be found in Table 1. Information from a publicly available database of remotely sensed data, some of which had records with different classifications, was used to build the independent variables. Numerous socioeconomic traits were constructed using remote sensing data. Table 1 includes links to the data sources and data catalogs in the Source of Data column. The imported data in this study were in the form of raster data.

The weather is one of the primary factors generating wildfires, according to numerous studies. The fire weather index (FWI) is a dataset provided by NASA that encompasses various weather information such as temperature, humidity, and transpiration [59]. It is employed widely outside of Canada [60–63]. Instead of merely using a few parameters for data models, in our study, we utilized the Terra Climate dataset [64] and the Global Land Data Assimilation System (GLDAS) [65], which offer a higher spatial resolution, as indicated in Table 1. We incorporated several climatic variables, both derived and primary, into our analysis. It is worth noting that Pakistan has not been extensively explored in terms of the FWI, and the global FWI only encompasses a limited number of factors. Given their superior spatial resolution, we opted for the GLDAS and Terra Climate datasets among the suggested datasets for FWI analysis.

To generate the training dataset, we established fire and no-fire training points along with corresponding conditional factors [66–75]. Subsequently, 34 independent variables were collected and resampled to match the spatial resolution. Information extraction techniques were employed to incorporate predictor values into the dataset. Training samples were then created for each latitude and longitude, resulting in a comprehensive dataset containing fire and non-fire sites, with associated driving variable data stored for each geographic location.

Table 1. Various datasets represent different variables, all obtained from reliable sources. These datasets have a spatial resolution of 1 km grid cells and cover the time period from November 2020 to October 2023.

S. No.	Parameter [Source]	Unit	Spatial Resolution	S. No.	Parameter [Source]	Unit	Spatial Resolution
1	Elevation [66]	Meters	90 m	18	Average surface skin temperature [68]	k	28 km × 28 km
2	Slope [66]	Degree		19	Soil moisture [68]	kg/m^2	
3	Aspect [66]			20	Actual evapotranspiration [67]	mm	4.5 km
4	Hill shadow [66]			21	Water deficit [67]	mm	
5	Population [76]	person/km (grid cell)	1 km	22	Downward surface shortwave radiation [67]	W/m^2	
6	Human modification [68]	Km^2		23	Precipitation accumulation [67]	mm	
7	Travel speed [69]	minutes/ meter		24	Minimum temperature [60]	°C	
8	Travel speed walk [69]			25	Maximum temperature [60]		
9	Settlement [73]	classes	100 m	26	Vapor pressure [60]	kPa	
10	Urban cover [67]	%	12 m	27	Tree cover [73]	%	100 m
11	Precipitation [71]	$kg/m^2/s$	28 km × 28 km	28	NDVI [74]	nm	500 m
12	Transpiration [71]	W/m^2		29	FPAR [75]		
13	Wind speed [71]	m/s		30	LAI [75]	m^2	
14	Soil temperature [71]	k		31	Land cover [73]	classes	100 m
15	Humidity [71]	kg/kg		32	Soil bulk density [70]	kg/m^3	250 m
16	Heat flux [71]	W/m^2		33	Soil taxonomy [72]	classes	
17	Albedo [71]	%		34	Soil texture [71]		

3.3. Convolutional Neural Network

Following the creation of the training set, the classification accuracy was evaluated. The supervised classifications frequently make use of the input training samples. It is essential to carry out multicollinearity in the creation of the training set, and the classification accuracy was evaluated, so it is essential to carry out a multicollinearity analysis before beginning fire modeling. In Figure 4 a multicollinearity analysis was conducted to check for any connected components. The link between the predicted wildland fire factors was then investigated using a multicollinearity analysis. The choice of features is extremely important in the prediction of fire susceptibility.

Multicollinearity analysis was conducted to assess the relationship between the climate, environment, socioeconomic, topography, and vegetation factors. The analysis aimed to determine whether there were high correlations or dependencies among these factors, which could impact the reliability and interpretability of the model. The results of the multicollinearity analysis provide insights into the extent to which these factors are correlated, helping to identify potential redundancies or overlapping information.

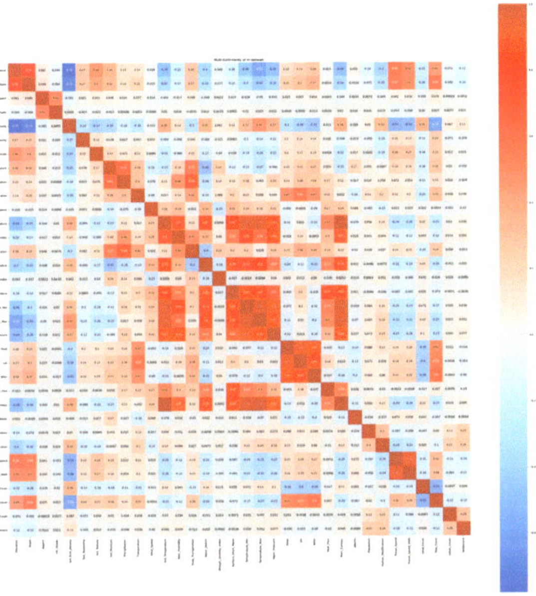

Figure 4. A thorough multicollinearity analysis examines the relationships between the climate, environment, socioeconomic, topography, and vegetation factors. This assessment aims to uncover high correlations that may affect the model reliability and interpretability.

Prior to implementing machine learning techniques, a comprehensive analysis of multicollinearity was conducted on the variables. The results indicated a strong correlation exceeding 0.8 among certain variables. Variables with correlation coefficients surpassing 0.8 or 0.9 were considered for removal. Consequently, these variables were excluded from the machine learning process based on the criterion of high correlation coefficients. This approach was adopted due to the understanding that strong correlations between variables imply the capture of similar information, thereby complicating the ability to discern their individual effects on the dependent variable. By prioritizing variables with lower correlations, the aim was to enhance the reliability and applicability of the models.

Feature Selection: Multicollinearity helped identify redundant features. In the multicollinearity analysis, we found that some of the variables were highly correlated, and they likely captured similar information. In such cases, in our feature importance analysis, we selected only one of these variables as it can be sufficient, reducing the dimensionality of the dataset and improving model performance.

Feature importance analysis: By assessing the importance of features, we gain insights into which variables have the most influence on the model's predictions. This analysis helps prioritize and focus on the most influential factors when interpreting and making decisions based on the model's outputs. In Figure 5 the results of the feature importance analysis provide a ranking or score for each feature, indicating their relative importance in predicting the target variable. This information can guide further analysis, model refinement, or decision-making processes.

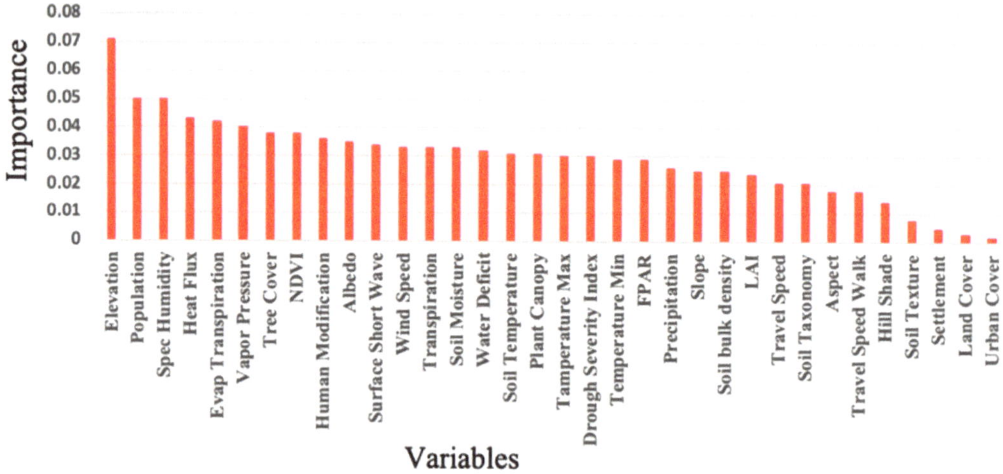

Figure 5. Feature importance analysis. Understanding the impact: unveiling the importance of features in our model aids in prioritizing influential variables for interpretation and decision-making. Each feature is assigned scores or rankings, providing guidance for further analysis and model refinement.

The model utilizes a CNN with a depth of 20 layers. The architecture of the CNN, consisting of 20 layers, is depicted in Figure 6. CNN has emerged as a fundamental concept in the field of deep learning. Its ability to leverage vast amounts of data in the era of big datasets is different from traditional methods, yielding promising results. Consequently, numerous applications have been developed. Furthermore, a CNN can be applied to process both 1D data and 2D images.

$$out(N, C_{out}) = bias(C_{out}) + \sum_{k=0}^{C_{in}-1} weight(C_{out}, k) * input(N, k) \quad (1)$$

In the 1D case, the output value of the layer with input size (N, C_{in}, L_{in}) and output (N, C_{out}, L_{out}) can be precisely described, where N is the batch size, C denotes the number of channels, and L is the length of the sequence (34 in our case). In the 2D case, the output value of the layer with input size $(N, C_{in}, H_{in}, W_{in})$ and output $(N, C_{out}, H_{out}, W_{out})$ can be precisely described, where N is the batch size, C denotes the number of channels, H is the height of the input planes in pixels, and W is the width in pixels.

The maps of wildland fire susceptibility were created using various classification methods including DT, SVM, RF, and KNN. Among these methods, RF employs an ensemble learning technique by combining the results of multiple classification trees. It achieves this by utilizing bootstrapping techniques and selecting subsets of observations to construct random binary trees. This approach involves taking random samples from the original training dataset. Using an RF offers several advantages over a single-participant method. It yields more accurate results due to the diversification and randomness introduced by the subset selection process, leading to enhanced generalization and robustness in the model.

The SVM is a popular and extensively employed classification method known for its non-parametric kernel-based approach. They are the most efficient at handling difficult, robust linear and non-linear classification and regression tasks.

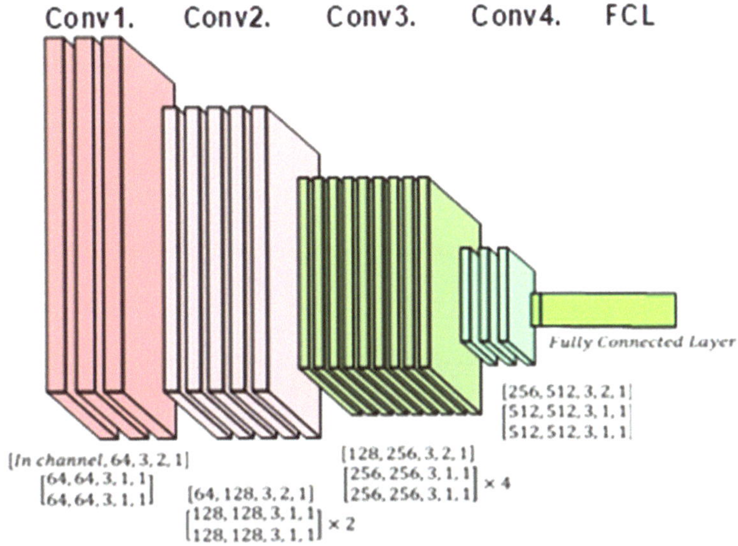

Figure 6. A 20-layer CNN architecture was used for this study with a fully connected layer of size 512 at the end, which will be fed to classifiers.

The KNN approach is a non-parametric model that is commonly used for classification and regression tasks. It adopts a lazy learning strategy, which means that it does not make any assumptions about the data. This makes the KNN model particularly valuable for predicting air pollution when there is a lack of general predictors. The KNN technique begins by randomly selecting a certain number of class centers (k) and then classifies the training data based on their proximity to these centers. The class centers are then iteratively adjusted to the middle of the training data, and the data are reclassified accordingly.

DTs are considered universal function approximators and belong to the category of supervised learning algorithms. However, achieving such universality in its basic form can be challenging. DTs are versatile and can be applied to both classification and regression problems. A DT is composed of a set of if-then-else rules organized as branches connected by decision nodes that ultimately lead to leaf nodes. The decision nodes represent the points at which the tree splits into different branches, indicating the specific decisions made by the algorithm. On the other hand, the leaf nodes represent the output of the model.

3.4. Validation and Evaluation Matrix

Once ML models have been trained to predict the origin of a fire, it becomes essential to evaluate their performance. To accomplish this, an accuracy evaluation was carried out. To validate the model, the sample dataset containing fire and non-fire zones was divided into training and testing datasets. The points were allocated in an 80:20 ratio, with 80% assigned for training and the remaining 20% for testing purposes.

Accuracy assessment: The performance of the ML model was evaluated using a widely recognized accuracy evaluation technique. This involved assessing characteristics such as overall accuracy, area under the curve score (AUC score), and precision score.

To conduct the accuracy evaluation, independent testing datasets were created by dividing the sample dataset at an 80:20 ratio, with 80% allocated for training and 20% reserved for testing.

The accuracy of a binary classification test is determined by its ability to correctly determine the presence or absence of a condition. It is measured as the percentage of accurate predictions, which includes both true positives (*TP*) and true negatives (*TN*), out of all the instances considered.

$$Accuracy = \frac{TP + TN}{TP + FP + FN + TN} \quad (2)$$

The F1 score combines precision and recall, focusing on the model's performance in binary classification tasks. It is particularly useful when there is class imbalance or varying impacts of false positives (*FP*) and false negatives (*FN*).

The F1 score combines two important metrics: precision and recall. Precision measures the accuracy of the model's positive predictions, while recall quantifies the model's ability to correctly identify positive instances. While TN are important for evaluating a classification model's overall performance, they do not directly factor into the F1 score calculation, which specifically focuses on the positive class.

Precision is a crucial metric in binary classification, emphasizing the quality of positive predictions made by the model. It is calculated by dividing the number of TN predictions by the sum of TP and FP predictions. Precision ensures that when the model claims something is positive, it is highly likely to be correct, minimizing FP. High precision indicates reliable and accurate positive predictions. Precision is often used alongside recall, which measures the model's ability to capture all positive instances. Precision quantifies the trustworthiness of the model's positive predictions. In summary, precision helps us understand the proportion of the true positive predictions among all instances predicted as positive by the model. It is an important metric, especially when FP has significant consequences.

$$Precision = \frac{TP}{TP + FP} \quad (3)$$

4. Results

In this section, the conclusions of the study are presented based on the methodology utilized. The first part reports the results of the fire occurrence sites obtained from the FIRMS dataset. The second half of this section reviews the findings of various machine learning methods. It is worth noting that the majority of the predictive variables used in this study were sourced from open-source platforms.

This study employed supervised ML methods (RF, DT, SVM, and KNN) with an 80–20 split for training and testing, respectively. The accuracy summary in Table 2 and Figure 7 shows that RF had the highest accuracy (96.56%), DT had the lowest (85.12%), and SVM and KNN both had 87.2%. Notably, all methods showed higher accuracy in predicting the absence of fire than in predicting its occurrence.

The evaluation script using RF algorithms is commonly used to determine the optimal number of maximum trees and max-depth for the RF model. These statistics directly influence the accuracy of the model, making this step crucial. Additionally, the script can provide insights into the number of leaf nodes needed for classifying two classes. Hyperparameter tuning involves selecting the best combination of hyperparameters to maximize model performance. Finding the perfect set of hyperparameters is essential for achieving optimal model performance. Manual tuning of hyperparameters involves manually setting and testing various combinations. However, this method can be time-consuming and impractical when there are numerous hyperparameters to test.

For a CNN 1D and a CNN 2D, the output of the completely connected layer is 512, which is fed directly into the RF. In the case of a CNN, the RF accuracy also significantly increased when the variable rose to 512. When the variable increased from 34 to 512, the RF accuracy rose from 86.48 to 91.34. A total of 512 variables were sent to the RF while using the picture data for a CNN 2D to maintain quick processing and achieve a 96.56 accuracy.

Table 2. An overview of the ML algorithm results for accuracy assessment including the RF, SVC, DT, and KNN models compared with the Conv1D, Conv2D, and no Conv approaches.

Classifier	Feature	Accuracy Score	AUC Score	Precision Score	F1-Score
RF	No-Conv	86.48	86.55	85.11	86.83
	Conv1D	91.34	90.90	90.04	91.71
	Conv2D	96.56	95.90	95.24	96.95
SVC	No-Conv	77.88	77.22	74.95	78.19
	Conv1D	82.19	82.58	80.02	82.52
	Conv2D	87.14	86.99	84.19	87.49
DT	No-Conv	76.41	76.40	70.91	76.72
	Conv1D	81.76	81.36	77.53	82.09
	Conv2D	85.13	82.37	80.03	85.47
KNN	No-Conv	77.18	77.10	71.99	77.49
	Conv1D	83.24	83.19	80.07	83.57
	Conv2D	87.19	88.24	82.27	87.54

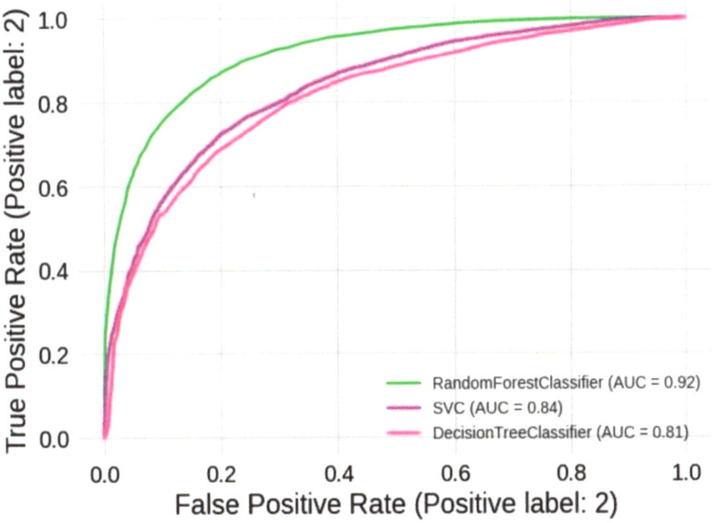

Figure 7. The ROC curves of RF, SVC, and DT.

The RF model's accuracy improved as the number of leaf nodes increased, reaching its peak at 110 leaf nodes and a maximum depth of 50. Beyond the 50 max-depth, the accuracy remained relatively stable. Based on these findings, it can be concluded that 110 leaf nodes were the optimal number of trees for the RF model in this experiment. The RF model with 110 trees and a maximum depth of 50 achieved an impressive accuracy of 96.91 percent, as demonstrated in Table 3.

Table 3. Hyperparameter (leaf node 110 and max-depth 50) tuning of the RF model and scores on the testing dataset.

Classifier	Feature	Accuracy Score	AUC Score	Precision Score	F1-Score
RF	No-Conv	86.73	86.53	85.29	87.13
	Conv1D	91.56	91.04	90.35	91.92
	Conv2D	96.91	95.94	95.54	97.29

It is essential to assess the performance of the RF model compared to other models under different temporal, geographical, and seasonal conditions to ensure its reliability. Extensive research has shown that the RF model consistently produces diverse results based on the specific location. Additionally, it is crucial to consider the temporal validity of the models. To enhance the model's applicability in real-time scenarios, the entire dataset was categorized into different subsets and subcategories. This approach aims to improve the model's adaptability in dynamic environments.

Temporal relatability of the model: One common issue in data-driven problems is the selection of the optimal time period for model training. The stochasticity or heterogeneity of the data presents a challenge, as using long-term data in model construction may result in a smoothing effect on behavior tendencies when applied to current data. Conversely, relying on short-term data could introduce higher levels of uncertainty. It is important to evaluate the durability of the model over an extended period to assess its reliability.

In Table 4 the dataset was divided into four groups with a six-year interval between them. The time periods from 2000 to 2005, 2006 to 2011, 2012 to 2017, and 2018 to 2023 were designated Y1, Y2, Y3, and Y4, respectively. It is important to note that there were fewer samples available in Y1 due to the data accessibility limitations for the year 2000. Across all time periods, the CNN 2D feature extraction technique consistently outperformed other feature techniques in terms of accuracy. Specifically, the highest accuracy score of 96.21 was achieved in Y1 using a CNN 2D feature extraction, while the lowest accuracy score of 86.01 was observed in Y1 with regular tabular data.

Table 4. The temporal relatabilities of the RF models from 2000 to 2005, 2006 to 2011, 2012 to 2017, and 2018 to 2023 were designated Y1, Y2, Y3, and Y4, respectively.

Division	Feature	Accuracy Score	AUC Score	Precision Score	F1-Score
Y1	No-Conv	86.01	83.47	83.87	86.35
	Conv1D	90.87	88.33	88.73	91.23
	Conv2D	96.21	93.67	94.07	96.59
Y2	No-Conv	83.52	83.35	83.20	83.85
	Conv1D	88.38	88.03	88.06	88.73
	Conv2D	93.72	93.55	93.40	94.09
Y3	No-Conv	84.23	84.10	85.74	84.56
	Conv1D	89.09	89.77	90.60	89.44
	Conv2D	94.43	95.02	95.94	94.80
Y4	No-Conv	83.67	83.59	82.44	84.00
	Conv1D	88.53	88.45	87.30	88.88
	Conv2D	93.87	93.79	92.64	94.24

Seasonal relatability of the model: Pakistan experiences four distinct seasons: S1 (mild, dry winter) from December to February, S2 (hot, dry spring), S3 (summer rainy season or southwest monsoon period), and S4 (receding monsoon period) from September to November. To analyze the data, it was divided into these seasonal categories (Table 5).

Notably, the CNN 2D approach consistently outperformed other feature techniques in each seasonal distribution. S3 with a CNN 2D feature extraction had the highest accuracy score of 98.71, whereas S3 with conventional tabular data had the highest accuracy score of 90.51.

Table 5. Seasonal relatability of the RF model.

Division	Feature	Accuracy Score	AUC Score	Precision Score	F1-Score
S1	No-Conv	82.75	80.63	82.86	83.16
	Conv1D	87.61	85.49	87.72	88.04
	Conv2D	92.95	90.83	93.06	93.41
S2	No-Conv	83.89	83.20	83.53	84.30
	Conv1D	88.75	88.06	88.39	89.19
	Conv2D	94.09	93.40	93.73	94.56
S3	No-Conv	90.51	87.43	89.28	90.96
	Conv1D	95.37	92.02	94.14	95.84
	Conv2D	98.71	97.63	99.48	99.20
S4	No-Conv	87.30	80.12	86.71	87.73
	Conv1D	92.16	84.98	91.57	92.62
	Conv2D	97.50	90.32	96.91	97.98

Recognizing the relationship between wildfires and precipitation during the summer rainy season is of utmost importance as it highlights the considerable influence that precipitation can exert on wildfire behavior and frequency. Rainfall plays a vital role in increasing the fuel moisture content, making it less susceptible to ignition and reducing the potential for fire spread. Moreover, higher humidity levels associated with rainy seasons can also influence the flammability of vegetation, further contributing to wildfire prevention. In our study, we found that the use of machine learning models including No-Conv, Conv1D, and Conv2D significantly contributed to the prediction of wildfires during the summer rainy season in Pakistan. Through extensive data collection on factors such as weather patterns, historical wildfire incidents, and fuel moisture content, we were able to train and evaluate our model accurately. Our findings support the notion that there is a strong relationship between wildfires and precipitation. Specifically, during the summer rainy season in Pakistan, increased rainfall leads to a higher moisture content in vegetation and fuels. This increased moisture makes them less susceptible to ignition and reduces the potential for fire occurrence and spread. Based on our analysis, we observed a significant reduction in wildfires during the summer rainy season, which was reflected in the high accuracy score of our prediction model. This aligns with our understanding of the connection between wildfires and precipitation, as rainfall plays a vital role in mitigating the risk of fire incidents.

Spatial assessment of the model: Although the data-driven approach in this study focused solely on a national perspective, the significance of distinctive attributes across various clusters within Pakistan is vividly illustrated in Figure 8. The data's relative characteristics separated in a pronounced manner, highlighting spatial variations. To delve into the impact of these variations on fire occurrences, the study opted for a selection of four distinct density-based clusters. For this investigation, the K-means clustering technique was employed, recognized for its simplicity and effectiveness. The primary objective of K-means clustering is to minimize the collective distance between each entity and its respective centroid, disregarding any predefined grouping. This process aims to identify the best arrangement of n entities into k clusters, assigning each object to the cluster with the nearest average, resulting in a partitioned data space known as Voronoi cells. Specifically, this study employed K-means spatial location clustering to segment the dataset into four clusters based on density. These clusters are denoted as C1, C2, C3, and C4, with corresponding colors of brown and pink for C1 and C2 and purple and green for C3 and C4, respectively.

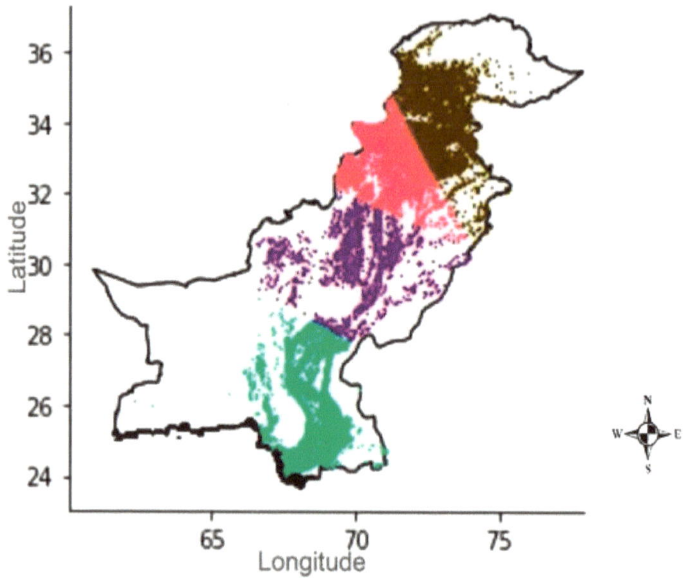

Figure 8. Cluster distribution of data from the north to the south of Pakistan (study region). C1 is represented by the color brown, C2 by pink, C3 by purple, and C4 by green.

The implementation of K-means clustering allowed for the categorization of spatially related data points into meaningful groupings, each representing a unique cluster. The selection of four clusters was based on the characteristics and distribution of the data. The distinctive attributes and patterns within each cluster provide insights into the spatial variations in fire occurrences. This approach enables a deeper exploration of the spatial dynamics of fires across different regions of Pakistan. By partitioning the data into these clusters, it becomes possible to identify areas with similar fire occurrence trends and understand the factors contributing to these patterns.

The resulting clusters, designated C1, C2, C3, and C4, offer a more refined understanding of the spatial distribution of fire incidents within Pakistan. The colors assigned to each cluster—brown and pink for C1 and C2 and purple and green for C3 and C4, respectively—further aid in visually distinguishing the distinct groupings. Through this K-means spatial clustering approach, the study gains valuable insights into how fire occurrences vary across the landscape. This information can be particularly useful for policymakers, land managers, and other stakeholders involved in fire management and prevention strategies. By considering the spatial patterns revealed by the clusters, more targeted and effective approaches to managing fire risks can be developed.

It was observed that in all cluster distributions, a CNN 2D outperformed other feature techniques. In Table 6 the highest accuracy score of 97.66 was observed in C4 with a CNN 2D feature extraction, and an 87.45 accuracy score was observed in C4 with normal tabular data.

Table 6. Spatial relatability of the RF model.

Division	Feature	Accuracy Score	AUC Score	Precision Score	F1-Score
C1	No-Conv	86.10	83.61	83.87	86.61
	Conv1D	90.96	88.47	88.73	91.50
	Conv2D	96.30	93.81	94.07	96.87
C2	No-Conv	82.71	81.75	82.85	83.20
	Conv1D	87.57	86.61	87.71	88.09
	Conv2D	92.91	91.95	93.05	93.46
C3	No-Conv	86.71	85.13	87.06	87.23
	Conv1D	91.57	89.99	91.92	92.11
	Conv2D	96.91	95.33	97.26	97.49
C4	No-Conv	87.46	87.09	82.90	87.98
	Conv1D	92.32	91.95	87.76	92.87
	Conv2D	97.66	97.29	93.10	98.24

To retain all of the available data, the researchers utilized one-month average values of daily datasets, considering that some figures were missing for specific dates. Consequently, they chose a minimum timeframe of one month for all driving-related components in their analysis.

The second objective of the study was to evaluate and compare various machine learning models to identify the most effective model for predicting fire occurrences. The test dataset was used to assess the performance of four different ML techniques, with the RF model demonstrating the most favorable outcomes. The RF model is characterized by multiple independently trained decision trees, and its classification decision is based on selecting the class with the highest number of votes from the individual trees. This ensemble approach contributes to the RF model's high precision. The researchers fine-tuned the RF model by adjusting parameters such as the maximum depth and the number of trees, which further improved its accuracy. The optimal configuration that yielded the best results consisted of 110 trees and a maximum depth of 50. Additionally, the study analyzed the importance of each variable to understand its contribution to the classifier's accuracy. Eliminating any of these variables led to a decrease in the classifier's accuracy rate.

Furthermore, researchers have explored the integration of a CNN model with RF to achieve even better prediction accuracy than other state-of-the art models [77]. In the case of a CNN 1D and a CNN 2D, the output of the completely connected layer was set to 512, which was directly fed into the RF. By increasing this variable from 34 to 512, the RF accuracy improved significantly, rising from 86.48% to 91.34%. This combination of a CNN and RF led to an overall accuracy of 96.56%, making it a successful approach for wildfire prediction.

Comparing the performance of the CNN alone with the ensemble approach, we observed a notable increase in precision and F1-score by incorporating traditional classifiers. This suggests that the ensemble approach better captures the intricate patterns in the wildfire data, resulting in more accurate predictions. In conclusion, while using a CNN directly for classification is a robust approach, the combination of a CNN with traditional classifiers offers a performance boost in terms of the key evaluation metrics. The ensemble approach's improved precision, F1-score, and accuracy highlight the benefits of synergizing deep learning and classical machine learning techniques in wildfire prediction.

The study proposes several succinct recommendations: first, the formulation of customized fire management plans for different sub-areas based on density clusters, utilizing appropriate datasets for training; second, the adoption of a seasonal approach to wildfire mitigation, incorporating insights into fire behavior across seasons for effective strategies; third, considering population and critical facility relocation from fire-prone regions to mitigate financial and public health risks; fourth, the enhancement of existing wildfire control decision support systems in Pakistan through the integration of the study's data, model insights, and platform development; finally, the utilization of freely available global databases to assess catastrophic risks and augment wildfire management approaches in emerging nations, fostering cross-border knowledge exchange for improved prevention and control efforts.

5. Discussion

The primary aim of this study was to improve the process of extracting data related to fire incidents. The aim of this study was to improve the process of extracting data related to fire incidents and no-fire conditions from historical inventories. These data were intended to be used in machine learning models for predicting wildfires and analyzing the factors that contribute to their occurrence. Here, we referenced previous novel studies that utilized both remotely sensed fire data and tree-based machine learning algorithms [78,79]. Our research goes beyond previous studies by incorporating a CNN, which is an advanced deep learning technique, to map and predict wildfires. Previous research often focused on identifying fire incidence sites using a 1 km-resolution dataset from FIRMS, but it neglected the inclusion of no-fire points. Additionally, previous research conducted on the wildland–urban interface (WUI) has revealed that wildfires have been observed to extend beyond the 1 km buffer zone surrounding the initial ignition points. This occurrence is attributed to the dispersion of firebrands over longer distances, leading to the ignition of spot fires. In our study, the classification of no-fire occurrence zones was based on a systematic approach. We defined a no-fire occurrence zone as areas that fell outside the firing zones within a 1 km buffer zone. The purpose of this classification was to differentiate between areas with documented fire incidents and areas that did not experience fires during the specified time period. The selection of the 1 km buffer zone as the basis for classifying no-fire occurrences was determined by several factors. First, it allowed us to establish a clear spatial boundary for analysis and comparison with fire occurrence zones. By defining a buffer zone around the fire zones, we can establish a contextually relevant reference area for identifying no-fire occurrences. This buffer zone allows us to capture the immediate vicinity surrounding the fire zones, enabling a focused analysis of the factors influencing fire occurrence and spread in nearby areas. The buffer zone helps in differentiating between areas with documented fire incidents and areas that remained unaffected by fires. It provides a clear spatial boundary within which we can classify and study the no-fire occurrence zones. By excluding areas within the 1 km buffer zone from the fire occurrence zones, we identified and analyzed the areas that fell outside this buffer zone as no-fire occurrence zones. Moreover, certain areas within the WUI exhibited an extended buffer zone of up to 5 km, which significantly increases the risk of wildfires encroaching upon residential properties [80,81]. While we acknowledge that wildfires can spread beyond a 1 km buffer zone, it is important to note that our study focused on analyzing the immediate impact and characteristics of fire incidents. The purpose was to identify factors contributing to fire susceptibility within a relatively close proximity to the fire zones. By examining the areas within this buffer zone, we aimed to gain insights into the factors influencing fire occurrence and spread in the immediate vicinity of fires.

Our study introduced advanced techniques such as the utilization of a CNN, which has demonstrated significant potential in image analysis and classification tasks. By applying this advanced technique to wildfire mapping and prediction assessment, we aimed to improve the accuracy and reliability of the results. Some of the new studies published recently in this field have aimed to contribute to the field of wildfire mapping and prediction assessment by exploring different approaches, metrics, and indices to improve our understanding and assessment of fire severity. These addressed specific regions and ecosystems and provide valuable insights into the application of remote sensing techniques and indices for assessing fire impacts [82–84]. Our research adopted a comprehensive approach by considering various contributing factors to wildfire occurrence and severity. We recognize that wildfires are complex events influenced by multiple factors such as weather conditions, vegetation type, terrain, and human activities. By taking into account these different factors, our study provides a more holistic understanding of the dynamics and patterns of wildfires. Moreover, our findings uncover novel relationships, patterns, or correlations that were not previously explored in the literature. The utilization of a CNN allows us to analyze and interpret large amounts of data, potentially revealing hidden insights and connections between contributing factors and wildfire occurrences. These novel findings contribute to the advancement of knowledge in wildfire mapping and prediction assessment, improving our ability to predict, prevent, and manage wildfires.

We build upon the literature by introducing advanced techniques, adopting a comprehensive approach, and potentially uncovering novel findings. Through our research, we aim to contribute to the advancement of wildfire mapping and prediction assessment, ultimately enhancing our understanding and ability to mitigate the devastating impacts of wildfires.

6. Conclusions with Constraints and Limitations

In the realm of data-driven approaches, the selection of an optimal time period for model training is a common challenge, influenced by data stochasticity and heterogeneity. Striking a balance between long-term data utilization, which may smooth behavior trends, and short-term data reliance, which introduces uncertainty, is pivotal. Evaluating a model's durability over an extended period is crucial to assessing its reliability. This study addressed this concern by dividing the dataset into four distinct time intervals, Y1 to Y4, each spanning six years. Notably, the CNN 2D feature extraction consistently showcased superior performance in accuracy across all time periods, with the highest accuracy score of 96.21 achieved in Y1.

Moving to seasonal applicability, Pakistan experiences four distinct seasons, and this study delved into the impact of seasonality on model performance. The CNN 2D approach consistently outperformed other feature techniques in each seasonal category. The highest accuracy score of 98.73 was achieved during S3, the summer rainy season, utilizing a CNN 2D feature extraction. These findings underscore the adaptability of a CNN 2D in capturing season-specific fire occurrence patterns, an essential aspect for accurate prediction.

Our study found that machine learning models (No-Conv, Conv1D, Conv2D) improved wildfire prediction in Pakistan's summer rainy season by considering weather patterns, historical incidents, and fuel moisture content, revealing a strong relationship between wildfires and precipitation. The prediction model achieved a high accuracy score of 98.71, confirming the effectiveness of considering wildfire–precipitation dynamics in minimizing fire risk and spread.

Moreover, the study investigated the spatial distribution of fire incidents within Pakistan, revealing significant attributes in various clusters through K-means clustering. This approach offered valuable insights into the diverse landscape of fire occurrences. The spatial clusters, labeled C1, C2, C3, and C4, enhanced our understanding of the fire dynamics in different regions, and the consistently superior performance of a CNN 2D was evident across all clusters. By utilizing these spatial patterns, policymakers and stakeholders can formulate targeted strategies for effective fire management and prevention.

Addressing the increasing wildfire risk in Pakistan, which is exacerbated by urban expansion encroaching upon wildland interfaces, demands immediate attention from policymakers. Mitigation strategies should encompass urban planning, zoning regulations, and community awareness initiatives to safeguard human settlements and infrastructure. This study highlights the potential of advanced machine learning models such as a CNN 2D and RF in predicting wildfires, emphasizing the role of technology in assessing fire risk. The integration of real-time remote sensing and weather monitoring data could further enhance prediction accuracy, empowering proactive wildfire management strategies.

In conclusion, this study's evaluation of historical fire events in Pakistan provides a solid foundation for constructing a comprehensive wildfire management framework.

By harnessing insights from data-driven analysis and advanced modeling techniques, policymakers can implement preemptive measures to protect communities and natural resources while mitigating the impacts of wildfires. Collaborative efforts among scientific communities, government bodies, and local stakeholders are pivotal in addressing the escalating wildfire threat in Pakistan. While data limitations have been acknowledged, refining the variable values, employing finer-resolution datasets, and aligning satellite data with real fire occurrences are crucial steps to enhance wildfire forecasting accuracy and management understanding. The alignment of remotely sensed data with forest department-maintained records promises further improvements in prediction reliability and overall management strategies.

Author Contributions: R.K.: Conceptualization, methodology, data curation and writing-original draft preparation. W.R.: Visualization, investigation, software and validation. M.I.: Writing-reviewing and editing. S.W.: Supervision. All authors have read and agreed to the published version of the manuscript.

Funding: The work was supported by the National Key R&D Program of China (2021YFC3000300).

Data Availability Statement: All the data that support this study are open access and can be accessed using websites or data repositories. The sources of the datasets are described accordingly.

Conflicts of Interest: The authors declare no conflict of interest.

References

1. Verde, J.C.; Zêzere, J.L. Assessment and validation of wildfire susceptibility and hazard in Portugal. *Nat. Hazards Earth Syst. Sci.* **2010**, *10*, 485–497. [CrossRef]
2. FerreirA-leiTe, F.; Lourenço, L.; Bento-Gonçalves, A. Large Forest fires in mainland Portugal, brief characterization. *Méditerranée. Rev. Géographique Pays Méditerranéens/J. Mediterr. Geogr.* **2013**, *121*, 53–65. [CrossRef]
3. Tedim, F.; Remelgado, R.; Borges, C.; Carvalho, S.; Martins, J. Exploring the occurrence of mega-fires in Portugal. *For. Ecol. Manag.* **2013**, *294*, 86–96. [CrossRef]
4. Brown, J.K.; Smith, J.K. *Wildland Fire in Ecosystems: Effects of Fire on Flora*; Gen. Tech. Rep. RMRS-GTR-42-vol. 2. Ogden, UT; U.S. Department of Agriculture, Forest Service, Rocky Mountain Research Station: Fort Collins, CO, USA, 2000; Volume 42, 257p.
5. Neary, D.G.; Ryan, K.C.; DeBano, L.F. *Wildland Fire in Ecosystems: Effects of Fire on Soils and Water*; Gen. Tech. Rep. RMRS-GTR-42-vol. 4. Ogden, UT; US Department of Agriculture, Forest Service, Rocky Mountain Research Station: Fort Collins, CO, USA, 2005; Volume 42, 250p.
6. Xu, R.; Yu, P.; Abramson, M.J.; Johnston, F.H.; Samet, J.M.; Bell, M.L.; Haines, A.; Ebi, K.L.; Li, S.; Guo, Y. Wildfires, global climate change, and human health. *N. Engl. J. Med.* **2020**, *383*, 2173–2181. [CrossRef] [PubMed]
7. Sandberg, D.V. *Wildland Fire in Ecosystems: Effects of Fire on Air*; US Department of Agriculture, Forest Service, Rocky Mountain Research Station: Fort Collins, CO, USA, 2003.
8. Johnston, L.M.; Wang, X.; Erni, S.; Taylor, S.W.; McFayden, C.B.; Oliver, J.A.; Stockdale, C.; Christianson, A.; Boulanger, Y.; Gauthier, S.; et al. Wildland fire risk research in Canada. *Environ. Rev.* **2020**, *28*, 164–186. [CrossRef]
9. Martell, D.L. *Forest Fire Management, in Handbook of Operations Research in Natural Resources*; Springer: Berlin/Heidelberg, Germany, 2007; pp. 489–509.
10. Shvidenko, A.Z.; Schepaschenko, D.G. Climate change and wildfires in Russia. *Contemp. Probl. Ecol.* **2013**, *6*, 683–692. [CrossRef]
11. Vadrevu, K.P.; Lasko, K.; Giglio, L.; Schroeder, W.; Biswas, S.; Justice, C. Trends in vegetation fires in south and southeast Asian countries. *Sci. Rep.* **2019**, *9*, 7422. [CrossRef] [PubMed]
12. Reddy, C.S.; Bird, N.G.; Sreelakshmi, S.; Manikandan, T.M.; Asra, M.; Krishna, P.H.; Jha, C.S.; Rao, P.V.N.; Diwakar, P.G. Identification and characterization of spatio-temporal hotspots of forest fires in South Asia. *Environ. Monit. Assess.* **2019**, *191*, 791. [CrossRef]
13. Hantson, S.; Pueyo, S.; Chuvieco, E. Global fire size distribution is driven by human impact and climate. *Glob. Ecol. Biogeogr.* **2015**, *24*, 77–86. [CrossRef]
14. Jolly, W.M.; Cochrane, M.A.; Freeborn, P.H.; Holden, Z.A.; Brown, T.J.; Williamson, G.J.; Bowman, D.M. Climate-induced variations in global wildfire danger from 1979 to 2013. *Nat. Commun.* **2015**, *6*, 7537. [CrossRef]
15. Barbero, R.; Abatzoglou, J.T.; Pimont, F.; Ruffault, J.; Curt, T. Attributing increases in fire weather to anthropogenic climate change over France. *Front. Earth Sci.* **2020**, *8*, 104. [CrossRef]
16. Oliveira, S.L.; Pereira, J.M.; Carreiras, J.M. Fire frequency analysis in Portugal (1975–2005), using Landsat-based burnt area maps. *Int. J. Wildland Fire* **2011**, *21*, 48–60. [CrossRef]
17. Stacey, R.; Gibson, S.; Hedley, P. *European Glossary for Wildfires and Forest Fires*; European Union-INTERREG IVC: Cham, Switzerland, 2012.

18. Álvarez-Díaz, M.; González-Gómez, M.; Otero-Giraldez, M.S. Detecting the socioeconomic driving forces of the fire catastrophe in NW Spain. *Eur. J. For. Res.* **2015**, *134*, 1087–1094. [CrossRef]
19. Flannigan, M.D.; Wotton, B.M. *Climate, Weather, and Area Burned, in Forest Fires*; Elsevier: Amsterdam, The Netherlands, 2001; pp. 351–373.
20. Tymstra, C.; Jain, P.; Flannigan, M.D. Characterisation of initial fire weather conditions for large spring wildfires in Alberta, Canada. *Int. J. Wildland Fire* **2021**, *30*, 823–835. [CrossRef]
21. Nunes, A.; Lourenço, L.; Meira, A.C. Exploring spatial patterns and drivers of forest fires in Portugal (1980–2014). *Sci. Total Environ.* **2016**, *573*, 1190–1202. [CrossRef] [PubMed]
22. Cao, X.; Cui, X.; Yue, M.; Chen, J.; Tanikawa, H.; Ye, Y. Evaluation of wildfire propagation susceptibility in grasslands using burned areas and multivariate logistic regression. *Int. J. Remote Sens.* **2013**, *34*, 6679–6700. [CrossRef]
23. Holsinger, L.; Parks, S.A.; Miller, C. Weather, fuels, and topography impede wildland fire spread in western US landscapes. *For. Ecol. Manag.* **2016**, *380*, 59–69. [CrossRef]
24. Calviño-Cancela, M.; Chas-Amil, M.L.; García-Martínez, E.D.; Touza, J. Interacting effects of topography, vegetation, human activities and wildland-urban interfaces on wildfire ignition risk. *For. Ecol. Manag.* **2017**, *397*, 10–17. [CrossRef]
25. Ghorbanzadeh, O.; Blaschke, T.; Gholamnia, K.; Aryal, J. Forest fire susceptibility and risk mapping using social/infrastructural vulnerability and environmental variables. *Fire* **2019**, *2*, 50. [CrossRef]
26. Gigović, L.; Pourghasemi, H.R.; Drobnjak, S.; Bai, S. Testing a new ensemble model based on SVM and random forest in forest fire susceptibility assessment and its mapping in Serbia's Tara National Park. *Forests* **2019**, *10*, 408.
27. Tehrany, M.S.; Jones, S.; Shabani, F.; Martínez-Álvarez, F.; Tien Bui, D. A novel ensemble modeling approach for the spatial prediction of tropical forest fire susceptibility using LogitBoost machine learning classifier and multi-source geospatial data. *Theor. Appl. Climatol.* **2019**, *137*, 637–653. [CrossRef]
28. Keeley, J.E.; Syphard, A.D. Historical patterns of wildfire ignition sources in California ecosystems. *Int. J. Wildland Fire* **2018**, *27*, 781–799. [CrossRef]
29. Rodrigues, M.; Jiménez, A.; de la Riva, J. Analysis of recent spatial–temporal evolution of human driving factors of wildfires in Spain. *Nat. Hazards* **2016**, *84*, 2049–2070. [CrossRef]
30. Choubin, B.; Abdolshahnejad, M.; Moradi, E.; Querol, X.; Mosavi, A.; Shamshirband, S.; Ghamisi, P. Spatial hazard assessment of the PM10 using machine learning models in Barcelona, Spain. *Sci. Total Environ.* **2020**, *701*, 134474. [CrossRef] [PubMed]
31. Ghorbanzadeh, O.; Meena, S.R.; Abadi HS, S.; Piralilou, S.T.; Zhiyong, L.; Blaschke, T. Landslide Mapping Using Two Main Deep-Learning Convolution Neural Network Streams Combined by the Dempster–Shafer Model. *IEEE J. Sel. Top. Appl. Earth Obs. Remote Sens.* **2020**, *14*, 452–463. [CrossRef]
32. Gholamnia, K.; Gudiyangada Nachappa, T.; Ghorbanzadeh, O.; Blaschke, T. Comparisons of diverse machine learning approaches for wildfire susceptibility mapping. *Symmetry* **2020**, *12*, 604. [CrossRef]
33. Iqbal, M.; Sameem MS, I.; Naqvi, N.; Kanwal, S.; Ye, Z. A deep learning approach for face recognition based on angularly discriminative features. *Pattern Recognit. Lett.* **2019**, *128*, 414–419. [CrossRef]
34. Iban, M.C.; Sekertekin, A. Machine learning based wildfire susceptibility mapping using remotely sensed fire data and GIS: A case study of Adana and Mersin provinces, Turkey. *Ecol. Inform.* **2022**, *69*, 101647. [CrossRef]
35. Yoon, H.J.; Voulgaris, P. Multi-time predictions of wildfire grid map using remote sensing local data. In Proceedings of the 2022 IEEE International Conference on Knowledge Graph (ICKG), Orlando, FL, USA, 30 November–1 December 2022; IEEE: Piscataway, NJ, USA, 2022; pp. 365–372.
36. Zhong, Y.; Ning, P.; Yan, S.; Zhang, C.; Xing, J.; Shi, J.; Hao, J. A machine-learning approach for identifying dense-fires and assessing atmospheric emissions on the Indochina Peninsula, 2010–2020. *Atmos. Res.* **2022**, *278*, 106325. [CrossRef]
37. Gould, J.S.; McCaw, W.L.; Cheney, N.P.; Ellis, P.F.; Knight, I.K.; Sullivan, A.L. *Project Vesta: Fire in Dry Eucalypt Forest: Fuel Structure, Fuel Dynamics and Fire Behaviour*; Csiro Publishing: Clayton, Australia, 2008.
38. Phelps, N.; Woolford, D.G. Comparing calibrated statistical and machine learning methods for wildland fire occurrence prediction: A case study of human-caused fires in Lac La Biche, Alberta, Canada. *Int. J. Wildland Fire* **2021**, *30*, 850–870. [CrossRef]
39. Cruz, M.G.; McCaw, W.L.; Anderson, W.R.; Gould, J.S. Fire behaviour modelling in semi-arid mallee-heath shrublands of southern Australia. *Environ. Model. Softw.* **2013**, *40*, 21–34. [CrossRef]
40. Nadeem, K.; Taylor, S.W.; Woolford, D.G.; Dean, C.B. Mesoscale spatiotemporal predictive models of daily human-and lightning-caused wildland fire occurrence in British Columbia. *Int. J. Wildland Fire* **2019**, *29*, 11–27. [CrossRef]
41. Woolford, D.G.; Martell, D.L.; McFayden, C.B.; Evens, J.; Stacey, A.; Wotton, B.M.; Boychuk, D. The development and implementation of a human-caused wildland fire occurrence prediction system for the province of Ontario, Canada. *Can. J. For. Res.* **2021**, *51*, 303–325. [CrossRef]
42. Rafaqat, W.; Iqbal, M.; Kanwal, R.; Song, W. Study of Driving Factors Using Machine Learning to Determine the Effect of Topography, Climate, and Fuel on Wildfire in Pakistan. *Remote Sens.* **2022**, *14*, 1918. [CrossRef]
43. Pérez-Porras, F.J.; Triviño-Tarradas, P.; Cima-Rodríguez, C.; Meroño-de-Larriva, J.E.; García-Ferrer, A.; Mesas-Carrascosa, F.J. Machine learning methods and synthetic data generation to predict large wildfires. *Sensors* **2021**, *21*, 3694. [CrossRef] [PubMed]
44. Cheng, S.; Prentice, I.C.; Huang, Y.; Jin, Y.; Guo, Y.K.; Arcucci, R. Data-driven surrogate model with latent data assimilation: Application to wildfire forecasting. *J. Comput. Phys.* **2022**, *464*, 111302. [CrossRef]

45. Rubí, J.N.S.; de Carvalho, P.H.P.; Gondim, P.R.L. Application of machine learning models in the behavioral study of forest fires in the Brazilian Federal District region. *Eng. Appl. Artif. Intell.* **2023**, *118*, 105649. [CrossRef]
46. Mahamed, M.; Wittenberg, L.; Kutiel, H.; Brook, A. Fire Risk Assessment on Wildland–Urban Interface and Adjoined Urban Areas: Estimation Vegetation Ignitability by Artificial Neural Network. *Fire* **2022**, *5*, 184. [CrossRef]
47. Mahamed, M.; Wittenberg, L.; Kutiel, H.; Brook, A. A novel urban vegetation mapping approach for fire risk assessment: A Mediterranean case study. *Urban Ecosyst.* **2023**, *26*, 1263–1274. [CrossRef]
48. Polinova, M.; Wittenberg, L.; Kutiel, H.; Brook, A. Reconstructing pre-fire vegetation condition in the wildland urban interface (WUI) using artificial neural network. *J. Environ. Manag.* **2019**, *238*, 224–234. [CrossRef]
49. Kattel, D.B.; Yao, T.; Ullah, K.; Rana, A.S. Seasonal near-surface air temperature dependence on elevation and geographical coordinates for Pakistan. *Theor. Appl. Climatol.* **2019**, *138*, 1591–1613. [CrossRef]
50. Begum, B.A.; Biswas, S.K.; Pandit, G.G.; Saradhi, I.V.; Waheed, S.; Siddique, N.; Seneviratne, M.S.; Cohen, D.D.; Markwitz, A.; Hopke, P.K. Long–range transport of soil dust and smoke pollution in the South Asian region. *Atmos. Pollut. Res.* **2011**, *2*, 151–157. [CrossRef]
51. Ghali, R.; Akhloufi, M.A. Deep Learning Approaches for Wildland Fires Using Satellite Remote Sensing Data: Detection, Mapping, and Prediction. *Fire* **2023**, *6*, 192. [CrossRef]
52. Nur, A.S.; Kim, Y.J.; Lee, C.-W. Creation of Wildfire Susceptibility Maps in Plumas National Forest Using InSAR Coherence, Deep Learning, and Metaheuristic Optimization Approaches. *Remote Sens.* **2022**, *14*, 4416. [CrossRef]
53. Lin, X.; Li, Z.; Chen, W.; Sun, X.; Gao, D. Forest Fire Prediction Based on Long- and Short-Term Time-Series Network. *Forests* **2023**, *14*, 778. [CrossRef]
54. Higa, L.; Marcato Junior, J.; Rodrigues, T.; Zamboni, P.; Silva, R.; Almeida, L.; Liesenberg, V.; Roque, F.; Libonati, R.; Gonçalves, W.N.; et al. Active Fire Mapping on Brazilian Pantanal Based on Deep Learning and CBERS 04A Imagery. *Remote Sens.* **2022**, *14*, 688. [CrossRef]
55. Meira Castro, A.C.; Nunes, A.; Sousa, A.; Lourenço, L. Mapping the causes of forest fires in portugal by clustering analysis. *Geosciences* **2020**, *10*, 53. [CrossRef]
56. Nolan, R.H.; Boer, M.M.; Collins, L.; Resco de Dios, V.; Clarke, H.; Jenkins, M.; Kenny, B.; Bradstock, R.A. Causes and consequences of eastern Australia's 2019-20 season of mega-fires. *Glob. Chang. Biol.* **2020**, *26*, 1039–1041. [CrossRef]
57. Edwards, R.B.; Naylor, R.L.; Higgins, M.M.; Falcon, W.P. Causes of Indonesia's forest fires. *World Dev.* **2020**, *127*, 104717. [CrossRef]
58. Tien Bui, D.; Tuan, T.A.; Hoang, N.D.; Thanh, N.Q.; Nguyen, D.B.; Van Liem, N.; Pradhan, B. Spatial prediction of rainfall-induced landslides for the Lao Cai area (Vietnam) using a hybrid intelligent approach of least squares support vector machines inference model and artificial bee colony optimization. *Landslides* **2017**, *14*, 447–458. [CrossRef]
59. Lawson, B.D.; Armitage, O. *Weather guide for the Canadian forest fire danger rating system*; Natural Resources Canada: Edmonton, AB, Canada, 2008; Available online: https://scf.rncan.gc.ca/pubwarehouse/pdfs/29152.pdf (accessed on 4 September 2023).
60. Horel, J.D.; Ziel, R.; Galli, C.; Pechmann, J.; Dong, X. An evaluation of fire danger and behaviour indices in the Great Lakes Region calculated from station and gridded weather information. *Int. J. Wildland Fire* **2014**, *23*, 202–214. [CrossRef]
61. de Jong, M.C.; Wooster, M.J.; Kitchen, K.; Manley, C.; Gazzard, R.; McCall, F.F. Calibration and evaluation of the Canadian Forest Fire Weather Index (FWI) System for improved wildland fire danger rating in the United Kingdom. *Nat. Hazards Earth Syst. Sci.* **2016**, *16*, 1217–1237. [CrossRef]
62. Romero, R.; Mestre, A.; Botey, R. A New Calibration for Fire Weather Index in Spain (AEMET). In *Advances in Forest Fire Research*; University of Coimbra: Coimbra, Portugal, 2014; p. 1044. Available online: http://hdl.handle.net/10316.2/34013 (accessed on 21 November 2014).
63. Tian, X.R.; Zhao, F.J.; Shu, L.F.; Wang, M.Y. Changes in forest fire danger for south-western China in the 21st century. *Int. J. Wildland Fire* **2014**, *23*, 185–195. [CrossRef]
64. Abatzoglou, J.T.; Dobrowski, S.Z.; Parks, S.A.; Hegewisch, K.C. TerraClimate, a high-resolution global dataset of monthly climate and climatic water balance from 1958–2015. *Sci. Data* **2018**, *5*, 170191. [CrossRef] [PubMed]
65. Rodell, M.; Houser, P.R.; Jambor, U.E.A.; Gottschalck, J.; Mitchell, K.; Meng, C.J.; Arsenault, K.; Cosgrove, B.; Radakovich, J.; Bosilovich, M.; et al. The global land data assimilation system. *Bull. Am. Meteorol. Soc.* **2004**, *85*, 381–394. [CrossRef]
66. Farr, T.G.; Rosen, P.A.; Caro, E.; Crippen, R.; Duren, R.; Hensley, S.; Kobrick, M.; Paller, M.; Rodriguez, E.; Roth, L.; et al. The Shuttle Radar Topography Mission. *Rev. Geophys.* **2007**, *45*. [CrossRef]
67. Pesaresi, M.; Huadong, G.; Blaes, X.; Ehrlich, D.; Ferri, S.; Gueguen, L.; Halkia, M.; Kauffmann, M.; Kemper, T.; Lu, L.; et al. A global human settlement layer from optical HR/VHR RS data: Concept and first results. *IEEE J. Sel. Top. Appl. Earth Obs. Remote Sens.* **2013**, *6*, 2102–2131. [CrossRef]
68. Kennedy, C.M.; Oakleaf, J.R.; Theobald, D.M.; Baruch-Mordo, S.; Kiesecker, J. Managing the middle: A shift in conservation priorities based on the global human modification gradient. *Glob. Chang. Biol.* **2019**, *25*, 811–826. [CrossRef]
69. Weiss, D.J.; Nelson, A.; Vargas-Ruiz, C.A.; Gligorić, K.; Bavadekar, S.; Gabrilovich, E.; Bertozzi-Villa, A.; Rozier, J.; Gibson, H.S.; Shekel, T.; et al. Global maps of travel time to healthcare facilities. *Nat. Med.* **2020**, *26*, 1835–1838. [CrossRef]
70. Hengl, T. *Soil Bulk Density (Fine Earth) 10 × kg/m-Cubic at 6 Standard Depths (0, 10, 30, 60, 100 and 200 cm) at 250 m Resolution*; Zenodo, Ed.; CERN: Geneva, Switzerland, 2018.

71. Hengl, T. *Soil Texture Classes (USDA System) for 6 Soil Depths (0, 10, 30, 60, 100 and 200 cm) at 250 m*; Zenodo, Ed.; CERN: Geneva, Switzerland, 2018.
72. Hengl, T.; Nauman, T. *Predicted USDA Soil Great Groups at 250 m (Probabilities)*; Zenodo, Ed.; CERN: Geneva, Switzerland, 2018.
73. Buchhorn, M.; Lesiv, M.; Tsendbazar, N.E.; Herold, M.; Bertels, L.; Smets, B. Copernicus global land cover layers—Collection 2. *Remote Sens.* **2020**, *12*, 1044. [CrossRef]
74. Didan, K. *MOD13A1 MODIS/Terra Vegetation Indices 16-Day L3 Global 500 m SIN Grid V006*; DAAC, Ed.; NASA: Washington, DC, USA, 2015.
75. Myneni, R.; Knyazikhin, Y.; Park, T. *MCD15A3H MODIS/Terra + Aqua Leaf Area Index/FPAR 4-Day L4 Global 500 m SIN Grid V006*; DAAC, Ed.; NASA: Washington, DC, USA, 2015.
76. Center for International Earth Science Information Network-CIESIN-Columbia University. *Gridded Population of the World, Version 4 (GPWv4): Population Count, Revision 11*; NASA Socioeconomic Data and Applications Center (SEDAC): Palisades, NY, USA, 2018.
77. Zhang, G.; Wang, M.; Liu, K. Deep neural networks for global wildfire susceptibility modelling. *Ecol. Indic.* **2021**, *127*, 107735. [CrossRef]
78. Mohajane, M.; Costache, R.; Karimi, F.; Pham, Q.B.; Essahlaoui, A.; Nguyen, H.; Laneve, G.; Oudija, F. Application of remote sensing and machine learning algorithms for forest fire mapping in a Mediterranean area. *Ecol. Indic.* **2021**, *129*, 107869. [CrossRef]
79. Florath, J.; Keller, S. Supervised machine learning approaches on multispectral remote sensing data for a combined detection of fire and burned area. *Remote Sens.* **2022**, *14*, 657. [CrossRef]
80. Kumar, M.; Li, S.; Nguyen, P.; Banerjee, T. Examining the existing definitions of wildland-urban interface for California. *Ecosphere* **2022**, *13*, e4306. [CrossRef]
81. Radeloff, V.C.; Hammer, R.B.; Stewart, S.I.; Fried, J.S.; Holcomb, S.S.; McKeefry, J.F. The wildland–urban interface in the United States. *Ecol. Appl.* **2005**, *15*, 799–805. [CrossRef]
82. Holsinger, L.M.; Parks, S.A.; Saperstein, L.B.; Loehman, R.A.; Whitman, E.; Barnes, J.; Parisien, M.A. Improved fire severity mapping in the North American boreal forest using a hybrid composite method. *Remote Sens. Ecol. Conserv.* **2022**, *8*, 222–235. [CrossRef]
83. Brook, A.; Hamzi, S.; Roberts, D.; Ichoku, C.; Shtober-Zisu, N.; Wittenberg, L. Total Carbon Content Assessed by UAS Near-Infrared Imagery as a New Fire Severity Metric. *Remote Sens.* **2022**, *14*, 3632. [CrossRef]
84. Giddey, B.L.; Baard, J.A.; Kraaij, T. Verification of the differenced Normalised Burn Ratio (dNBR) as an index of fire severity in Afrotemperate Forest. *S. Afr. J. Bot.* **2022**, *146*, 348–353. [CrossRef]

MDPI AG
Grosspeteranlage 5
4052 Basel
Switzerland
Tel.: +41 61 683 77 34

Remote Sensing Editorial Office
E-mail: remotesensing@mdpi.com
www.mdpi.com/journal/remotesensing

www.ingramcontent.com/pod-product-compliance
Lightning Source LLC
Chambersburg PA
CBHW041407210926
51850CB00013B/3197
* 9 7 8 3 7 2 5 8 3 9 9 2 6 *